NATO ASI Series

Advanced Science Institutes Series

A series presenting the results of activities sponsored by the NATO Science Committee, which aims at the dissemination of advanced scientific and technological knowledge, with a view to strengthening links between scientific communities.

The Series is published by an international board of publishers in conjunction with the NATO Scientific Affairs Division

A Life Sciences B Physics	Plenum Publishing Corporation London and New York
C Mathematical and Physical Sciences D Behavioural and Social Sciences E Applied Sciences	Kluwer Academic Publishers Dordrecht, Boston and London
F Computer and Systems Sciences G Ecological Sciences H Cell Biology I Global Environmental Change	Springer-Verlag Berlin Heidelberg New York London Paris Tokyo Hong Kong Barcelona Budapest

NATO-PCO DATABASE

The electronic index to the NATO ASI Series provides full bibliographical references (with keywords and/or abstracts) to more than 30000 contributions from international scientists published in all sections of the NATO ASI Series. Access to the NATO-PCO DATABASE is possible in two ways:

– via online FILE 128 (NATO-PCO DATABASE) hosted by ESRIN, Via Galileo Galilei, I-00044 Frascati, Italy.

– via **CD-ROM** "NATO-PCO DATABASE" with user-friendly retrieval software in English, French and German (© WTV GmbH and DATAWARE Technologies Inc. 1989).

The CD-ROM can be ordered through any member of the Board of Publishers or through NATO-PCO, Overijse, Belgium.

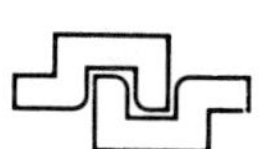

Series H: Cell Biology Vol. 50

The ASI Series Books Published as a Result of Activities of the Special Programme on CELL TO CELL SIGNALS IN PLANTS AND ANIMALS

The books published as a result of the activities of the Special Programme are:

Vol. 1: **Biology and Molecular Biology of Plant-Pathogen Interactions.** Edited by J.A. Bailey. 1986.

Vol. 2: **Glial-Neuronal Communication in Development and Regeneration.** Edited by H.H. Althaus and W. Seifert. 1987.

Vol. 3: **Nicotinic Acetylcholine Receptor: Structure and Function.** Edited by A. Maelicke. 1986.

Vol. 4: **Recognition in Microbe-Plant Symbiotic and Pathogenic Interactions.** Edited by B. Lugtenberg. 1986.

Vol. 5: **Mesenchymal-Epithelial Interactions in Neural Development.** Edited by J.R. Wolff, J. Sievers, and M. Berry. 1987.

Vol. 6: **Molecular Mechanisms of Desensitization to Signal Molecules.** Edited by T.M. Konjin, P.J.M. Van Haastert, H. Van der Starre, H. Van der Wel, and M.D. Houslay. 1987.

Vol. 7: **Gangliosides and Modulation of Neuronal Functions.** Edited by H. Rahmann. 1987.

Vol. 9: **Modification of Cell to Cell Signals During Normal and Pathological Aging.** Edited By S. Govoni and F. Battaini. 1987.

Vol. 10: **Plant Hormone Receptors.** Edited by D. Klämbt. 1987.

Vol. 11: **Host-Parasite Cellular and Molecular Interactions in Protozoal Infections.** Edited by K.-P. Chang and D. Snary. 1987.

Vol. 12: **The Cell Surface in Signal Transduction.** Edited by E. Wagner, H. Greppin, and B. Millet. 1987.

Vol. 19: **Modulation of Synaptic Transmission and Plasticity in Nervous Systems.** Edited by G. Hertting and H.-C. Spatz. 1988.

Vol. 20: **Amino Acid Availability and Brain Function in Health and Disease.** Edited by G. Huether. 1988.

Vol. 21: **Cellular and Molecular Basis of Synaptic Transmission.** Edited by H. Zimmermann. 1988.

Vol. 23: **The Semiotics of Cellular Communication in the Immune System.** Edited by E.E. Sercarz, F. Celada, N.A. Mitchison, and T. Tada. 1988.

Vol. 24: **Bacteria, Complement and the Phagocytic Cell.** Edited by F.C. Cabello and C. Pruzzo. 1988.

Vol. 25: **Nicotinic Acetylcholine Receptors in the Nervous System.** Edited by F. Celementi, C. Gotti, and E. Sher. 1988.

Vol. 26: **Cell to Cell Signals in Mammalian Development.** Edited by S.W. de Laat, J.G. Bluemink, and C.L. Mummery. 1989.

Vol. 27: **Phytotoxins and Plant Pathogenesis.** Edited by A. Graniti, R.D. Durbin, and A. Ballio. 1989.

Vol. 31: **Neurobiology of the Inner Retina.** Edited by R. Weiler and N.N. Osborne. 1989.

Vol. 32: **Molecular Biology of Neuroreceptors and Ion Channels.** Edited by A. Maelicke. 1989.

Vol. 33: **Regulatory Mechanisms of Neuron to Vessel Communication in the Brain.** Edited by F. Battaini, S. Govoni, M.S. Magnoni, and M. Trabucchi. 1989.

Vol. 35: **Cell Separation in Plants: Physiology, Biochemistry and Molecular Biology.** Edited by D.J. Osborne and M.B. Jackson. 1989.

Vol. 36: **Signal Molecules in Plants and Plant-Microbe Interactions.** Edited by B.J.J. Lugtenberg. 1989.

Vol. 39: **Chemosensory Information Processing.** Edited by D. Schild. 1990.

Vol. 41: **Recognition and Response in Plant-Virus Interactions.** Edited by R.S.S. Fraser. 1990.

Vol. 43: **Cellular and Molecular Biology of Myelination.** Edited by G. Jeserich, H. H. Althaus, and T. V. Waehneldt. 1990.

Vol. 44: **Activation and Desensitization of Transducing Pathways.** Edited by T. M. Konijn, M. D. Houslay, and P. J. M. Van Haastert. 1990.

Vol. 45: **Mechanism of Fertilization: Plants to Humans.** Edited by B. Dale. 1990.

Vol. 46: **Parallels in Cell to Cell Junctions in Plants and Animals.** Edited by A. W. Robards, W. J. Lucas, J. D. Pitts, H. J. Jongsma, and D. C. Spray, 1990.

Vol. 50: **Phytochrome Properties and Biological Action.** Edited by B. Thomas and C. B. Johnson, 1991.

Vol. 51: **Cell to Cell Signals in Plants and Animals.** Edited by V. Neuhoff and J. Friend, 1991.

Phytochrome Properties and Biological Action

Edited by

Brian Thomas

Horticulture Research International
Worthington Road
Littlehampton, West Sussex BN17 6LP
U.K.

Christopher B. Johnson

Department of Botany
University of Reading
Whiteknights
Reading RG6 2AS
U.K.

Springer-Verlag
Berlin Heidelberg New York London Paris Tokyo
Hong Kong Barcelona Budapest
Published in cooperation with NATO Scientific Affairs Division

Proceedings of the NATO Advanced Research Workshop on Phytochrome Properties and Biological Action held in Chichester, U.K., July 22–27, 1990.

ISBN 3-540-51770-7 Springer-Verlag Berlin Heidelberg New York
ISBN 0-387-51770-7 Springer-Verlag New York Berlin Heidelberg

Library of Congress Cataloging-in-Publication Data
NATO Advanced Research Workshop on Phytochrome Properties and Biological Action (1990 : Chichester, England) Phytochrome properties and biological action / edited by Brian Thomas, Christopher B. Johnson. (NATO ASI series. Series H, Cell biology ; vol. 50) "Proceedings of the NATO Advanced Research Workshop on Phytochrome Properties and Biological Action, held in Chichester, U.K., July 22–27, 1990"—T.p. verso
Includes bibliographical references and index.
ISBN 3-540-51770-7.—ISBN 0-387-51770-7
1. Phytochrome—Congresses. I. Thomas, Brian, II. Johnson, Christopher B. III. Title. IV. Series. QK898.P67N38 1990 581.19'218—dc20 91-17411

Printed in Germany

Typesetting: camera ready by authors

31/3140-543210 – Printed on acid-free-paper

PREFACE

International meetings devoted exclusively to phytochrome are relatively infrequent. The Advanced Research Workshop on Phytochrome Properties and Biological Action held in July 1990 at Bishop Otter College in Chichester, an old English cathedral town situated between the south downs and the sea, provided a rare opportunity for those working directly on the mechanism of phytochrome action to get together for a focused and intensive period of discussion on the subject. The number of delegates was limited to fifty but contained representatives from almost all the major laboratories engaged in this area of science. In a five day meeting there were twenty invited lectures, four round tables and three poster sessions which allowed all delegates to present and discuss their work.

The purpose of the meeting was to review, in depth, the current state of knowledge concerning the properties of the plant photoreceptor, phytochrome, in relation to its mechanism of biological action. The central themes of the meeting were the application of molecular biology to provide major new insights into the phytochrome system and the integration of molecular, genetic and biochemical information to generate an understanding of photoreceptor action at the whole organism level. As a record of the meeting each invited speaker prepared a chapter on the subject of his or her lecture and these represent the main body of this book. During the meeting four round table sessions were held on specific topics and the summaries, prepared by the chairmen, are also included. In this way some of the flavour of the discussions, which were a major component of the meeting, can be discerned by the reader.

The book has been prepared in camera-ready format by the editors from manuscripts produced by the participants at the meeting and have been reformatted to a relatively uniform style of presentation. The editors are grateful to the authors for the prompt provision of their manuscripts. Any delay in preparing the book comes from the editorial stage of preparation. One of the topics for discussion at the meeting was the nomenclature to be used for phytochrome proteins and genes. This is not yet standardised although recommendations are made in this book. We did not think it possible to impose any standard use of terminology for phytochromes while editing the book, as authors use different operational criteria for defining their terms, but rather have accepted those in use by the different authors. Although this may cause some confusion to the casual reader it closely reflects the situation as it currently exists in the subject.

The meeting and book would not have been possible without the help of a number of people. The editors would like to thank Eberhard Schafer and Peter Quail for their help in drawing up the scientific programme. Ruthanne Smith from the liaison section at Horticulture Research International provided valuable administrative assistance during the organisation and running of the meeting , as did Dennis Hart, who kept an eye on the finances. At the meeting Martin Hughes, David Mozley, Nick Douthwaite and Natalie Wright helped to ensure that there were no technical or organisational hitches. The editors would like to thank the PVA section at HRI for help in preparation of some of the figures. Special thanks to Mrs M. Barratt at Reading University and in particular to Mrs A. Morley at HRI for word processing and final preparation of the manuscript.

Brian Thomas

Christopher B. Johnson

CONTENTS

Section A. PHYTOCHROME PROPERTIES

B Thomas The properties and biological action of phytochrome: prologue 1

P H Quail, H P Hershey, K B Idler, RR A Sharrock, A H Christensen, B M Parks, D Somers, J Tepperman, W B Bruce and K Dehesh Phy-gene structure, evolution and expression 13

L H Pratt, M-M Cordonnier, Y-C Wang, S J Stewart and M Moyer Evidence for three phytochromes in *Avena* 39

W Rüdiger and F Thümmler Phytochrome in lower plants 57

M Furuya, K Tomizawa, N Ito, D Sommer, L Deforce, K Konomi, D Farrens and P-S Song Biogenesis of phytochrome apoprotein in transgenic organisms and its assembly to the chromophore 71

W Parker, M Romanowski and P-S Song Conformation and its functional implications in phytochrome 85

J R Cherry, D Hondred, J M Keller, H P Hershey and R D Vierstra The use of transgenic plants to study phytochrome domains involved in structure and function 113

Section B: PHYTOCHROME ACTION - MOLECULAR BIOLOGY

S A Kay *In vitro* protein-DNA interactions in the rice phytochrome promoter 129

P M Gilmartin, J Memelink and N-H Chua Dissection of the light-responsive elements of pea *RBCS3A* 141

U Schindler, J R Ecker and A R Cashmore An *Arabidopsis thaliana* leucine zipper protein that binds to G-box promoter sequences 157

E M Tobin, J A Brusslan, J S Buzby, G A Karlin-Neumann, D M Kehoe, P A Okubara, S A Rolfe, L Sun and T Yamada Phytochrome regulation of transcription: biochemical and genetic approaches 167

A Batschauer, B Ehmann, M Furuya, R Grimm, E Hoffmann, K Harter, T Kretsch, A Nagatani, B Ocker, V Speth and E Schäfer Light and cell specific gene expression 181

E Adam, M Szell, A Pay, E Fejes and F Nagy Circadian clock and light regulated transcription of the wheat CAB-1 gene in wheat and in transgenic tobacco plants 191

W F Thompson, R C Elliott, L F Dickey, M Gallo, T J Pedersen and D A Sowinski Unusual features of the light response system regulating ferredoxin gene expression 201

Section C: PHYTOCHROME ACTION - DEVELOPMENTAL PHYSIOLOGY

H Smith, G C Whitelam and A C McCormac Do the members of the phytochrome family have different roles? Physiological evidence from wild-type, mutant and transgenic plants 217

J L Peters, J C Wesselius, K C Georghiou, R E Kendrick, A van Tuinen and M Koornneef The physiology of photomorphogenetic tomato mutants 237

C Hanstein, U Lange, H A W Schneider-Poetsch, F Grolig and G Wagner Immunolocalization of phytochrome and immunodetection of actin in *Mougeotia* 249

H Mohr Integration of phytochrome response 259

C B Johnson, S M Allsebrook, H Carr-Smith and B Thomas A quantitative approach to the molecular biology of phytochrome action 273

W R Briggs and T W Short The transduction of light signals in plants: responses to blue light 289

Appendix: ROUND TABLE REPORTS

G C Whitelam Post-translational modification of phytochrome 303

B Thomas Phytochrome destruction 307

D Vince-Prue Phytochrome action under natural conditions 313

G I Jenkins *Cis*-acting DNA sequence elements and *trans*-acting factors of photoregulated genes 321

Index 331

LIST OF PARTICIPANTS

Director

Dr B. Thomas HRI, Worthing Road, Littlehampton, West Sussex, BN17 6LP, UK

Lecturers

Dr E Adam Institut of Plant Physiology, Biological Research Center, Hungarian Academy of Sciences, 6701 Szeged, POB 521, HUNGARY

Prof W R Briggs Carnegie Institution of Washington, Department of Plant Biology, 290 Panama Street, Stanford, California, 94305, USA

Dr A R Cashmore Plant Science Institute, Department of Biology, University of Pennsylvania, Philadelphia, PA 19104, USA

Prof M Furuya Frontier Research Program, RIKEN, Wako, Saitamo 351-01, JAPAN

Dr P M Gilmartin Lab of Plant Molecular Biology, The Rockerfeller University, 1230 York Avenue, New York, NY 10021-6399, USA

Dr C Johnson Plant Science Laboratories, Reading University, Whiteknights, Reading RG6 2AS, UK

Dr S A Kay Laboratory of Molecular Biology, The Rockerfeller University, 1230 York Avenue, New York, NY 10021-6399 USA

Prof R E Kendrick Plant Physiological Research, Agricultural University, General Foulkesweg 72/6703 BW, Wageningen, THE NETHERLANDS

Prof Dr H Mohr Institut fur Biologie II/Botanik, Schanzlestrasse 1, University of Freiburg, D-7800 Freiburg, FRG

Prof L H Pratt Department of Botany, The University of Georgia, Athens, Georgia 30602, USA

Prof P H Quail UC Berkeley/USDA Plant Gene Exp Centre, 800 Buchanan Street, Albany, CA 94710, USA

Prof Dr W Rudiger Botanisches Institut Der Universitat, Munchen, Menzinger Strasse 67, D-8000 Munchen 19, FRG

Prof Dr E Schafer Institut fur Biologie II/Botanik, Schanzlestrasse 1, University of Freiburg, D-7800 Freiburg, FRG

Prof H Smith Department of Botany, University of Leicester, University Road, Leicester, LE1 7RH, UK

Prof P-S Song Department of Chemistry, University of Nebraska, Lincoln, NE 68588 USA

Prof W F Thompson Department of Botany, North Carolina State University, Box 7612, Raleigh, NC 27695, USA

Dr E M Tobin Department of Biology, University of California, 405 Hilgard Avenue, Los Angeles, California 90024, USA

Dr R D Vierstra Department of Horticulture, University of Wisconsin, 1575 Linden Drive, Madison, WI 53706, USA

Prof Dr G Wagner Institute of Botany I, Justus-Liebig-Universitat, Senckenbergstr 17 , D-6300 Giessen, FRG

Other participants

Dr A Batschauer Institut fur Biologie II/Botanik, Schanzlestrasse 1, D-7800 Freiburg, FRG

Dr T Becker INRA Laboratoire de Biologie Cellulaire, Route de Saint-Cyr, F-78026, Versailles Cedex, FRANCE

Dr P Bonner Dept of Applied Biology, Lancashire Polytechnic, Preston, PR1 2TQ UK

Dr J Cherry University of Wisconsin-Madison, Department of Horticulture, 1575 Linden Drive, Madison WI 53706, USA

Dr K Dehesh Plant Gene Expression Center, 800 Buchanan Street, Albany, California 94710, USA

Dr N Douthwaite School of Biological Sciences, Queen Mary & Westfield College, University of London, Mile End Road, London E1 4NS, UK

Dr E Fernbach Institut fur Biologie II/Botanik, Schanzlestrasse 1, D-7800 Freiburg, FRG

Dr R Grimm Institut fur Biologie II/Botanik Schanzlestrasse 1, D-7800 Freiburg, FRG

Dr N Harberd John Innes Centre for Plant Science Research, Colney Lane, Norwich, Norfolk, NR4 7UJ, UK

Dr G Jenkins Dept of Biochemistry, University of Glasgow, Glasgow, G12 8QQ, UK

Dr K Kazan (Turkey) Dept of Agronomy & Soils, Washington State University, Pullman, WA 99164, USA

Dr D Kristie Department of Biology , Acadia University, Wolfville, Nova Scotia, CANADA, BOP 1X0

Dr B Langer Institut fur Biologie II/Botanik Schanzlestrasse 1, D-7800 Freiburg, FRG

Dr L Leurentop University of Antwerpen, Dept of Biology, Universiteitsplein 1, B-2610 Wilrijk, Antwerpen, BELGIUM

Dr P Lumsden School of Applied Biology, Lancashire Polytechnic, Preston PR1 2TQ, UK

Dr M Partis Biochemistry and Molecular Biology Dept, IHR Littlehampton, Worthing Road, Littlehampton, West Sussex, BN17 6LP, UK

Dr Y Pauncz Israel Inst of Technology, Dept of Biology, Technion City, Haifa 32000, ISRAEL

Dr J Peters Dept of Plant Physiological Res, Wageningen Ag University, Gen Foulkesweg 72/6703 BW, Wageningen, THE NETHERLANDS

Dr J Silverthorne 424 Sinsheimer Labs, University of California, Santa Cruz, California 95064, USA

Dr M Terry Dept of Biochem and Biophys, University of California, Davis California 95616 USA.

Prof G Towers Dept of Botany, University of British Columbia, 3529-6270 University Boulevard, Vancouver, BC, CANADA V6T 2B1

Dr N Urwin Department of Biochemistry & Botany, University of Glasgow, University Avenue, Glasgow, G12 8QQ, UK.

Dr O van Kooten ATO Agrotechnologie, Haagsteeg 6, Postbus 17, 6700 AA Wageningen, THE NETHERLANDS

Dr W VanDerWoude Plant Photobiology Lab, USDA-ARS, Bldg 046A, Ag Research Center, Beltsville, Maryland 20705, USA

Prof D Vince-Prue 2 Maple Court, Goring-on-Thames, READING, RG8 9BQ, UK

Prof M Wada Department of Biology, Tokyo Metropolitan University, Fukazawa 2-1-1, Setagaya-ku, Tokyo 158, JAPAN

Dr G Whitelam Department of Botany, University of Leicester, University Road, Leicester, LE1 7RH, UK

Section A.

PHYTOCHROME PROPERTIES

THE PROPERTIES AND BIOLOGICAL ACTION OF PHYTOCHROME: PROLOGUE

B. Thomas
Horticulture Research International,
Littlehampton,
West Sussex, BN17 6LP,
UK.

INTRODUCTION

Plants possess sophisticated mechanisms for detecting light quality and quantity and adapting their patterns of growth and development depending upon their environment. This process of photomorphogenesis requires photoreceptor pigments which absorb light and provide the initial biochemical signal to trigger the plant's photoresponse. The major photoreceptor in plants is phytochrome. This chromoprotein is found in all green plants including algae, mosses and ferns. It exists, characteristically, as two photoisomers called Pr and Pfr, each with a distinct absorbance spectrum. Typically, Pr has a major absorbance maximum (λmax) at about 660 nm and a secondary maximum at about 380 nm. Pfr on the other hand has a λmax at about 730 nm and a secondary peak at about 400 nm. When either form absorbs light a series of photochemical and protein conformational changes is initiated leading within milliseconds to the formation of the other photoisomer. For many responses red light at about 660 nm is the most effective part of the spectrum and the effect of red can be prevented by a subsequent far-red irradiation at about 730 nm. In such cases sequentially antagonistic actions of red and far-red can be seen over several cycles of irradiation. The effectiveness of Pfr-forming irradiation and reversal by Pr-forming irradiation leads inevitably to the conclusion that Pfr is the biologically active isomer and Pr an inactive form of phytochrome. It is worth bearing in mind that short alternating treatments with narrow waveband light, as used in these experiments is not necessarily comparable with natural irradiation. Nevertheless, the distinction between the biological activity of Pr and Pfr has provided the rationale underlying four decades of phytochrome research. The mechanisms by which phytochrome regulates changes in plant development still remain elusive. Progress towards elucidating these processes is the subject of this book. The aim of this introductory chapter is to provide a contextual outline within which the remaining

NATO ASI Series, Vol. H 50
Phytochrome Properties and Biological Action
Edited by B. Thomas and C. B. Johnson

chapters, which deal in detail with recent advances in highly active laboratories, can be placed. The work falls in three separate areas, the properties of phytochrome, the cellular and molecular physiology of its action and the mechanisms by which phytochrome regulates gene expression.

PHYTOCHROME PROPERTIES

Phytochrome is a large soluble chromoprotein, containing for example 1127 amino acids in *Avena*. It normally occurs as a dimer, the molecular weight of the monomer being in the range 120-127 kDa (Thomas, 1991). Comparison of nucleic acid and deduced amino acid sequences and the reactivity of monoclonal antibodies indicates a significant degree of sequence and structural conservation in higher plant phytochromes.

Phytochrome genes. The cloning of *Avena* phytochrome cDNA by Peter Quail's group (Hershey *et al.*, 1984) stands out as the most prominent landmark in the landscape of recent phytochrome research. Besides being a technical achievement in that phytochrome was one of the earliest large plant genes to be fully sequenced it also provided a tranche of absolute biochemical information on the molecule. The amino acid composition and sequence was established unequivocally along with the size of the primary gene product. The cloned genes provided probes to clarify the transcriptional regulation of phytochrome synthesis by light and for the isolation of genes from other species. Analysis of sequences of phytochrome genes from a range of species in turn is beginning to identify conserved regions which are potentially important for phytochrome action. A second quantum leap in our understanding of the phytochrome system arises from recent work, also from Quail's laboratory. This reveals that phytochrome is coded for by a complex family of genes. In *Arabidopsis* three distinct cDNA cones with less than 70% sequence similarity were isolated and two further unassigned bands were observed on Southern Blots (Sharrock and Quail, 1989). The structure and organisation of phytochrome genes is discussed in Quail's chapter (page 13) and further information on phytochrome sequences in lower plants is dealt with in the chapter by Rüdiger (page 57) in this volume.

Types of phytochrome. Almost everything we know about the properties of phytochrome molecules derives from studies on phytochrome extracted from dark-grown tissues.

However, phytochrome extracted from light-grown plants differs in several respects from that in etiolated plants. Tokuhisa *et al.* (1985) purified "green" phytochrome from light-grown *Avena* and found a photoreversible species with an altered difference spectrum, molecular weight, peptide map and immunoreactivity from "dark" phytochrome. Similar results were obtained by Shimazaki and Pratt (1985) in both green and norflurazon-bleached *Avena* and by Abe *et al.* (1985) with *Pisum*. The phytochrome present in etiolated plants at high levels is destroyed rapidly on transfer to light leaving a residual pool which is more resistant to destruction as Pfr. These two pools of differing stability correlate with two immunochemically distinguishable pools sometimes called Types I and II phytochrome. Types I and II phytochrome have been detected in germinating embryos of *Avena* (Thomas and Hilton, 1985; Tokuhisa and Quail, 1987) and in embryo tissue from *Pisum* (Konomi *et al.*, 1987). That immunochemically distinct pools of phytochrome might represent different gene products is consistent with the family of genes characterized by Sharrock and Quail (1989). An underlying theme to much ongoing research is: how can the properties and action of phytochrome in whole plants be linked assigned to different gene products. The inadequacy of the current nomenclature for phytochrome proteins is highlighted in the chapter by Pratt (page 39) who can distinguish at least three distinct phytochrome proteins in *Avena* and further addressed in the round table report by Vince-Prue (page 313) which proposes guidelines for the use of phytochrome terminology.

Chromophore biosynthesis and holoprotein assembly. Each phytochrome monomer contains a single, linear tetrapyrrole chromophore linked to a cysteine residue in a highly conserved region of the apoprotein *via* a thioether linkage. The chromophore in Pr is thought to be protonated and in an extended conformation (Rüdiger, 1986). Spectral differences between Pr and Pfr arise, in part, from a *cis-trans* isomerization at the C15 bridge between rings C and D with the Pr chromophore being the *cis* isomer. However, this is not sufficient in itself to cause the wavelength shift to 730 nm; chromophore-protein interactions are also required. Loss of the N-terminal 6-10 kDa of the protein results in a shift of λmax for Pfr from 730 to 720 nm. This N-terminal domain appears to interact with the chromophore and take part in a conformational change on phototransformation.

In an attempt to clarify the mechanism of chromophore attachment and the nature of its interaction with the protein during phototransformation expression systems for producing apoprotein for *in vitro* biogenesis studies are being used. In his chapter (page 71) Furuya describes the expression of pea apoprotein in *E. Coli* and Yeast and the autocatalytic attachment of the chromophore as originally proposed by Lagarias and Lagarias, 1989). As well as chromophore attachment, phytochrome may undergo other post-translational modifications as outlined in the round table report by Whitelam (page 303).

Phytochrome destruction. Phytochrome is synthesised as Pr which, in darkness, normally accumulates. Pfr, on the other hand has a much shorter half life and following its formation when plants are transferred from dark to light phytochrome is degraded rapidly within the cell. Pfr destruction in dark grown seedlings which receive a brief irradiation is rapid with a half-life of 1-2 hours. Destruction is a complex process involving biochemical modification and subcellular relocation. Immunostained phytochrome becomes redistributed following a red light treatment. From being cytoplasmically dispersed phytochrome rapidly becomes aggregated in discrete areas of the cell. This is followed by loss of immunostaining as phytochrome is broken down. The biochemical mechanism involved in phytochrome destruction apparently involves covalent binding of ubiquitin to the molecule. The process is discussed more fully in the chapters by Schäfer (page 181) and Vierstra (page 113) and the round table report by Thomas (page 307).

Phytochrome activation. The mechanisms by which phytochrome becomes activated by light and by which this information is transduced within the cell to bring about the photomorphogenetic response are poorly understood. Responses are diverse, ranging from short term changes in membrane properties to the activation or repression of specific genes. The initial step in the chain is a modification of the properties of phytochrome when it is converted from Pr to Pfr. Although several physicochemical changes have been described and analyzed (see Thomas, 1991) definitive information awaits successful crystallisation of phytochrome. The factors which have thwarted attempts at crystallisation include microheterogeneity of the protein due to the co-expression of members of multigene families and the formation of non-homogeneous mixtures of Pr and Pfr in light. Another approach is to use structural predictions based on sequence analysis as described in Song's chapter (page 85). By using a correlation function and secondary structure predictions sequences can be divided into eleven structural domains which, in some cases, correspond with known functional regions on the molecule. One of the domains is the N-terminal region of phytochrome which is apparently the only domain to undergo significant changes in secondary structure during Pr-Pfr photoisomerisation.

PHYTOCHROME AND PLANT DEVELOPMENT

Genetic approaches. The problem of analyzing a system in which a family of similar photoreceptors may be operating simultaneously demands approaches in which the contribution from specific genes can be isolated. In practice this means the use of defined genetic mutants which are valuable tools in unravelling complex developmental mechanisms. A major advance in the last few years has been the identification of plant mutants with altered photomorphogenetic responses. In particular, Koornneef, Kendrick and their colleagues have identified a range of mutants in *Arabidopsis*, Cucumber and tomato characterised by a partially etiolated appearance e.g. long hypocotyls, small leaves and pale green leaves, when grown in light (Adamse *et al.*, 1988). Some of these have reduced or non-functional phytochrome, while others are transduction or response mutants. Some recent progress in the analysis of a range of mutants, including both insensitive and hypersensitive to light is described in the chapter by Kendrick (page 237).

Transgenic plants. A complementary approach to the use of genetic lines with defined alterations in photoresponse is to introduce phytochrome genes into plants by transformation and evaluate their effect. Initially three groups independently transformed plants with phytochrome genes under the control of a strong constitutive promoter. Rick Vierstra's group (Madison Wi, USA) expressed oat phytochrome in tobacco (Keller *et al.*, 1989). Kay *et al.* (1989) also used tobacco but with the rice gene and Boylan and Quail (1989) expressed the oat gene in tomato. Both Vierstra and Quail found that the transgenic plants showed marked phenotypic changes, such as dwarfed size and dark green colour. The original transformants from Kay did not show this phenotype, despite over-expressing both mRNA and apoprotein, although small changes in growth could be demonstrated with appropriate irradiation protocols. Strong phenotypic effects have, however been observed when the same genes were used to transform a different tobacco cultivar. Interpretation of results obtained using transgenic, over-expressing plants, therefore needs to be made with extreme caution and include an assessment of effects which may be species or even cultivar-specific.

Work with transgenic plants has involved using Type I phytochrome from a monocot and overexpressing it in light grown dicots. The appearance of spectrally active phytochrome indicates a well preserved mechanism for chromophore attachment in species which are not closely related. This might be in part explained by the ability of the phytochrome molecule to catalyse chromophore attachment to itself. However the availability of precursor requires correct intracellular localisation or targeting of the apoprotein. It is not yet known whether other post-translational modifications occur correctly in the transgenic tissues although presumably any essential for phytochrome function must take place.

Experiments with transgenic plants have already demonstrated that monocot phytochrome is active in dicot plants, indicating that functional parts of the molecule have been retained during recent evolution. This offers a promising experimental approach to identifying which phytochrome sequence elements are essential for action. Such experiments are still in their infancy but initial progress is reported in the chapter by Vierstra's group (page 113).

Molecular physiology. A promising approach for understanding phytochrome action is transgenic plants with altered phytochrome content, which provide excellent models for physiological studies. Superficially, the observation that over-expressing plants produce phenotypes consistent with an increased phytochrome "input" would seem to support the belief that the decisive component is the absolute level of Pfr rather than the ratio between Pr and Pfr levels. However, the effect of spectral distribution and irradiance and response needs to be analyzed in detail in such plants before this is confirmed. The prospects for such work are highlighted in the chapter by Smith's group (page 217). Taken with the pattern of response observed with mutants it is clear that the quantity of phytochrome in plant tissues has a significant effect on plant phenotype. The reduced response to light when (Type I) phytochrome levels are reduced in mutants such as the *aurea* mutant of tomato compared an exaggerated light-grown phenotype associated with elevated levels of phytochrome in transgenic plants argues strongly that photomorphogenesis is a positive consequence of Pfr formation, rather than a negative reaction to the removal of Pr. Furthermore, aberrant phenotypes are seen in plants grown in light. This argues strongly that the function of Type I phytochrome is not only to act as the photoreceptor for de-etiolation and greening, but also to regulate development in light-grown plants. This is also supported by work with genetically unmodified plants. Physiological experiments show High Irradiance Responses, usually assumed to be attributable to Type I phytochrome, can also be found in light-grown plants (Carr-Smith *et al.*, 1989). By using quantitative immunoassays these responses can be correlated with changes in the quantity of Type I phytochrome apoprotein as described in the chapter by Johnson (page 273). A wider understanding of the roles of different phytochromes can be gained by extending investigations of its properties and action to lower plants. Wagner, in his chapter (page 249), reports on possible interactions between phytochrome and cytoskeletal comonents in *Mougeotia.*

It is tempting to see photoregulated systems in terms of a photoreceptor and a response with a direct linear causality between the activation of the former and the expression of the latter. In practice, phytochrome provides one input, albeit sometimes decisive, into a complex regulatory network. Against this background understanding the role of individual phytochrome genes requires both that the quantitative relationship be established between the photoreceptor and response and also that the interaction between phytochrome and other regulatory factors be understood. The chapter by Mohr (page 259) addresses this problem.

PHYTOCHROME AND GENE EXPRESSION

The expression of a number of genes has been shown to be up or down regulated by phytochrome at the transcriptional or post-transcriptional level (Tobin and Silverthorne, 1985). The convenience of using defined light treatments to regulate these genes selectively, makes them excellent models for the regulation of higher plant gene expression. Genes are sometimes classified according to the amount of light required for their activation. Very low fluence (VLF) responses are so sensitive to Pfr that they are induced by far-red alone and therefore do not show red/far red reversibility. For the majority of genes, reversibility can be demonstrated, placing them in the low fluence (LF) group. Some examples of VLF genes are Cab and phy, while ferredoxin and rbcs fall into the LF category (Nagy *et al.*, 1988). Different genes also show complexity in the time courses of their response to light. Where mRNA abundance for a number of genes has been measured following a single light pulse a complex pattern has been obtained, with almost every gene having a distinct pattern of response (Kaufmann *et al.*, 1985, 1986). Furthermore, complexity is also exhibited in variation in the time course of escape from reversibility by far red in LF genes, by dependence on the developmental state of the tissues or interaction with other photoreceptors (Jenkins, 1988). One additional complication is that photoregulated genes may also be co-regulated by endogenous circadian rhythms.

Phytochrome autoregulation. As luck would have it, one of the best characterized phytochrome-regulated genes is phytochrome itself. That light reduced the ability of corn seedlings to synthesise phytochrome was first observed by Duke *et al.* (1977). Subsequent *in vitro* translation and cDNA-RNA hybridization experiments confirmed that the response was due at least in part to reduction in phytochrome mRNA abundance in both *Avena* and pea seedlings. The effect on mRNA abundance showed a lag of about 15 minutes but assay of transcriptional activity by "run-on" experiments indicated repression of *phy* transcription within 5 minutes (Lissemore and Quail, 1988). The response was unimpaired by treatments which inhibited protein synthesis by over 90%, indicating that all the components of the transduction chain are present in the cell prior to irradiation. Red/far-red photoreversibility can be demonstrated if appropriate protocols are used, showing that phytochrome is the photoreceptor for its own regulation (Colbert *et al.*, 1985). In germinating *Avena* seeds and in pea seedlings photoregulation of phytochrome synthesis is by a stable Pfr population (Hilton and Thomas, 1987; Thomas *et al.*, 1989). Thus Type II phytochrome may be capable of regulating Type I phytochrome synthesis. Down regulation of the phytochrome genes by light is not a universal feature in plants. Tomato shows little effect of light on phytochrome mRNA levels while several other species have a partial response to light (Colbert, 1988; Sharrock *et al.*, 1988).

Other photoregulated genes. Several groups are actively analyzing the *cis*-acting elements and *trans*-acting factors which confer light-responsiveness to phytochrome-regulated genes.

These studies have identified a number of upstream regions that mediate light-responsive transcription and discrete sequence elements within them that bind nuclear proteins. Chapters in this book by Quail, Gilmartin, Kay, Tobin, Adam and Cashmore all deal in detail with the analysis of *cis*-acting elements and the identification and cloning of putative regulatory DNA-binding proteins. The round table report by Jenkins (page 321) pulls together the current activity and nomenclature in this area.

While attention inevitably focuses on transcriptional control the possibility of post-transcriptional control should not be ignored and this has been explored by Thompson's group, using ferredoxin as a model and is described in his chapter (page 201). One feature of certain photoregulated genes is that they are, in addition, under the control of a circadian rhythm. In the chapter by Adam (page 191), the analysis of promoter elements in *Cab-1* clearly shows that the light and rhythmic components of the control mechanism involve distinct regulatory DNA sequences.

The characterisation of *cis*-acting elements and *trans*-acting factors represents major progress towards understanding the regulatory mechanisms involved in the regulation of gene transcription by light. The number of regulatory sequences and DNA binding proteins identified so far is relatively small. Even with these, however, the potential interactions possible when sequence elements and protein factors are in combination are numerous. Many of these interactions are probably subtle and this may be a major factor in establishing the complexity of expression patterns in response to light. A further factor is almost certainly the cellular background within which the regulatory elements exist. One can ask whether all cells of a certain tissue type respond in the same way to light, or whether the same gene responds in the same way to light stimuli when in different cellular backgrounds. These questions have been asked by Schäfer's group and his chapter (page 181) further underlines the subtlety of the plant's response to light.

CONCLUDING REMARKS

This book addresses the question of how phytochrome regulates plant development. The content of the book largely reflects the distribution of effort on, and our current understanding of, various components of the overall mechanism. What is almost completely absent is high quality information on the early stages of the signal completely absent is high quality information on the early stages of the signal transduction chain between the phytochrome molecule and the target response. Phytochrome is not the only plant photoreceptor and Briggs, in his chapter (page 289) provides an overview of research on another, namely the blue-light photoreceptor. The contrast with knowlege of the phytochrome system is striking. There are now good indications as to an early step in the transduction chain for blue-light regulation, where protein phosphorylation can clearly be demonstrated. On the other hand the molecular identity of the blue-light photoreceptor still remains cryptic. Maybe progress in such parallel systems will provide the inspiration for unearthing the mechanisms involved in phytochrome signal transduction.

REFERENCES

Abe H, Yamamoto KT, Nagatani A, Furuya M (1985) Characterization of green tissue-specific phytochrome isolated immunochemically from pea seedlings. Plant Cell Physiol. 26:1387-1399

Adamse P, Kendrick RE, Koorneef M (1988) Photomorphogenetic mutants of higher plants. Photochem. Photobiol. 48:833-841

Boylan MT, Quail PH (1989) Oat phytochrome is biologically active in transgenic tomatoes. The Plant Cell 1:765-773

Carr-Smith H, Thomas B, Johnson CB (1989) An action spectrum for the effects of continuous light on flowering in wheat. Planta 179:428-432

Colbert JT (1988) Molecular biology of phytochrome. Plant Cell Environ. 11:305-318

Colbert JT, Hershey HP, Quail PH (1985) Phytochrome regulation of phytochrome mRNA abundance. Plant Mol. Biol. 5:91-101.

Duke SO, Naylor AW, Wickliff JL (1977) Phytochrome control of longitudinal growth and phytochrome synthesis in maize seedlings. Physiol. Plant 40:59-68

Hershey HP, Colbert JT, Lissemore JL, Barker RF, Quail PH (1984) Molecular cloning of cDNA for *Avena* phytochrome. Proc. Nat. Acad. Sci. U.S.A. 81:2332-2336

Hilton JR, Thomas B (1985) A comparison of seed and seedling phytochrome in *Avena sativa* using monoclonal antibodies. J. Exp. Bot. 36:1937-1946

Hilton JR, Thomas B (1987) Photoregulation of phytochrome synthesis in germinating embryos of *Avena sativa* L. J. Expt Bot 38:1704-1712

Jenkins GI (1988) Photoregulation of gene expression in plants. Photochem. Photobiol. 48:821-832

Kaufman L, Briggs WR, Thompson W (1985) Phytochrome control of specific mRNA levels in pea buds: the presence of both very low and low fluence responses. Plant Physiol. 78:388-393

Kaufman L, Roberts LL, Briggs WR, Thompson W (1986) Phytochrome control of specific mRNA levels in developing pea buds: kinetics of accumulation, reciprocity and and escape kinetics of the low fluence response. Plant Physiol. 81:1033-1038

Kay SA, Nagatani A, Keith B, Deak M, Furuya M, Chua N-H (1989) Rice phytochrome is biologically active in transgenic tobacco. The Plant Cell 1:775-782

Keller JM, Shanklin J, Vierstra RD, Hershey HP (1989) expression of a fuctional monocotyledonous phytochrome in transgenic tobacco. EMBO J. 8:1005-1012

Konomi K, Abe H, Furuya M (1987) Changes in the content of phytochrome I and II apoproteins in embryonic axes of pea seeds during imbibition. Plant Cell Physiol. 28:1443-1451

Lagarias JC, Lagarias DM (1989) Self-assembly of synthetic phytochrome holoprotein *in vitro*. Proc. Nat. Acad. Sci. USA 86:5778-5780

Lissemore JL, Quail PH (1988) Rapid transcriptional regulation by phytochrome of the genes for phytochrome and chlorophyll a/b-binding protein in *Avena sativa*. Mol. Cell. Biol. 8:4840-4850

Nagy F, Kay SA, Chua N-H (1988) Gene regulation by phytochrome. Trends Genet. 4:37-42

Rüdiger W (1986) The Chromophore. In: Kendrick RE, Kronenberg GHM (eds) Photomorphogenesis in Plants. Martinus Nihoff Publ, Dordrecht, The Netherlands. p17

Sharrock RA, Parks BM, Koornneef M, Quail PH (1988) Molecular analysis of the phytochrome deficiency in an *aurea* mutant of tomato. Molec. Gen. Genet. 213:9-14

Sharrock RA, Quail PH (1989) Novel phytochrome sequences in *Arabidopsis thaliana*: structure, evolution and differential expression of a plant regulatory photoreceptor family Gene. Develop. 3:1745-1757

Shimazaki Y, Pratt LH (1985) Immunochemical detection with rabbit polyclonal and mouse monoclonal antibodies of different pools of phytochrome from etiolated and green *Avena* shoots. Planta 164:333-344

Thomas B (1991) Phytochrome properties and biological action. In: V Neuhoff (ed) Recent advances in cell to cell signalling. Springer-Verlag in press

Thomas B, Penn SE, Jordan BR (1989) Factors affecting phytochrome transcripts and apoprotein synthesis in germinating embryos of *Avena sativa* L. J. Expt. Bot. 40:1299-1304

Tobin EM, Silverthorne J (1985) Light regulation of gene expression in higher plants. Annu. Rev. Plant Phys. 36:569-593

Tokuhisa JG, Daniels SM, Quail PH (1985) Phytochrome in green tissue: spectral and immunochemical evidence for two distinct molecular species of phytochrome in light-grown *Avena sativa*. Planta 64:321-332

Tokuhisa JG, Quail PH (1987) The levels of two distinct species of phytochrome are regulated differently during germination in *Avena sativa* L. Planta 172:371-377

Phy-GENE STRUCTURE, EVOLUTION, AND EXPRESSION

P.H. Quail, H.P. Hershey[1], K.B. Idler[2], R.A. Sharrock[3], A.H. Christensen[4],
B.M. Parks, D. Somers, J. Tepperman, W.B. Bruce[5] and K. Dehesh
U.C. Berkeley/USDA Plant Gene Expression Center,
800 Buchanan Street,
Albany,
CA, 94710,
USA.

INTRODUCTION

The fundamental notion that phytochrome controls plant development through differential regulation of gene expression (Mohr, 1966) is now well supported by direct experimental evidence (Benfey and Chua, 1989; Gilmartin *et al.*, 1990; Kuhlemeier *et al.*, 1987; Nagy *et al.*, 1988; Tobin and Silverthorne, 1985). However, the molecular mechanism by which the photoreceptor transduces its regulatory signal to genes under its control remains unknown. For some time we have approached this question by simultaneously studying the properties of the photoreceptor molecule and the negative autoregulation of phytochrome (phy) genes as a paradigm of phytochrome-regulated gene expression (Colbert, 1988; Lissemore and Quail, 1988; Quail *et al.*, 1987b, 1990). Recent molecular-genetic studies have revealed that phytochrome is encoded by a small family of divergent and differentially regulated genes (Dehesh *et al.*, 1990b; Sharrock and Quail,

---

[1]Present address; E.I. du Pont Nemours and Co., Central Research & Development Dept., Experimental Station, Wilmington, DE 19898, USA.

[2]Present address; Abbott Laboratories, Dept. 93-D, AP-9A/3, Abbott Park, IL 60064, USA.

[3]Present address; Dept. of Biology, Montana State University, Bozeman, MT59717, USA.

[4]Present address; DepT. of Biology, George Mason University, 4400 University Drive, Fairfax, VA 22030-4444, USA.

[5]Present address; Plant Molecular Biology Center, Montgomery Hall 325, Northern Illinois University, De Kalb, IL 60115-2861, USA.

NATO ASI Series, Vol. H 50
Phytochrome Properties and Biological Action
Edited by B. Thomas and C. B. Johnson

1989); have shown that overexpression of the photoreceptor in heterologous, transgenic plants provides a system for directed mutational analysis of functional regions of the polypeptide (Boylan and Quail, 1989; Kay *et al.*, 1989c; Keller *et al.*, 1989); and have begun to provide insight into the *cis*-regulatory elements and *trans*-acting factors involved in phy gene transcription (Bruce *et al.*, 1989, 1990; Dehesh *et al.*, 1990a).

PHYTOCHROME STRUCTURE AND EVOLUTION

Early physiological (Briggs and Chon, 1966; Hillman, 1967) and later spectroscopic (Brockman and Schäfer, 1982; Jabben and Deitzer, 1978; Jabben and Holmes, 1983) studies suggested the presence of more than one pool of phytochrome in plant tissue. However, when multiple phytochrome cDNAs were first isolated from oats (Hershey *et al.*, 1984, 1985), the encoded polypeptides were found to be 98% identical, thus providing no evidence of significant heterogeneity. Indeed, the high level of nucleotide sequence identity, even in the introns of the two oat phy genes that have thus far been sequenced (Fig. 1), was consistent with the notion that the multiple phy sequences in this species represented recent duplication events. Moreover, because oats is hexaploid it was uncertain whether or not these multiple genes simply reflected the multiple genome copies in this species. The first direct evidence for structurally distinct phytochrome molecules, designated Type 1 or etiolated-tissue phytochrome, and Type 2 or green-tissue phytochrome, respectively, came from immunochemical and spectroscopic studies with plant extracts (Cordonnier *et al.*, 1986a; Pratt and Cordonnier, 1987; Shimazaki and Pratt, 1985, 1986; Shimazaki *et al.*, 1983; Tokuhisa and Quail, 1983, 1987, 1989; Tokuhisa *et al.*, 1985). However, the data did not definitively establish whether or not these molecules were encoded by different structural genes.

Evidence for the presence of multiple, divergent phy genes in the same genome came initially from studies with *Arabidopsis* (Sharrock and Quail, 1989). Genomic Southern hybridization analysis indicates that *Arabidopsis* contains four (and possibly five) phytochrome genes. cDNA clones for three of these genes, designated *phyA*, *phyB*, and *phyC*, have been sequenced, and the genes mapped to chromosomes 1, 2, and 5, respectively (E. Meyerowitz, pers. comm.). The predicted amino acid sequences of the three polypeptides are equally divergent from each other with only 50% identity for each pairwise comparison. In consequence, the *Arabidopsis phyA* sequence is more closely related (65-80% identity) to all other published sequences from monocots and dicots (Christensen and Quail, 1989; Hershey *et al.*, 1985; Kay *et al.*, 1989a,b; Sato, 1988; Sharrock *et al.*, 1986) than it is to the *phyB* and *phyC* sequences in the same genome in which it resides (Sharrock and Quail, 1989). We have therefore designated these other published sequences as members of the *phyA* subfamily.

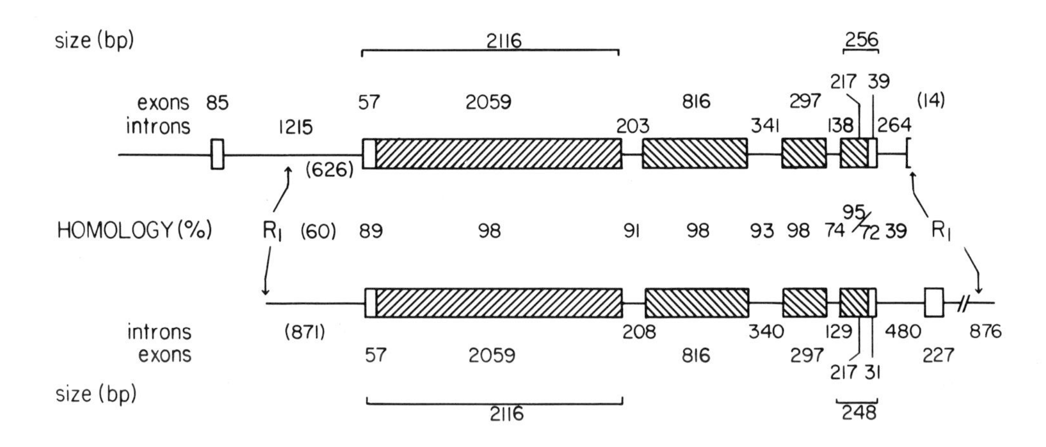

Figure 1. Schematic representation of oat *phyA3* (top) and *phyA4* (bottom) genes. Boxes = exons; lines = introns and flanking DNA; shaded = protein coding sequence; open = untranslated sequence. Sizes of exons and introns are indicated in bp. Homology (%) = percent identical nucleotides between *phyA3* and *phyA4* in the regions of the genes indicated. R1 = *Eco* RI sites. *phyA3* is truncated at the R1 site at the 3′ end, and *phyA4* at the R1 site at the 5′ end, these two sites being the ends of the respective genomic clones. The schematic of *phyA3* is modified from Hershey *et al.* (1987). The *phyA4* sequence is from K. Idler, H. Hershey, and P. Quail (unpublished).

Recently, we have obtained evidence that rice contains single-copy homologs of the *Arabidopsis* genes (K. Dehesh, unpublished). We have determined the amino acid sequence encoded by rice *phyB* and found that it is more closely related to *Arabidopsis phyB* (73% identity) than to existing *phyA* and *phyC* sequences (50% identity) (Dehesh *et al.*, 1990b). Together with the *Arabidopsis* data, these results indicate that the three-way divergence of the *phyA*, *B*, and *C* subfamilies was an ancient evolutionary event preceding the divergence of monocots and dicots, and predict that all angiosperms possess the three subfamilies. On the other hand, since the two *phyB* sequences are more closely related to each other (73% amino acid identity) than are monocot and dicot *phyA* sequences (65% identity), the *phyA* sequences may be evolving more rapidly than the *phyB* sequences.

All available evidence strongly indicates that *phyA* genes encode the abundant Type 1 phytochrome that predominates in etiolated tissue. This evidence includes the observation that *phyA* transcripts likewise predominate in etiolated tissue (Dehesh *et al.*, 1990b; Sharrock and Quail, 1989). More compelling, however, are microsequencing data for phytochrome purified from etiolated oat (Grimm *et al.*, 1988; Jones and Quail, 1989; Lagarias and Rapoport, 1980) which indicate identity with the *phyA* nucleotide-predicted amino acid sequence in regions of divergence from *phyB* and *phyC* sequences. The low abundance and constitutive levels of *phyB* and *phyC* transcripts (Dehesh *et al.*, 1990b; Sharrock and Quail, 1989) are consistent with the possibility that they encode Type 2 phytochrome polypeptides, which are also apparently expressed at low, constitutive levels (Tokuhisa and Quail, 1987). Comparison of recently reported preliminary micro-sequencing data for purified Type 2 pea phytochrome (Abe *et al.*, 1985; Furuya, 1989) with *Arabidopsis* (Sharrock and Quail, 1989) and rice (Dehesh *et al.*, 1990b) sequences indicates two regions that are more closely related to *phyB* than to *phyA* or *phyC* (data not shown). However, definitive data on this issue are still lacking because of the limited extent of the sequenced regions available for comparison.

All full-length phytochrome polypeptide sequences currently available have been compiled and aligned to maximize sequence identity in Figure 2. (Note that the numbering here starts with the initiator methionine in the *phyB* sequences. Thus, *phyA* and *phyC* sequences begin at position 42). The distribution of invariant amino acids along the nine aligned sequences is shown graphically in Figure 3 together with the hydropathy profiles of *Arabidopsis phyA*, *B*, and *C* for comparison. Despite the apparent lack of extensive regions of absolute amino acid sequence identity, the general similarity of the hydropathy profiles indicates that the overall three-dimensional structure of the molecule has been conserved. This observation is reflected at the sequence level by frequent conservative amino acid substitutions (Fig. 2). The most dramatic departures from this pattern are the NH_2- and COOH-terminal extensions of the *phyB* polypeptides which overall are 40-60 amino acids longer than *phyA* and *C*. It will be of interest to determine whether these extensions have a role in potential functional and/or locational differences between these different phytochromes.

Table 1 Amino Acid Residues Conserved Among Sequenced Phytochromes

Residue	No. Conserved Residues in Ara B	Total Residues	Percent Conserved Residues*
A	25	88	28
R	12	65	19
N	9	35	26
D	20	62	32
C	10	25	40
Q	14	57	25
E	20	80	25
G	31	90	34
H	9	22	41
I	15	72	21
L	44	111	40
K	9	55	16
M	14	42	33
F	22	47	47
P	19	51	37
S	12	100	12
T	7	47	15
W	7	11	64
Y	12	26	46
V	19	86	22
Total	330	1172	28
A+G	56	178	31
S+T	19	147	13
D+E	40	142	28
D+E+N+Q	63	234	27
H+K+R	30	142	21
D+E+H+K+R	70	284	25
I+L+M+V	92	311	30
F+W+Y	41	84	49
I+L+M+V+F+W+Y	133	395	34

*Percent residues conserved is the number of each amino acid identical in all phytochrome sequences expressed as a percentage of the total number of that residue in *Arabidopsis phyB (Ara B)*, the longest phytochrome polypeptide.

In general, the extent of sequence identity is greatest in the central core region of the NH_2-terminal domain surrounding the chromophore-attachment site (Fig. 3). This observation may reflect the need for stricter structural constraints on polypeptide segments involved in protein-chromophore interactions. The single, longest, conserved stretch of residues is in this region and has the sequence PFPLRYACEFL (positions 432 to 442) in a relatively hydrophobic segment of the protein (Figs. 2,3). Figure 3 shows that in fact this coincidence of peaks of amino acid sequence conservation with hydrophobic segments of the polypeptide is a general feature of the comparison. This coincidence is quantified in Table 1 which shows that the hydrophobic residues in the phytochrome polypeptide are conserved to a significantly greater extent (34%) than the hydrophilic residues (25%). Overall, this general analysis shows that phytochrome fits the pattern observed for other globular proteins (Bajaj and Blundell, 1984) whereby: (a) invariant functionally important amino acids may comprise only a small percentage of the total; and (b) residues constituting the hydrophobic core are relatively more conserved than hydrophilic external residues.

During the period since the first *phyA* sequences were determined, there have been a number of suggested functions for various amino acid sequence motifs within the polypeptide, based either on similarities to sequences of established or proposed function in other proteins or on various biochemical analyses of phytochrome. With the availability of an expanded group of *phyA* sequences and the recognition of the divergent *phyB* and *phyC* subfamilies, it is of interest to review to what extent these proposed functional motifs are conserved across all these phy sequences.

Figure 2 (opposite and following pages). Aligned amino acid sequences of all currently available phy polypeptides. The letters A, B, C after each plant name indicates the *phyA*, *phyB*, and *phyC* sequence, respectively, from that plant. The numbers after the letter A indicate individual members of the *phyA* family in maize and oats where multiple copies of *phyA* genes have been identified. Zuc = zucchini; Ara = *Arabidopsis*. Consensus = residues invariant in all sequences represented by a letter; positions where one or more residues are different among sequences represented by a dash. Some of the names given to clones by the original authors have been changed here as indicated below in an attempt to introduce uniform nomenclature. Published sequences: rice *phyA* (originally *phy18*; Kay *et al.*, 1989a,b); maize *phyA1* (Christensen and Quail, 1989); oat *phyA3* (originally Type 3 or *phy3*; Hershey *et al.*, 1985, 1987); pea *phyA* (originally *phy*; Sato, 1988); zuc *phyA* (originally pFMD1; Lissemore *et al.*, 1987; Sharrock *et al.*, 1986); rice *phyB* (Dehesh *et al.*, 1990a); Ara *phyA*, *phyB*, *phyC* (Sharrock and Quail, 1989).

```
           1                                                      50
   Rice A  .......... .......... .......... .......... .MSSSRPTQC
 Maize A1  .......... .......... .......... .......... .MSSSRPAHS
   Oat A3  .......... .......... .......... .......... .MSSSRPA..
    Pea A  .......... .......... .......... .......... .MSTTRPS..
    Zuc A  .......... .......... .......... .......... .MSTSRPS..
    Ara A  .......... .......... .......... .......... .MSGSRPT..
   Rice B  MASGSRATPT RSPSSARPAA PRHQHHHSQS SGGSTSRAGG GGGGGGGGGG
    Ara B  MVSGV..... .GGSGGGRGG GRGGEEEPSS SHTPNNRRGG EQAQSSGT..
    Ara C  .......... .......... .......... .......... .MSSN.....
Consensus  ---------- ---------- ---------- ---------- ----------

           51                                                    100
   Rice A  SSSSSRTRQS SRARILAQTT LDAELNAEYE ...EYGDSFD YSKLVEAQRT
 Maize A1  SSSSSRTRQS SRARILAQTT LDAELNAEYE ...ESGDSFD YSKLVEAQRS
   Oat A3  SSSSSRNRQS SQARVLAQTT LDAELNAEYE ...ESGDSFD YSKLVEAQRD
    Pea A  QSSNNSGRSR NSARIIAQTT VDAKLHATFE ...ESGSSFD YSSSVRVSGS
    Zuc A  QSSSNSGRSR HSTRIIAQTS VDANVQADFE ...ESGNSFD YSSSVRVTSD
    Ara A  QSSEGSRRSR HSARIIAQTT VDAKLHADFE ...ESGSSFD YSTSVRVTGP
   Rice B  ....GAAAAE SVSKAVAQYT LDARLHAVFE QSGASGRSFD YTQSLRASPT
    Ara B  KSLRPRSNTE SMSKAIQQYT VDARLHAVFE QSGESGKSFD YSQSLKTTTY
    Ara C  TSRSCSTRSR QNSRVSSQVL VDAKLHGNFE ...ESERLFD YSASINLNMP
Consensus  ---------- -------Q-- -DA------E --------FD Y---------

           101                                                   150
   Rice A  TGPEQQARSE KVI.AYLHHI QRAKLIQPFG CLLAL.DEKT FNVIALSENA
 Maize A1  TPPEQQGRSG KVI.AYLQHI QRGKLIQPFG CLLAL.DEKS FRVIAFSENA
   Oat A3  GPPVQQGRSE KVI.AYLQHI QKGKLIQTFG CLLAL.DEKS FNVIAFSENA
    Pea A  VDGDQQPRSN KVTTAYLNHI QRGKQIQPFG CLLAL.DEKT CKVVAYSENA
    Zuc A  VSGDQQPRSD KVTTAYLHHI QKGKLIQPFG CLLAL.DDKT FKVIAYSENA
    Ara A  VVENQPPRSD KVTTTYLHHI QKGKLIQPFG CLLAL.DEKT FKVIAYSENA
   Rice B  PSS....EQQ ..IAAYLSRI QRGGHIQPFG CTLAVADDSS FRLLAYSENT
    Ara B  GSS..VPEQQ ..ITAYLSRI QRGGYIQPFG CMIAV.DESS FRIIGYSENA
    Ara C  SSSCEIPSSA ..VSTYLQKI QRGMLIQPFG CLIVV.DEKN LKVIAFSENT
Consensus  ---------- -----YL--I Q----IQ-FG C-----D--- ------SEN-

           151                                                   200
   Rice A  PEMLTTVS.H AVPSVDDPPK ...LRIGTNV RSLFTDPGTT ALQKALGFAD
 Maize A1  PEMLTTVS.H AVPNVDDPPK ...LGIGTNV RSLFTDPGAT ALQKALGFAD
   Oat A3  PEMLTTVS.H AVPSVDDPPR ...LGIGTNV RSLFSDQGAT ALHKALGFAD
    Pea A  PEMLTMVS.H AVPSVGDHPA ...LGIGTDI RTVFTAPSAS ALQKALGFAE
    Zuc A  PEMLTMVS.H AVPSMGDYPV ...LGIGTDV RTIFTAPSAS ALLKALGFGE
    Ara A  SELLTMAS.H AVPSVGEHPV ...LGIGTDI RSLFTAPSAS ALQKALGFGD
   Rice B  ADLLDLSPHH SVPSLDSSAV PPPVSLGADA RLLFAPSSAV LLERAFAARE
    Ara B  REMLGIMP.Q SVPTLEKPEI ...LAMGTDV RSLFTSSSSI LLERAFVARE
    Ara C  QEMLGLIP.H TVPSMEQREA ...LTIGTDV KSLFLSPGCS ALEKAVDFGE
Consensus  ---L------ -VP------- ------G--- ---F------ -L--A-----

           201                                                   250
   Rice A  VSLLNPILVQ CKTSGKPFYA IVHRATGCLV VDFEPVKPTE FPATAAGALQ
 Maize A1  VSLLNPILVQ CKTSGKPFYA IVHRATGCLV VDFEPVKPTE FPATAAGALQ
   Oat A3  VSLLNPILVQ CKTSGKPFYA IVHRATGCLV VDFEPVKPTE FPATAAGALQ
    Pea A  VSLLNPILVH CKTSGKPFYA IIHRVTGSLI IDFEPVKPYE VPMTAAGALQ
    Zuc A  VTLLNPILVH CKTSGKPFYA IVHRVTGSLI IDFEPVKPYE GPVTAAGALQ
    Ara A  VSLLNPILVH CRTSAKPFYA IIHRVTGSII IDFEPVKPYE VPMTAAGALQ
   Rice B  ISLLNPLWIH SRVSSNPFYA ILHRIDVGVV IDLEPARTED PALSIAGAVQ
    Ara B  ITLLNPVWIH SKNTGKPFYA ILHRIDVGVV IDLEPARTED PALSIAGAVQ
    Ara C  ISILNPITLH CRSSSKPFYA ILHRIEEGLV IDLEPVSPDE VPVTAAGALR
Consensus  ---LNP---- ------PFYA I-HR------ -D-EP----- -----AGA--
```

```
             251                                                     300
  Rice A     SYKLAAKAIS KIQSLPGGSM EVLCNTVVKE LFDLTGYDRV MAYKFHEDDH
 Maize A1    SYKLAAKAIS KIQSLPGGSM EALCNTVVKE VFDLTGYDRV MAYKFHEDEH
   Oat A3    SYKLAAKAIS KIQSLPGGSM EVLCNTVVKE VFDLTGYDRV MAYKFHEDDH
   Pea A     SYKLAAKAIT RLQSLASGSM ERLCDTMVQE VFELTGYDRV MAYKFHEDDH
   Zuc A     SYKLAAKAIT RLQSLPSGSM ARLCDTMVQE VFELTGYDRV MAYKFHDDDH
   Ara A     SYKLAAKAIT RLQSLPSGSM ERLCDTMVQE VFELTGYDRV MAYKFHEDDH
  Rice B     SQKLVVRAIS RLQALPGGDV KLLCDTVVEH VRELTGYDRV MVYRFHEDEH
   Ara B     SQKLAVRAIS QLQALPGGDI KLLCDTVVES VRDLTGYDRV MVYKFHEDEH
   Ara C     SYKLAAKSIS RLQALPSGNM LLLCDALVKE VSELTGYDRV MVYKFHEDGH
Consensus    S-KL----I- --Q-L--G-- --LC---V-- ---LTGYDRV M-Y-FH-D-H

             301                                                     350
  Rice A     GEVFAEITKP GLEPYLGLHY PATDIPQAAR FLFMKNKVRM ICDCRARSIK
 Maize A1    GEVFAEITKP GIEPYIGLHY PATDIPQAAR FLFMKNKVRM ICDCRARSVK
   Oat A3    GEVFSEITKP GLEPYLGLHY PATDIPQAAR LLFMKNKVRM ICDCRARSIK
   Pea A     GEVIAEIAKP GLEPYLGLHY PATDIPQAAR FLFMKNKVRM IVDCNAKHVK
   Zuc A     GEVISEVAKP GLQPYLGLHY PATDIPQAAR FLFMKNKVRM IVDCRAKHLK
   Ara A     GEVVSEVTKP GLEPYLGLHY PATDIPQAAR FLFMKNKVRM IVDCNAKHAR
  Rice B     GEVVAESRRS NLEPYIGLHY PATDIPQASR FLFRQNRVRM IADCHAAPVR
   Ara B     GEVVAESKRD DLEPYIGLHY PATDIPQASR FLFKQNRVRM IVDCNATPVL
   Ara C     GEVIAECCRE DMEPYLGLHY SATDIPQASR FLFMRNKVRM ICDCSAVPVK
Consensus    GEV--E---- ---PY-GLHY -ATDIPQA-R -LF--N-VRM I-DC-A----

             351                                                     400
  Rice A     IIEDESLHLD ISLCGSTLRA PHSCHLQYME NMNSIASLVM AVVVNENEDD
 Maize A1    IIEDEALSID ISLCGSTLRA PHSCHLKYME NMNSIASLVM AVVVNENEED
   Oat A3    VIEAEALPFD ISLCGSALRA PHSCHLQYME NMNSIASLVM AVVVNENEED
   Pea A     VLQDEKLPFD LTLCGSTLRA PHSCHLQYMA NMDSIASLVM AVVVNDSDED
   Zuc A     VLQDEKLQFD LTLCGSTLRA PHSCHLQYME NMNSIASLVM AVVVNEGDEE
   Ara A     VLQDEKLSFD LTLCGSTLRA PHSCHLQYMA NMDSIASLVM AVVVNEEDGE
  Rice B     VIQDPALTQP LCLVGSTLRS PHGCHGQYMA NMGSIASLVM AVIISSGGDD
   Ara B     VVQDDRLTQS MCLVGSTLRA PHGCHSQYMA NMGSIASLAM AVIINGNEDD
   Ara C     VVQDKSLSQP ISLSGSTLRA PHGCHAQYMS NMGSVASLVM SVTINGSDSD
Consensus    ------L--- --L-GS-LR- PH-CH--YM- NM-S-ASL-M -V--------

             401                                                     450
  Rice A     DEVGADQPAQ QQKRKKLWGL LVCHHESPRY VPFPLRYACE FLAQVFAVHV
 Maize A1    DEPEPEQPPQ QQKKKRLWGL IVCHHESPRY VPFPLRYACE FLAQVFAVHV
   Oat A3    DEAESEQPAQ QQKKKKLWGL LVCHHESPRY VPFPLRYACE FLAQVFAVHV
   Pea A     GDS.ADA.VL PQKKKRLWGL VVCHNTTPRF VPFPLRYACE FLAQVFAIHV
   Zuc A     NEG.PAL..Q QQKRKRLWGL VVCHNSSPRF VPFPLRYACE FLAQVFAIHV
   Ara A     GDA.PDATTQ PQKRKRLWGL VVCHNTTPRF VPFPLRYACE FLAQVFAIHV
  Rice B     DHN.IARG.S IPSAMKLWGL VVCHHTSPRC IPFPLRYACE FLMQAFGLQL
   Ara B     GSN.VASG.. .RSSMRLWGL VVCHHTSSRC IPFPLRYACE FLMQAFGLQL
   Ara C     EMN.RDL... .QTGRHLWGL VVCHHASPRF VPFPLRYACE FLTQVFGVQI
Consensus    ---------- ------LWGL -VCH----R- -PFPLRYACE FL-Q-F----

             451                                                     500
  Rice A     NKEFELERQV REKSILRMQT MLSDMLLRES SPLSIVSGTP NIMDLVKCDG
 Maize A1    NKEFELEKQI REKNILRMQT MLSDMLFKES SPLSIVSGSP NIMDLVKCDG
   Oat A3    NREFELEKQL REKNILKMQT MLSDMLFREA SPLTIVSGTP NIMDLVKCDG
   Pea A     NKEIELEYQI LEKNILRTQT LLCDMLMRDA .PLGIVSQSP NIMDLVKCDG
   Zuc A     NKELELENQI IEKNILRTQT LLCDMLMRDA .PLGIVSRSP NIMDLVKSDG
   Ara A     NKEVELDNQM VEKNILRTQT LLCDMLMRDA .PLGIVSQSP NIMDLVKCDG
  Rice B     NMELQLAHQL SEKHILRTGT LLCDMLLRDS .PTGIVTQSP SIMDLVKCDG
   Ara B     NMELQLALQM SEKRVLRTQT LLCDMLLRDS .PAGIVTQSP SIMDLVKCDG
   Ara C     NKEAESAVLL KEKRILQTQS VLCDMLFRNA .PIGIVTQSP NIMDLVKCDG
Consensus    N-E------- -EK--L---- -L-DML---- -P--IV---P -IMDLVK-DG
```

	501				550
Rice A	AALLYGGKVW	RLQNAPTESQ	IRDIAFWLSD	VHRDSTGLST	DSLHDAGYPG
Maize A1	AALLYGDKVW	RLQTAPTESQ	IRDIAFWLSE	VHGDSTGLST	DSLQDAGYPG
Oat A3	AALLYGGKVW	RLRNAPTESQ	IHDIAFWLSD	VHRDSTGLST	DSLHDAGYPG
Pea A	AALFYRNKLW	LLGATPTESQ	LREIALWMSE	YHTDSTGLST	DSLSDAGFPG
Zuc A	AALLYKKKIW	RLGLTPNDFQ	LLDIASWLSE	YHMDSTGLST	DSLYDAGYPG
Ara A	AALLYKDKIW	KLGTTPSEFH	LQEIASWLCE	YHMDSTGLST	DSLHDAGFPR
Rice B	AALYYHGKYY	PLGVTPTEVQ	IKDIIEWLTM	CHGDSTGLST	DSLADAGYSG
Ara B	AAFLYHGKYY	PLGVAPSEVQ	IKDVVEWLLA	NHADSTGLST	DSLGDAGYPG
Ara C	AALYYRDNLW	SLGVTPTETQ	IRDLIDWVLK	SHGGNTGFTT	ESLMESGYPD
Consensus	AA--Y-----	-L---P----	------W---	-H---TG--T	-SL---G---

	551				600
Rice A	AAALGDMICG	MAVAKINSKD	ILFWFRSHTA	AEIRWGGAKH	DPSDKDDSRR
Maize A1	AASLGDMICG	MAVAKITSKD	ILFWFRSHTA	AEIKWGGAKH	DPSDKDDNRR
Oat A3	AAALGDMICG	MAVAKINSKD	ILFWFRSHTA	AEIRWGGAKN	DPSDMDDSRR
Pea A	ALSLSDTVCG	MAAVRITSKD	IVFWFRSHTA	AEIRWGGAKH	EPGDQDDGRK
Zuc A	AIALGDEVCG	MAAVRITNND	MIFWFRSHTA	SEIRWGGAKH	EHGQKDDARK
Ara A	ALSLGDSVCG	MAAVRISSKD	MIFWFRSHTA	GEVRWGGAKH	DPDDRDDARR
Rice B	AADLGDAVSG	MAVAYITPSD	YLFWFRSHTA	KEIKWGGAKH	HPEDKDDGQR
Ara B	AAALGDAVCG	MAVAYITKRD	FLFWFRSHTA	KEIKWGGAKH	HPEDKDDGQR
Ara C	ASVLGESICG	MAAVYISEKD	FLFWFRSSTA	KQIKWGGARH	DPNDR.DGKR
Consensus	A--L-----G	MA---I---D	--FWFRS-TA	----WGGA--	------D---

	601				650
Rice A	MHPRLSFKAF	LEVVKMKSLP	WNDYEMDAIH	SLQLILRGTL	NDD.IKP...
Maize A1	MHPRLSFKAF	LEVVKTKSLP	WSDYEMDAIH	SLQLILRGTL	NDA.SKP...
Oat A3	MHPRLSFKAF	LEVVKMKSLP	WSDYEMDAIH	SLQLILRGTL	NDA.SKP...
Pea A	MHPRSSFKAF	LEVVKARSVP	WKDFEMDAIH	SLQLILRNAS	KDT.DII...
Zuc A	MHPRSSFKAF	LEVVKTRSLP	WKDYEMDAIH	SLQLILRNTF	KDT.DAT...
Ara A	MHPRSSFKAF	LEVVKTRSLP	WKDYEMDAIH	SLQLILRNAF	KDS.ETT...
Rice B	MHPRSSFKAF	LEVVKSRSLP	WENAEMDAIH	SLQLILRDSF	RDSAEGTSNS
Ara B	MHPRSSFQAF	LEVVKSRSQP	WETAEMDAIH	SLQLILRDSF	KES.EAAMNS
Ara C	MHPRSSFKAF	MEIVRWKSVP	WDDMEMDAIN	SLQLIIKGSL	QE........
Consensus	MHPR-SF-AF	-E-V---S-P	W---EMDAI-	SLQLI-----	----------

	651				700
Rice A	..TRAASLDN	QVGDLKLDGL	AELQAVTSEM	VRLMETATVP	ILAVDSNGLV
Maize A1	..AQASGLDN	QIGDLKLDGL	AELQAVTSEM	VRLMETATVP	ILAVDGNGLV
Oat A3	..KREASLDN	QIGDLKLDGL	AELQAVTSEM	VRLMETATVP	ILAVDGNGLV
Pea A	.DLNTKAINT	RLNDLKIEGM	QELEAVTSEM	VRLIETATVP	ILAVDVDGTV
Zuc A	.EINRKSIQT	TLGDLKIEGR	QELESVTSEM	VRLIETATVP	ILAVDLDGLI
Ara A	.DVNTKVIYS	KLNDLKIDGI	QELEAVTSEM	VRLIETATVP	ILAVDSDGLV
Rice B	KAIVNGQVQL	..GELELRGI	DELSSVAREM	VRLIETATVP	IFAVDTDGCI
Ara B	K.VVDGVVQP	CRDMAGEQGI	DELGAVAREM	VRLIETATVP	IFAVDAGGCI
Ara C	.EHSKTVVDV	PLVDNRVQKV	DELCVIVNEM	VRLIDTAAVP	IFAVDASGVI
Consensus	----------	----------	-EL-----EM	VRL--TA-VP	I-AVD--G--

	701				750
Rice A	NGWNQKVAEL	TGLRVDEAIG	RHILT.VVEE	SSVPVVQRML	YLALQGKEEK
Maize A1	NGWNQKVAEL	SGLRVDEAIG	RHILT.LVED	SSVSLVQRML	YLALQGREEK
Oat A3	NGWNQKAAEL	TGLRVDDAIG	RHILT.LVED	SSVPVVQRML	YLALQGKEEK
Pea A	NGWNIKIAEL	TGLPVGEAIG	KHLLT.LVED	SSTDIVKKML	NLALQGEEEK
Zuc A	NGWNTKIAEL	TGLPVDKAIG	KHLLT.LVED	SSVEVVRKML	FLALQGQEEQ
Ara A	NGWNTKIAEL	TGLSVDEAIG	KHFLT.LVED	SSVEIVKRML	ENALEGTEEQ
Rice B	NGWNAKVAEL	TGLSVEEAMG	KSLVNDLIFK	ESEETVNKLL	SRALRGDEDK
Ara B	NGWNAKIAEL	TGLSVEEAMG	KSLVSDLIYK	ENEATVNKLL	SRALRGDEEK
Ara C	NGWNSKAAEV	TGLAVEQAIG	KP.VSDLVED	DSVETVKNML	ALALEGSEER
Consensus	NGWN-K-AE-	-GL-V--A-G	----------	-----V---L	--AL-G-E--

```
             751                                                      800
   Rice A    EVKFEVKTHG SKRDDGPVIL VVNACASRDL HDHVVGVCFV AQDMTVHKLV
 Maize A1    EVRFELKTHG SKRDDGPVIL VVNACASRDL HDHVVGVCFV AQDMTVHKLV
   Oat A3    EVRFEVKTHG PKRDDGPVIL VVNACASRDL HDHVVGVCFV AQDMTVHKLV
    Pea A    NVQFEIKTHG DQVESGPISL IVNACASKDL RENVVGVCFV AQDITAQKTV
    Zuc A    NVQFEIKTHG SHIEVGSISL VVNACASRDL RENVVGVFFV AQDITGQKMV
    Ara A    NVQFEIKTHL SRADAGPISL VVNACASRDL HENVVGVCFV AHDLTGQKTV
   Rice B    NVEIKLKTFG PEQSKGPIFV IVNACSTRDY TKNIVGVCFV GQDVTGQKVV
    Ara B    NVEVKLKTFS PELQGKAVFV VVNACSSKDY LNNIVGVCFV GQDVTSQKIV
    Ara C    GAEIRIRAFG PKRKSSPVEL VVNTCCSRDM TNNVLGVCFI GQDVTGQKTL
Consensus    ---------- ---------- -VN-C---D- -----GV-F- --D-T--K--

             801                                                      850
   Rice A    MDKFTRVEGD YKAIIHNPSP LIPPIFGADE FGWCSEWNAA MTKLTGWHRD
 Maize A1    MDKFTRVEGD YKAIIHNPNP LIPPIFGADQ FGWCSEWNAA MTKLTGWHRD
   Oat A3    MDKFTRVEGD YKAIIHNPNP LIPPIFGADE FGWCSEWNAA MTKLTGWNRD
    Pea A    MDKFTRIEGD YKAIVQNPNQ LIPPIFGTDE FGWCCEWNAA MIKLTGWKRE
    Zuc A    MDKFTRLEGD YKAIVQNPNP LIPPIFGSDE FCGWSEWNPA MAKLTGWSRE
    Ara A    MDKFTRIEGD YKAIIQNPNP LIPPIFGTDE FGWCTEWNPA MSKLTGLKRE
   Rice B    MDKFINIQGD YKAIVHNPNP LIPPIFASDE NTCCSEWNTA MEKLTGWSRG
    Ara B    MDKFINIQGD YKAIVHSPNP LIPPIFAADE NTCCLEWNMA MEKLTGWSRS
    Ara C    TENYSRVKGD YARIMWSPST LIPPIFITNE NGVCSEWNNA MQKLSGIKRE
Consensus    --------GD Y--I---P-- LIPPIF---- -----EWN-A M-KL-G--R-

             851                                                      900
   Rice A    EVINKMLLGE VFDSTNASCL VKNKDAFVSL CILINSALAG D.ETEKAPFS
 Maize A1    EVVDKMLLGE VFNSSNASCL LKSKDAFVRL CIVINSALAG E.EAEKASFG
   Oat A3    EVLDKMLLGE VFDSSNASCP LKNRDAFVSL CVLINSALAG E.ETEKAPFG
    Pea A    EVMDKMLLGE VFGTQMSCCR LKNQEAFVNF GIVLNKAMTG L.ETEKVPFG
    Zuc A    EVIDKMLLGE VFGVHKSCCR LKNQEAFVNL GIVLNNAMCG Q.DPEKASFG
    Ara A    EVIDKMLLGE VFGTQKSCCR LKNQEAFVNL GIVLNNAVTS Q.DPDKVSFA
   Rice B    EVVGKLLVGE VFG...NCCR LKGPDALTKF MIVLHNAIGG Q.DCEKFPFS
    Ara B    EVIGKMIVGE VFG...SCCM LKGPDALTKF MIVLHNAIGG Q.DTDKFPFP
    Ara C    EVVNKILLGE VFTTDDYGCC LKDHDTLTKL RIGFNAVISG QKNIEKLLFG
Consensus    EV--K---GE VF------C- -K-------- ---------- -----K--F-

             901                                                      950
   Rice A    FFDRNGKYIE CLLSVNRKVN ADGVITGVFC FIQVPSHELQ HALHVQQASQ
 Maize A1    FFDRNEKYVE CLLSVNRKVN ADGVVTGVFC FIHVPSDDLQ HALHVQQASE
   Oat A3    FFDRSGKYIE CLLSANRKEN EGGLITGVFC FIHVASHELQ HALQVQQASE
    Pea A    FFSRKGKYVE CLLSVSKKID AEGLVTGVFC FLQLASPELQ QALHIQRLSE
    Zuc A    FLARNGMYVE CLLCVNKILD KDGAVTGFFC FLQLPSHELQ QALNIQRLCE
    Ara A    FFTRGGKYVE CLLCVSKKLD RKGVVTGVFC FLQLASHELQ QALHVQRLAE
   Rice B    FFDKNGKYVQ ALLTANTRSR MDGEAIGAFC FLQIASPELQ QAFEIQRHHE
    Ara B    FFDRNGKFVQ ALLTANKRVS LEGKVIGAFC FLQIPSPELQ QALAVQRRQD
    Ara C    FYHRDGSFIE ALLSANKRTD IEGKVTGVLC FLQVPSPELQ YALQVQQISE
Consensus    F--------- -LL------- --G---G--C F----S--LQ -A---Q----

             951                                                     1000
   Rice A    QNALTKLKAY SYMRHAINNP LSGMLYSRKA LKNTGLNEEQ MKEVNVADSC
 Maize A1    QTAQRKLKAF SYMRHAINKP LSGMLYSRET LKSTGLNEEQ MRQVRVGDNC
   Oat A3    QTSLKRLKAF SYMRHAINNP LSGMLYSRKA LKNTDLNEEQ MKQIHVGDNC
    Pea A    QTALKRLKVL TYMKRQIRNP LAGIVFSSKM LEGTDLETEQ KRIVNTSSQC
    Zuc A    QTALKRLRAL GYIKRQIQNP LSGIIFSRRL LERTELGVEQ KELLRTSGLC
    Ara A    RTAVKRLKAL AYIKRQIRNP LSGIMFTRKM IEGTELGPEQ RRILQTSALC
   Rice B    KKCYARMKEL AYIYQEIKNP LNGIRFTNSL LEMTDLKDDQ RQFLETSTAC
    Ara B    TECFTKAKEL AYICQVIKNP LSGMRFANSL LEATDLNEDQ KQLLETSVSC
    Ara C    HAIACALNKL AYLRHEVKDP EKAISFLQDL LHSSGLSEDQ KRLLRTSVLC
Consensus    ---------- -Y-------P ---------- -----L---Q ---------C
```

```
           1001                                                   1050
   Rice A  HRQLNKILSD LDQDSVMNKS SCLDLEMVEF VLQDVFVAAV SQVLITCQGK
 Maize A1  HRQLNKILAD LDQDNITDKS SCLDLDMAEF VLQDVVVSAV SQVLIGCQAK
   Oat A3  HHQINKILAD LDQDSITEKS SCLDLEMAEF LLQDVVVAAV SQVLITCQGK
    Pea A  QRQLSKILDD SDLDGIID.. GYLDLEMAEF TLHEVLVTSL SQVMNRSNTK
    Zuc A  QKQISKVLDE SDIDKIID.. GFIDLEMDEF TLHEVLMVSI SQVMLKIKGK
    Ara A  QKQLSKILDD SDLESIIE.. GCLDLEMKEF TLNEVLTAST SQVMMKSNGK
   Rice B  EKQMSKIVKD ASLQSIED.. GSLVLEKGEF SLGSVMNAVV SQVMIQLRER
    Ara B  EKQISRIVGD MDLESIED.. GSFVLKREEF FLGSVINAIV SQAMFLLRDR
    Ara C  REQLAKVISD SDIEGIEE.. GYVELDCSEF GLQESLEAVV KQVMELSIER
Consensus  --Q------- ---------- ----L---EF -L-------- -Q--------

           1051                                                   1100
   Rice A  GIRVSCNLPE RYMKQTVYGD GVRLQQILSD FLFVSVKFSP VG.GSVEISC
 Maize A1  GIRVACNLPE RSMKQKVYGD GIRLQQIVSD FLFVSVKFSP AG.GSVDISS
   Oat A3  GIRISCNLPE RFMKQSVYGD GVRLQQILSD FLFISVKFSP VG.GSVEISS
    Pea A  GIRIANDVAE HIARETLYGD SLRLQQVLAD FLLISINSTP NG.GQVVIAA
    Zuc A  GIQIVNETPE EAMSETLYGD SLRLQQVLAD FLLISVSYAP SG.GQLTIST
    Ara A  SVRITNETGE EVMSDTLYGD SIRLQQVLAD FMLMAVNFTP SG.GQLTVSA
   Rice B  DLQLIRDIPD EIKEASAYGD QYRIQQVLCD FLLSMVRFAP AENGWVEIQV
    Ara B  GLQLIRDIPE EIKSIEVFGD QIRIQQLLAE FLLSIIRYAP SQ.EWVEIHL
    Ara C  KVQISCDYPQ EVSSMRLYGD NLRLQQILSE TLLSSIRFTP AL.RGLCVSF
Consensus  ---------- --------GD --R-QQ---- ---------- ----------

           1101                                                   1150
   Rice A  SL..TKNSIG ENLHLIDLEL RIKHQGKGVP ADLLSQMYED DNKEQSDEGM
 Maize A1  KL..TKNSIG ENLHLIDFEL RIKHRGAGVP AEILSQMYEE DNKEQSEEGF
   Oat A3  KL..TKNSIG ENLHLIDLEL RIKHQGLGVP AELMAQMFEE DNKEQSEEGL
    Pea A  SL..TKEQLG KSVHLVNLEL SITHGGSGVP EAALNQMFGN .NVLESEEGI
    Zuc A  DV..TKNQLG KSVHLVHLEF RITYAGGGIP ESLLNEMFGS .EEDASEEGF
    Ara A  SL..RKDQLG RSVHLANLEI RLTHTGAGIP EFLLNQMFGT .EEDVSEEGL
   Rice B  RPNIKQNSDG ..TDTMLFPF RFACPGEGLP PEIVQDMFSN SRW.TTQEGI
    Ara B  SQLSKQMADG ..FAAIRTEF RMACPGEGLP PELVRDMFHS SRW.TSPEGL
    Ara C  KVIARIEAIG KRMKRVELEF RIIHPAPGLP EDLVREMFQP LRKGTSREGL
Consensus  ---------G ---------- -------G-P ------M--- -------EG-

           1151                                                   1200
   Rice A  SLAVSRNLLR LMN.GDVRHM REAGMSTFIL SVELASAPAK ..........
 Maize A1  SLAVSRNLLR LMN.GDIRHL REAGMSTFIL TAELAAAPSA VGR.......
   Oat A3  SLLVSRNLLR LMN.GDVRHL REAGVSTFII TAELASAPTA MGQ.......
    Pea A  SLHISRKLLK LMN.GDVRYL KEAGKSSFIL SVELAAAHKL KG........
    Zuc A  SLLISRKLVK LMN.GDVRYM REAGKSSFII TVELAAAHKS RTT.......
    Ara A  SLMVSRKLVK LMN.GDVQYL RQAGKSSFII TAELAAANK. ..........
   Rice B  GLSICRKILK LMG.GEVQYI RESERSFFHI VLELPQPQQA ASRGTS....
    Ara B  GLSVCRKILK LMN.GEVQYI RESERSYFLI ILELPVPRKR PLSTASGSGD
    Ara C  GLHITQKLVK LMERGTLRYL RESEMSAFVI LTEFPLI... ..........
Consensus  -L-------- LM--G----- -----S-F-- --E------- ----------

           1201
   Rice A  .......
 Maize A1  .......
   Oat A3  .......
    Pea A  .......
    Zuc A  .......
    Ara A  .......
   Rice B  .......
    Ara B  MMLMMPY
    Ara C  .......
Consensus  -------
```

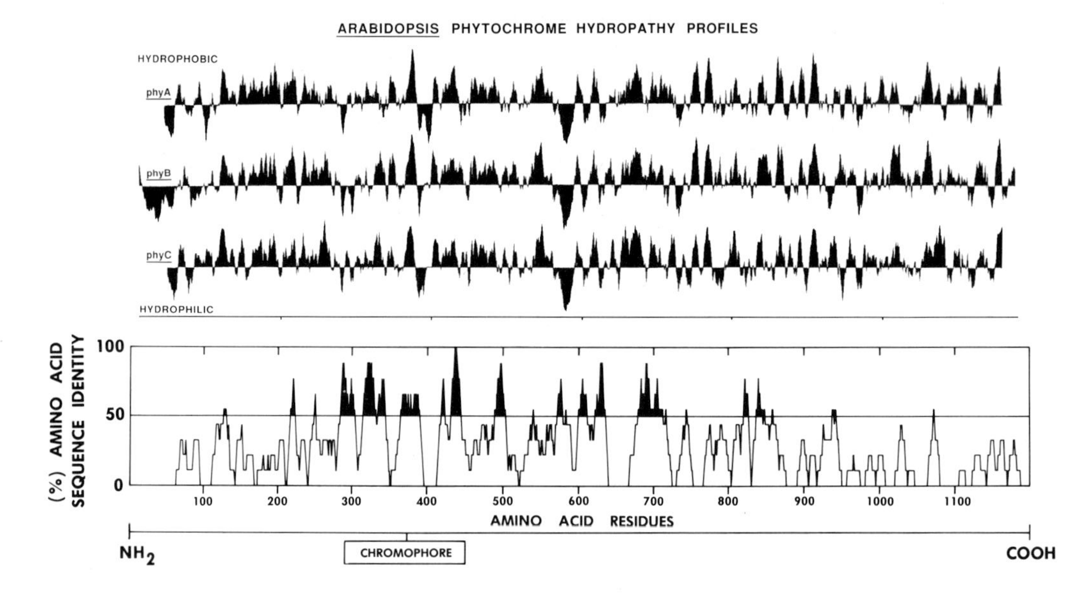

Figure 3. (Top) Hydropathy profiles of *Arabidopsis phyA*, *phyB*, and *phyC* (modified after Sharrock and Quail, 1989). (Bottom) Distribution of invariant amino acid residues along the aligned polypeptides of all phy sequences currently available (see Fig. 2). The number of invariant residues (consensus line in Fig. 2) within a moving window of 9 amino acids is expressed as percent identity and plotted at the middle position of the window. Shaded portions indicate regions of >50% identity. A schematic diagram of the longest polypeptide, Ara *phyB*, with chromophore attachment site is indicated below the plots.

W. Rüdiger (pers. comm.) has noted that the conserved R at position 369 in *phyA* sequences, just upstream of the conserved C-374, the chromophore attachment residue (Fig. 2), is positionally identical to a structurally critical R at the chromophore attachment site of α-C-phycocyanin (Fig. 4). Crystallographic data show that this R provides a point charge which interacts with the chromophore in α-C-phycocyanin. R-369 is conserved in all nine phy sequences consistent with any proposed role in protein-chromophore interaction.

CHROMOPHORE ATTACHMENT SITE

α-C-phycocyanin	A	R	G	K	S	K	C	A	R	D	I
phyA con.	L	R	A	P	H	S	C	H	L	–	Y
all *phy* con.	L	R	–	P	H	–	C	H	–	–	Y
							374				

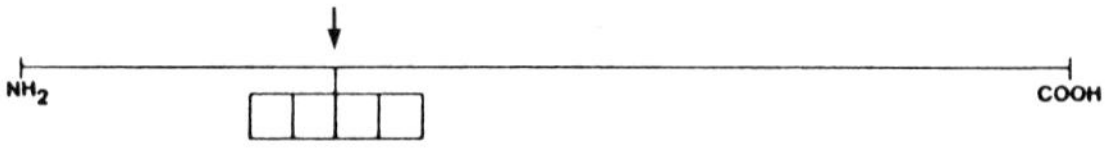

NH2 COOH

Figure 4. Comparison of sequences surrounding chromophore covalent attachment site in phytochrome (Fig. 2) and α-C-phycocyanin (Glazer, 1980). *phyA* con = consensus sequence for all *phyA* polypeptides; all phy con = consensus sequence for all phy polypeptides. Boxed C residues = chromophore attachment site (C-374 in Fig. 2); boxed R's = critical R residue. Arrow indicates sequence position in polypeptide.

In connection with studies demonstrating the presence of protein kinase activity in purified phytochrome preparations, Lagarias and coworkers (1987) indicated that a segment of the oat *phyA3* polypeptide has sequence similarity to mammalian protein kinases (Fig. 5). However, because this region of the protein kinases is not one of the invariant motifs diagnostic of these enzymes (Leader, 1988), and because this region is not well-conserved among the phy sequences (Fig. 5), the likelihood has decreased that this similarity may indicate that phytochrome itself is a protein kinase.

Phytochrome was proposed by Rechsteiner and colleagues (Rogers *et al.*, 1986) on the basis of the oat *phyA3* sequence to contain a so-called "PEST" sequence which was suggested to be involved in the rapid turnover of the photoreceptor. All the essential features of this sequence present in the *phyA3* polypeptide - i.e., positively charged residues at the beginning and end of the sequence, no internal R, K, or H residues, a high local concentration of P, E, D, S and/or T residues, and a clustering of negatively

PROTEIN KINASE HOMOLOGY

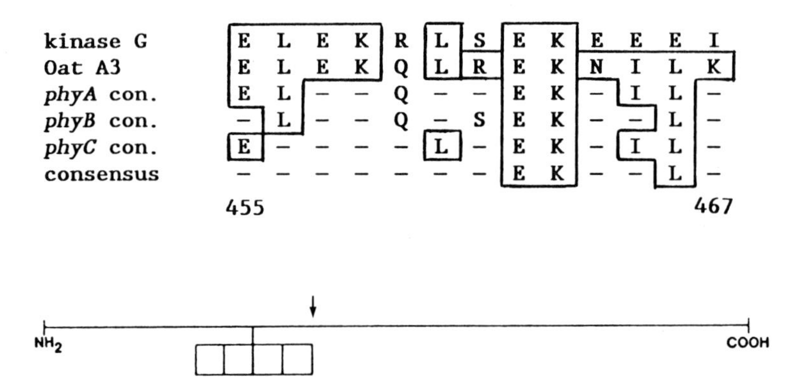

Figure 5. Lack of conservation of phytochrome homology to protein kinases. Oat A3 = region of oat *phyA3* sequence designated as homologous to protein kinases by Lagarias *et al.* (1987). Boxed residues in Oat A3 were identified as being identical to residues in one or more of the kinases used in the original comparison. *phyA* con, *phyB* con, and *phyC* con = consensus sequence for each of the phy subfamilies (from Fig. 2). Consensus = invariant residues for all phy sequences. Boxed residues are those identical to oat *phyA3*. Residue positions are indicated below. Arrow indicates sequence position in the polypeptide.

charged residues - are conserved among all *phyA* sequences (Fig. 6). In contrast, the *phyB* and *phyC* sequences either do not end with a positive residue, do contain internal H or R residues, and/or lack P residues in this region (Figs. 2,6). In addition, the cluster of negatively-charged D and E residues is less extensive than for *phyA*, with the result that the local net negative charge for *phyB* and *phyC* is -3 versus -5 to -7 for the *phyA* sequences (Fig. 7). Grimm *et al.*, (1988) have shown that E-404 here (E-354 their numbering) is preferentially cleaved by endoproteinase Glu-C in the Pfr form, suggesting that this segment of the polypeptide moves closer to the surface of the molecule upon photoconversion to the active form. It has been proposed that this movement might account for the observation that Pfr is more negatively charged than Pr. Taken together, these data are consistent with the notion that *phyA* gene products (Type 1 phytochrome) are preferentially and rapidly degraded in the Pfr form *in vivo* as a result of conformational rearrangement of the PEST sequence upon photoconversion; and that the *phyB* and *phyC* gene products, in contrast, may prove not to be light-labile because they lack

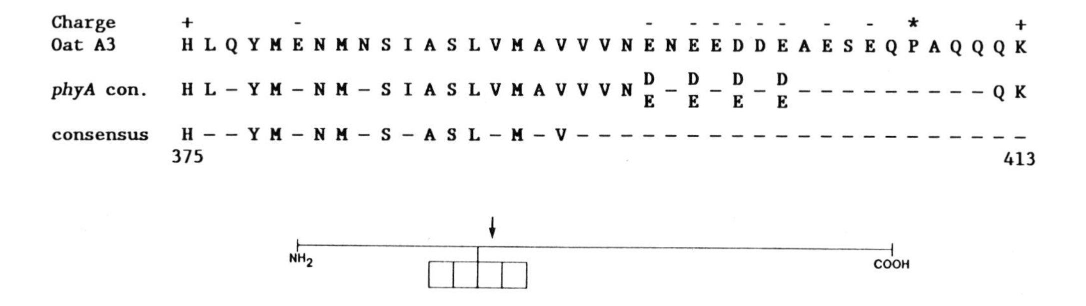

Figure 6. PEST region in phytochrome polypeptides. Oat A3 = oat *phyA3* sequence designated as a PEST region by Rogers *et al.* (1986). *phyA* con = consensus for all *phyA* sequences (Fig. 2). Consensus = invariant residues for all phy sequences. Charge = charged residues in oat A3. Asterisk indicates P in oat A3. All *phyA* sequences contain a P in this region although the position is not conserved. Residue positions are indicated below. Arrow indicates sequence position within the polypeptide.

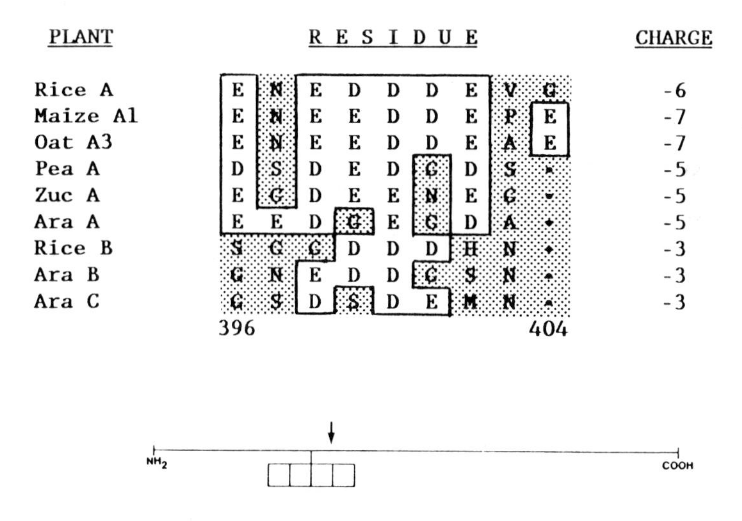

Figure 7. Acidic residue cluster in phytochrome polypeptides. Sequences are as designated in Fig. 2 with residue positions indicated below. Acidic residues are boxed, unshaded. Net negative charge for this region in each sequence is indicated to the right. Arrow indicates sequence position in the polypeptide.

Pr-CHROMOPHORE INTERACTION SITE

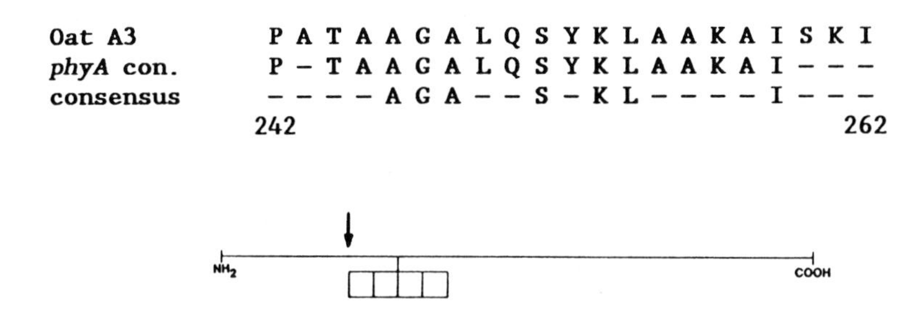

Figure 8. Proposed Pr-chromophore interaction site (Quail *et al.*, 1987a). Sequence designations as in Fig. 6. Residue positions indicated below. Arrow indicates sequence position within the polypeptide.

a functional PEST sequence. The *phyB* and *phyC* gene products would then have one of the properties of Type 2 phytochrome.

The ubiquitin pathway has been proposed as an alternative mechanism for selective Pfr degradation and evidence of Pfr-dependent ubiquitination of phytochrome has been presented (Shanklin *et al.*, 1987, 1989). The region of the oat *phyA* polypeptide between positions 806 and 889 (Fig. 2) has been identified as containing ubiquitinated residues by interference with the binding of specific monoclonal antibodies to epitopes that have been mapped to that region (Shanklin *et al.*, 1989). Since K residues are the target amino acids for ubiquitin conjugation, it is of interest to compare the phy subfamilies for differences in K distribution. All *phyA* sequences contain four K residues in this region, and these residues are all invariant in the *phyB* sequences (Fig. 2). Thus, if *phyB* encodes Type 2 phytochrome, differences in target-residues for ubiquitination in this region of the polypeptide cannot account for the instability of Type 1 Pfr relative to Type 2 Pfr. On the other hand, the *Arabidopsis phyC* sequence contains one substitution in this region at position 812, leaving open the possibility that this polypeptide might be stable as Pfr due to lack of these ubiquitination target sites.

Based on proteolytic and spectroscopic studies, it has been proposed (Jones and Quail, 1989; Quail *et al.*, 1987a) that a 20-amino acid segment in the oat *phyA3* polypeptide is necessary for correct interaction of the Pr form of the chromophore with the protein. This segment is strongly conserved among the *phyA* sequences, but diverges in the *phyB* and *phyC* sequences (Fig. 8). It will be of interest to determine whether these differences in the *phyB* and *phyC* sequences result in a shift in the Pr spectrum peak to shorter wavelengths as observed for Type 2 phytochrome (Tokuhisa *et al.*, 1985).

Pratt and colleagues (Thompson *et al.*, 1989) have determined the amino acid sequence of the epitope in the oat *phyA* polypeptide recognized by a monoclonal antibody designated P-25. This antibody recognizes apparent phytochrome polypeptides in a broad range of plant species across the phylogenetic spectrum (Cordonnier *et al.*,

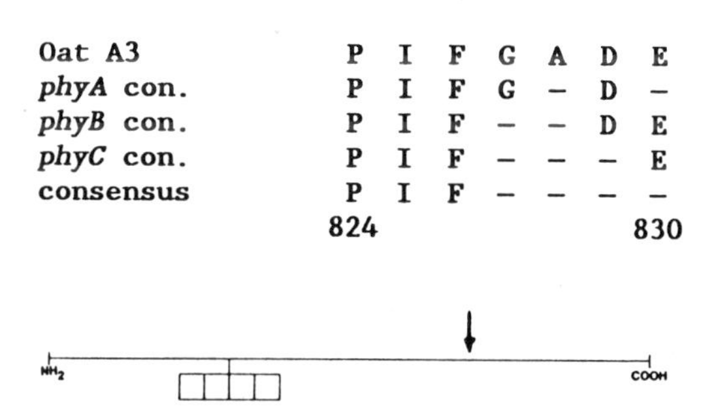

Figure 9. Lack of conservation of P-25 epitope among sequenced phytochromes. Oat A3 is the amino acid sequence identified as the epitope for monoclonal antibody P-25 in the oat *phyA3* sequence (Thompson *et al.*, 1989). Other sequences indicate residues conserved within the *phyA*, *phyB*, and *phyC* subfamily sequences, as well as among all sequenced phytochromes ("consensus"). Arrow indicates sequence position within the polypeptide.

1986b). Some of the *phyA* sequences which are recognized by P-25 have a single amino acid change in the epitope (Fig. 9), apparently in non-critical positions. Since the *phyB* and *phyC* sequences have 2 and 3 substitutions, respectively, it will be of interest to determine whether P-25 is capable of recognizing either of these polypeptides.

PHYTOCHROME-DEFICIENT MUTANTS AND OVEREXPRESSION IN TRANSGENIC PLANTS

The question of the potential functional significance of various regions and sequence motifs in the phytochrome polypeptide can be addressed using targeted *in vitro* mutagenesis of phy sequences followed by examination of the effects of the mutagenesis in transgenic plants. Ideally, one would wish to use a null or phytochrome-deficient mutant as a recipient for such mutagenized sequences. To this end, we initially investigated the molecular basis for the *aurea* (*au*) and long-hypocotyl (*hy*) mutants of tomato and *Arabidopsis*, respectively (Koornneef and Kendrick, 1986). Although it was found that *au*, *hy1* and *hy2* are clearly deficient in spectrally- and biologically-active phytochrome (Chory *et al.*, 1989; Koornneef and Kendrick, 1986; Parks *et al.*, 1987, 1989; Sharrock *et al.*, 1988), all available evidence indicates that the genetic lesions are not in the genes that encode the phytochrome polypeptides. This evidence includes a lack of coincidence in the chromosomal locations of the phenotypic loci and the phy genes (Chang *et al.*, 1988; Meyerowitz, pers. comm.; Sharrock *et al.*, 1988). More directly, however, recent feeding experiments have shown that biliverdin, a precursor to the phytochrome chromophore, rescues the wild-type phenotype and reinstates wild-type levels of spectrally active phytochrome in the *hy1* and *hy2* mutants (B. M. Parks, unpublished). These data indicate that at least these *Arabidopsis* mutants have lesions in the chromophore biosynthetic pathway. We speculate that our failure thus far to identify phytochrome-polypeptide mutants may be because the members of the multigene phy family, although divergent, may have overlapping or redundant biological activities. In this case, only lesions with global effects on all members of the phy family, such as disrupted chromophore synthesis, would be expected to result in phytochrome-deficient phenotypes.

One approach to overcoming this problem is to overexpress a heterologous phytochrome polypeptide in transgenic plants. In this way, it is hoped either that overexpression of the wild-type sequence itself will produce a phenotype, or that overexpression of mutagenized sequences will interfere with endogenous phytochrome function to produce a dominant-negative phenotype (Herskowitz, 1987). It has now been shown that overexpression of oat or rice wild-type *phyA* sequences in tomato and tobacco results in the formation of spectrally and biologically active monocot phytochrome in the dicot cells, and that high levels of overexpression can lead to a dwarf phenotype (Boylan and Quail, 1989; Kay *et al.*, 1989c; Keller *et al.*, 1989). This result thus provides a potential system for examining mutagenized phy sequences. A major shortcoming of this strategy, however, is that the assay for disruption of function in a given mutagenized sequence is the failure of that sequence to induce the dwarf phenotype upon over-expression. Unfortunately, a substantial proportion of the transgenic plants that overexpress the wild-type, heterologous phy sequence also fail to exhibit a detectable phenotype either in tomato (Boylan and Quail, 1989) or tobacco (S. Kay, pers. comm.). It is difficult, therefore, to provide definitive evidence that the failure of any given mutagenized sequence to produce a phenotype results from disruption of functional activity, rather than from insufficient levels of the introduced polypeptide or some other unknown factor. On the other hand, transgenic plants do provide the opportunity to examine the effects of mutagenizing such motifs as the PEST sequence on the structural properties and intracellular behavior of the phytochrome molecule itself.

Phy-GENE EXPRESSION

The availability of gene specific probes has enabled the relative levels of expression and mode of regulation of individual members of the phy gene family to be examined. In *Arabidopsis*, *phyA* mRNA constitutes ~90% of the total detectable phy transcripts, with *phyB* and *C* being ~5% each (Sharrock and Quail, 1989). *phyA* is down-regulated in light, whereas *phyB* and *C* are constitutively expressed regardless of light treatment. Similar data have recently been obtained for rice except that in dark-grown tissue, *phyA* is only ~5-fold higher than *phyB* (Dehesh *et al.*, 1990b). In consequence of these contrasting expression patterns, *phyA*, *B*, and *C* transcripts appear to be of roughly equal abundance in fully green tissue. As mentioned above, the constitutive expression of *phyB* and *C* is reminiscent of the behavior of the Type 2 phytochrome detected by immunoblot analysis in germinating oat seedlings (Tokuhisa and Quail, 1987). The observation that *phyA* is light-regulated and *phyB* is constitutively expressed in both *Arabidopsis* and rice indicates that divergence in the mode of regulation of these genes was an ancient evolutionary event accompanying divergence of the structural genes.

An additional level of complexity in *phyA* gene structure and expression was reported for pea by Sato (1988). Three different transcription start sites were detected for pea *phyA* by S1 nuclease protection assay. Moreover, the abundance of the three transcripts was found to differ and to respond differentially to light, with the shortest transcript being initially the most abundant and down-regulated, whereas the others were more or less unaffected by light treatment (Sato, 1988; Tomizawa *et al.*, 1989). Northern blot analysis of *Arabidopsis phyA* expression resulted in the detection in light-treated tissue of transcripts that are less abundant and larger in size (4.4 kb) than the major transcript (4.0 kb) present in etiolated tissue (Sharrock and Quail, 1989). These data suggested that *Arabidopsis phyA* may be similar in this respect to pea *phyA*. We have now performed S1 nuclease protection assays with the *Arabidopsis phyA* gene and do indeed detect three different transcription start sites (D. Somers and R. Sharrock, unpublished) consistent with the different sized transcripts detected on Northern blots. The relative positions of these start sites are compared with those of pea in Figure 10. Whether light changes the relative abundance of the transcripts detected by S1 nuclease analysis remains to be determined. These results indicate a difference between monocot and dicot *phyA* gene structure and expression since in all three monocot genes thus far examined, oat (Hershey *et al.*, 1987), maize (Christensen and Quail, 1989), and rice (Kay *et al.*, 1989a,b), only a single transcription start site has been detected. It will be of interest to determine whether this difference is related to the generally stronger down-regulation of *phyA* expression observed in monocots than in dicots (Christensen and Quail, 1989; Kay *et al.*, 1989a,b; Lissemore and Quail, 1988; Quail, *et al.*, 1987a; Sharrock and Quail, 1989).

The negative autoregulation that phytochrome exerts over the expression of its own *phyA* genes in monocots provides an excellent model system for investigating the molecular basis of phytochrome-controlled gene expression (Colbert, 1988). The photoreceptor initiates repression of *phyA* transcription within 5 min of Pfr formation via a mechanism that is independent of new protein synthesis (Lissemore and Quail, 1988). These results indicate that all necessary components of the transduction chain pre-exist in the cell before light-signal perception.

The recent advent of the microprojectile-mediated gene transfer procedure (Klein *et al.*, 1987, 1988, 1989) has opened up the opportunity for the first time to dissect light-regulated promoters in monocot tissue for cis-acting elements involved in control of

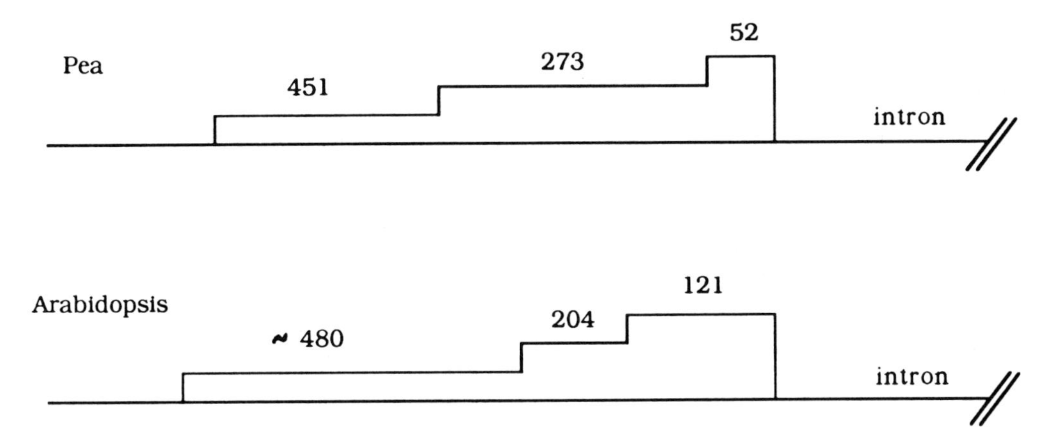

Figure 10. Multiple transcription start-sites in dicot *phyA* promoters. The locations of the three transcription start sites are indicated as steps and the sizes of the three 5′ noncoding exons contained in these alternate transcripts are indicated in bp above each step. Data for pea are modified from Sato (1988). Data for *Arabidopsis* are from D. Somers, R. Sharrock, and P. Quail (unpublished).

expression. We have shown that oat and rice *phyA* promoters transferred into etiolated rice seedlings drive phytochrome-regulated expression of chimaeric genes in a manner parallel to the endogenous *phyA* genes (Bruce *et al.*, 1989; Bruce and Quail, 1990; Dehesh *et al.*, 1990a). Deletion and sequence substitution analysis has led to the identification of three positive elements in the oat *phyA3* promoter (Bruce and Quail, 1990), and an unrelated GT-containing positive element in the rice *phyA* promoter (Dehesh *et al.*, 1990a). The data suggest that these *phyA* promoters are modular in structure with the oat and rice genes having different cis-regulatory elements which perform the same activator function (Quail, *et al.*, 1990).

Recently, we have cloned a transcription factor, designated GT-2, that binds in a highly sequence-specific fashion to the functionally critical GT-element in the rice *phyA* promoter (Dehesh *et al.*, 1990a). This GT-motif is related to, but different from, a motif described in positively light-regulated *rbcS* genes, that interacts with a factor designated GT-1 present in crude nuclear extracts (Gilmartin and Chua, 1990; Green *et al.*, 1987, 1988; Kuhlemeier *et al.*, 1988). The data suggest that there may be a family of functionally distinct trans-acting factors that interact with related GT-containing motifs in a highly sequence specific manner.

CONCLUSIONS

The recognition that plants contain multiple phytochromes encoded by divergent genes has, on the one hand, increased the complexity of the analysis of this photoreceptor system. On the other hand, it has provided the opportunity to determine whether the diversity of functional roles attributed to phytochrome might be accounted for by the structural diversity, differential expression, and/or potential differences in spatial location of the different family members. Moreover, the existence of at least one, and possibly two, more phy genes yet to be characterized (K. Dehesh, unpublished; Sharrock and Quail, 1989) indicates that a rich store of information about this unique regulatory system remains to be uncovered. The advent of routine gene-transfer procedures has provided us with powerful new tools for dissecting both the functional domains of the phytochrome molecule and the transcriptional machinery involved in the terminal step of signal transduction to phytochrome-regulated genes. The gap between the two ends of this signal transduction chain appears to be steadily narrowing.

ACKNOWLEDGEMENTS

We thank R. A. Wells for preparation and editing of this manuscript. Supported by grants from the National Science Foundation (DCB-8796344), the Department of Energy (Division of Energy Biosciences no. DE-FG03-87ER13742), the U. S. Department of Agriculture (Competitive Research Grants Program no. 89-37280-4800), and the National Institutes of Health (RO1-GM36361).

REFERENCES

Abe H, Yamamoto KT, Nagatani A, Furuya M (1985) Characterization of green tissue-specific phytochrome isolated immunochemically from pea seedlings. Plant Cell Physiol 326:1387-1399

Bajaj M, Blundell T (1984) Evolution and the tertiary structure of proteins. Ann Rev Biophys Bioeng 13:453-492

Benfey PN, Chua N-H (1989) Regulated genes in transgenic plants. Science 244:174-181

Boylan MT, Quail PH (1989) Oat phytochrome is biologically active in transgenic tomatoes. Plant Cell 1:765-773

Briggs WR, Chon HP (1966) The physiological versus spectro-photometric status of phytochrome in corn coleoptiles. Plant Physiol 41:1159-1166

Brockmann J, Schäfer E (1982) Analysis of Pfr destruction in *Amaranthus caudatus* L.: Evidence for two pools of phytochrome. Photochem Photobiol 35:555-558

Bruce WB, Christensen AH, Klein T, Fromm M, Quail PH (1989) Photoregulation of a phytochrome gene promoter from oat transferred into rice by particle bombardment. Proc Natl Acad Sci USA 86:9692-9696

Bruce WB, Quail PH (1990) Cis-acting elements involved in photoregulation of an oat phytochrome promoter in rice. Plant Cell submitted

Chang C, Bowman JL, DeJohn AW, Lander ES (1988) Restriction fragment length polymorphism linkage map for *Arabidopsis thaliana*. Proc Natl Acad Sci USA 85:6856-6860

Chory J, Peto CA, Ashbaugh M, Saganich R, Pratt L, Ausubel F (1989) Different roles for phytochrome in etiolated and green plants deduced from characterization of *Arabidopsis thaliana* mutants. Plant Cell 1:867-880

Christensen AH, Quail PH (1989) Structure and expression of a maize phytochrome--encoding gene. Gene 85:381-390

Colbert JT (1988) Molecular Biology of Phytochrome. Plant, Cell & Environ 11:305-318

Cordonnier M-M, Greppin H, Pratt L (1986a) Phytochrome from green *Avena* shoots characterized with a monoclonal antibody to phytochrome from etiolated *Pisum* shoots. Biochemistry 25:7657-7666

Cordonnier M-M, Greppin H, Pratt LH (1986b) Identification of a highly conserved domain on phytochrome from angiosperms to algae. Plant Physiol 80:982-987

Dehesh K, Bruce WB, Quail PH (1990a) Light-regulated expression of a trans-acting factor that binds to a GT-motif in a phytochrome gene promoter. Science submitted

Dehesh K, Tepperman J, Christensen AH, Quail PH (1990b) *phyB* is evolutionarily conserved and constitutively expressed in rice-seedling shoots. Mol Gen Genetics submitted

Furuya M (1989) Molecular properties and biogenesis of phytochrome I and II Adv Biophys 25:133-167

Gilmartin PM, Chua N-H (1990) Spacing between GT-1 binding sites within a light-responsive element is critical for transcriptional activity. Plant Cell 2:447-455

Gilmartin PM, Sarokin L, Memelink J, Chua N-H (1990) Molecular light switches for plant genes. Plant Cell 2:369-378

Glazer AN (1980) Structure and evolution of photosynthetic accessory pigment systems with special reference to phycobiliproteins. In: Sigman D, Brazier MAB (eds) The Evolution of Protein Structure and Function. Academic Press, New York, pp. 221-244

Green P, Kay S, Chua N-H (1987) Sequence-specific interactions of a pea nuclear factor with light responsive elements upstream of the rbcS-3A gene. EMBO J 6:2543-2549

Green PJ, Yong M-H, Cuozzo M, Kano-Murakami Y, Silverstein P, Chua N-H (1988) Binding site requirements for pea nuclear protein factor GT-1 correlate with sequences required for light-dependent transcriptional activation of the *rbcS-3A* gene. EMBO J 7:4035-4044

Grimm R, Eckerskorn C, Lottspeich F, Zenger C, Rüdiger W (1988) Sequence analysis of proteolytic fragments of 124-kilodalton phytochrome from etiolated *Avena sativa* L.: Conclusions on the conformation of the native protein. Planta 174:396-401

Hershey HP, Barker RF, Idler KB, Lissemore JL, Quail PH (1985) Analysis of cloned cDNA and genomic sequences for phytochrome: Complete amino acid sequence for two gene products expressed in etiolated *Avena*. Nucleic Acids Res 13:8543-8559

Hershey, HP, Barker, RF, Idler, KB, Murray, MG, Quail, PH (1987) Nucleotide sequence and characterization of a gene encoding the phytochrome polypeptide from *Avena*. Gene 61:339-348

Hershey HP, Colbert JT, Lissemore JL, Barker RF, Quail PH (1984) Molecular cloning of cDNA for *Avena* phytochrome. Proc Natl Acad Sci USA 81:2332-2336

Herskowitz I (1987) Functional inactivation of genes by dominant negative mutations. Nature 329:219-222

Hillman WS (1967) The physiology of phytochrome. Ann Rev Plant Physiol 18:301-324

Jabben M, Deitzer GF (1978) Spectrophotometric phytochrome measurements in light-grown *Avena sativa* L. Planta 143:309-313

Jabben M, Holmes MG (1983) Phytochrome in light-grown plants. In: Shropshire W, Mohr H (eds) Photomorphogenesis. Springer- Verlag, Berlin, pp 704-722

Jones AM, Quail PH (1989) Phytochrome structure: Peptide fragments from the amino-terminal domain involved in protein-chromophore interactions. Planta 178:147-156

Kay SA, Keith B, Shinozaki K, Chye M-L, Chua N-H (1989a) The rice phytochrome gene: Structure, autoregulated expression, and binding of GT-1 to a conserved site in the 5′ upstream region. Plant Cell 1:351-360

Kay SA, Keith B, Shinozaki K, Chua N-H (1989b) The sequence of the rice phytochrome gene. Nucl Acids Res 17:2865-2866

Kay SA, Nagatani A, Keith B, Deak M, Furuya M, Chua N-H (1989c) Rice phytochrome is biologically active in transgenic tobacco. Plant Cell 1:775-782

Keller JM, Shanklin J, Vierstra RD, Hershey HP (1989) Expression of a functional monocotyledonous phytochrome in transgenic tobacco. EMBO J 8:1005-1012

Klein TM, Fromm ME, Weissinger A, Tomes D, Schaaf S, Sletten M, Sanford JC (1988) Transfer of foreign genes into intact maize cells with high-velocity microprojectiles. Proc Natl Acad Sci USA 85:4305-4309

Klein TM, Roth BA, Fromm ME (1989) Regulation of anthocyanin biosynthetic genes introduced into intact maize tissues by microprojectiles. Proc Natl Acad Sci USA 86:6681-6685

Klein TM, Wolf ED, Wu R, Sanford JC (1987) High-velocity microprojectiles for delivering nucleic acids into living cells. Nature 327:70-73

Koornneef M, Kendrick RE (1986) A genetic approach to photomorphogenesis. In: Kendrick RE, Kronenburg GHM (eds) Photomorphogenesis in Plants. Martinus Nijhoff, Dordrecht, pp 521-546

Kuhlemeier C, Cuozzo M, Green PJ, Goyvaerts E, Ward K, Chua N-H (1988) Localization and conditional redundancy of regulatory elements in rbcS-3A, a pea gene encoding the small subunit of ribulose-bisphosphate carboxylase. Proc Natl Acad Sci USA 85:4662-4666

Kuhlemeier C, Green PJ, Chua N-H (1987) Regulation of gene expression in higher plants. Ann Rev Plant Physiol 38:221-257

Lagarias JC, Rapoport H (1980) Chromopeptides from phytochrome. The structure and linkage of the Pr form of the phytochrome chromophore. J Am Chem Soc 102:4821-4828

Lagarias JC, Wong Y-S, Berkelman TR, Kidd DG, McMichael Jr. RW (1987) Structure--function studies on *Avena* phytochrome In: Furuya M (ed) Phytochrome and Photoregulation in Plants. Academic Press, Tokyo, pp 51-61

Leader DP (1988) Identification of protein kinases by computer. Nature 332:208

Lissemore JL, Colbert JT, Quail PH (1987) Cloning of cDNA for phytochrome from etiolated *Cucurbita* and coordinate photoregulation of the abundance of two distinct phytochrome transcripts. Plant Mol Biol 8:485-496

Lissemore J, Quail PH (1988) Rapid transcriptional regulation by phytochrome of the genes for phytochrome and chlorophyll a/b-binding protein in *Avena sativa*. Mol Cell Biol 8:4840-4850

Mohr H (1966) Differential gene activation as a mode of action of phytochrome 730. Photochem. Photobiol 5:469-483

Nagy F, Kay SA, Chua N-H (1988) Gene regulation by phytochrome. Trends in Genetics 4:37-41

Parks BM, Jones AM, Adamse P, Koornneef M, Kendrick RE, Quail PH (1987) The *aurea* mutant of tomato is deficient in spectrophotometrically and immunochemically detectable phytochrome. Plant Mol Biol 9:97-107

Parks BM, Shanklin J, Koornneef M, Kendrick RE, Quail PH (1989) Immunochemically detectable phytochrome is present at normal levels but is photochemically nonfunctional in the *hy 1* and *hy 2* long hypocotyl mutants of *Arabidopsis*. Plant Mol Biol 12:425-437

Pratt LH, Cordonnier M-M (1987) Phytochrome from green *Avena*. In: Furuya M (ed) Phytochrome and Photoregulation in Plants. Academic Press, Tokyo, pp 83-94

Quail PH, Christensen AH, Jones AM, Lissemore JL, Parks BM, Sharrock RA (1987a) The phytochrome molecule and the regulation of its genes. In: Kon OL (ed) Integration and Control of Metabolic Processes. Cambridge University Press, Cambridge (UK), pp 41-54

Quail PH, Gatz C, Hershey HP, Jones AM, Lissemore JL, Parks BM, Sharrock RE, Barker RF, Idler K, Murray MG, Koornneef M, Kendrick RE (1987b) Molecular biology of phytochrome. In: Furuya M (ed) Phytochrome and Photoregulation in Plants. Academic Press, New York, pp 23-27

Quail PH, Bruce WB, Dehesh K, Dulson J (1990) *phyA* gene promoter analysis. In: Herrmann RG (ed) Plant Molecular Biology. Plenum Press, London, in press

Rogers S, Wells R, Rechsteiner M (1986) Amino acid sequences common to rapidly degraded proteins: The PEST hypothesis. Science 234:364-368

Sato N (1988) Nucleotide sequence and expression of the phytochrome gene in *Pisum sativum*: Differential regulation by light of multiple transcripts. Plant Mol Biol 11:697-710

Shanklin J, Jabben M, Vierstra RD (1989) Partial purification and peptide mapping of ubiquitin-phytochrome conjugates from oat. Biochemistry 28:6028-6034

Shanklin J, Jabben M, Vierstra RD (1987) Red light-induced formation of ubiquitin--phytochrome conjugates: Identification of possible intermediates of phytochrome degradation. Proc Natl Acad Sci USA 84:359-363

Sharrock RA, Lissemore JL, Quail PH (1986) Nucleotide and derived amino acid sequence of a *Cucurbita* phytochrome cDNA clone: Identification of conserved features by comparison with *Avena* phytochrome. Gene 47:287-295

Sharrock RA, Parks BM, Koornneef M, Quail PH (1988) Molecular analysis of the phytochrome deficiency in an *aurea* mutant of tomato. Mol Gen Genetics 213:9-14

Sharrock RA, Quail PH (1989) Novel phytochrome sequences in *Arabidopsis thaliana*: Structure, evolution, and differential expression of a plant regulatory photoreceptor family. Genes Develop 3:1745-1757

Shimazaki Y, Cordonnier M-M, Pratt LH (1983) Phytochrome quantitation in crude extracts of *Avena* by enzyme-linked immunosorbent assay with monoclonal antibodies. Planta 159:534-544

Shimazaki Y, Pratt LH (1985) Immunochemical detection with rabbit polyclonal and mouse monoclonal antibodies of different pools of phytochrome from etiolated and green *Avena* shoots. Planta 164:333-344

Shimazaki Y, Pratt LH (1986) Immunoprecipitation of phytochrome from green *Avena* by rabbit antisera to phytochrome from etiolated *Avena*. Planta 168:512-515

Thompson LK, Pratt LH, Cordonnier M-M, Kadwell S, Darlix J-L, Crossland L (1989) Fusion protein-based epitope mapping of phytochrome. J Biol Chem 264:12426--12431

Tobin E, Silverthorne J (1985) Light Regulation of Gene Expression in Higher Plants. Ann Rev Plant Physiol 36:569-593

Tokuhisa JG, Daniels SM, Quail PH (1985) Phytochrome in green tissue: Spectral and immunochemical evidence for two distinct molecular species of phytochrome in light grown *Avena sativa*. Planta 164:321-332

Tokuhisa JG, Quail PH (1983) Spectral and immunochemical characterization of phytochrome isolated from light-grown *Avena sativa*. (Abstr) Plant Physiol (Suppl) 72:85

Tokuhisa JG, Quail PH (1987) The levels of two distinct species of phytochrome are regulated differently during germination in *Avena*. Planta 172:371-377

Tokuhisa JG, Quail PH (1989) Phytochrome in green tissue: Partial purification and characterization of the 118-kiloDalton phytochrome species from light-grown *Avena sativa* L. Photochem Photobiol 50:143-152

Tomizawa K-I, Sato N, Furuya M (1989) Phytochrome control of multiple transcripts of the phytochrome gene in *Pisum sativum*. Plant Mol Biol 12:295-299

EVIDENCE FOR THREE PHYTOCHROMES IN *AVENA*

L.H. Pratt[1], M-M. Cordonnier[2*], Y-C. Wang[1], S.J. Stewart[2] and M. Moyer[2]
[1]Department of Botany,
University of Georgia,
Athens, GA 30602, USA.

INTRODUCTION

Shortly after the identification of phytochrome as a chromoprotein, Butler and Lane (1965) raised the question of whether the 'bulk' phytochrome detected in etiolated seedlings was physiologically functional, or whether only the much smaller amount of phytochrome that remains in continuous illumination might be active. The implication was that these two pools of phytochrome differed from one another, either in their physicochemical properties or in their cellular distribution and/or subcellular localization. Subsequently, Hillman (1967) in a seminal review suggested by analogy to other pigment-protein complexes "...the likelihood that several phytochromes may exist, differing only slightly in structure but significantly in biological activity."

While the results of transgenic experiments indicate that 'bulk' phytochrome is physiologically active (Keller *et al.*, 1989; Boylan and Quail, 1989; Kay *et al.*, 1989), it has also become apparent that there is more than one phytochrome within the same plant, as suggested by Butler and Lane (1965) and Hillman (1967). Whether these different phytochromes have different biological activities, however, is as yet unknown. The goals of this contribution are (i) to summarize data documenting the existence of two different gene products, each of which is identifiable as phytochrome, and (ii) to introduce new data indicating that there is at least one more type of phytochrome in an oat (*Avena sativa* L., cv. Garry) shoot, yielding a minimum of three phytochromes.

Of these three phytochromes, one is present at a high level in etiolated tissue (etiolated-oat phytochrome), while the other two are present at a low level in light-grown tissue. At least one of these latter two is also present in comparable amounts in etiolated oat shoots, although at a level only ~2% that of etiolated-oat phytochrome (Tokuhisa and Quail, 1987). These latter two will be referred to as green-oat phytochromes, because

[2]Ciba-Geigy Biotechnology, Research Triangle Park, NC 27709-2257, USA.

NATO ASI Series, Vol. H 50
Phytochrome Properties and Biological Action
Edited by B. Thomas and C. B. Johnson

they predominate in light-grown shoots. Use of the terms 'etiolated-' and 'green-oat phytochrome' should not be taken to mean that they exist in only etiolated or light-grown tissue, respectively. Etiolated- and green-oat phytochromes are equivalent to type I and type II phytochromes as defined by Abe *et al.* (1985) with reference to pea phytochromes. The descriptive terms 'green-' and 'etiolated-plant phytochrome' will be used here both because they are less abstract and because their use will prevent possible confusion with the nomenclature (types 3, 4, 5 and 6) used by Hershey *et al.* (1985) to refer to genes encoding etiolated-oat phytochrome.

ETIOLATED- *VERSUS* GREEN-PLANT PHYTOCHROMES

Repeated attempts were made to test the early suggestions that multiple phytochromes exist, and in particular the hypothesis that the phytochrome predominating in a light-grown plant might differ from that which was being characterized in and from etiolated tissue (unpublished data and personal communications). It was not until 1983, however, that the first report documenting an immunochemical difference between phytochrome from dark- and light-grown plants appeared (Tokuhisa and Quail, 1983). Because of the necessity to probe exhaustively for possible experimental artifacts that might account for the immunochemical difference observed by Tokuhisa and Quail, formal reports did not appear for another two years. These reports (Tokuhisa *et al.*, 1985; Shimazaki and Pratt, 1985) established that phytochrome from green and etiolated oats differ not only immunochemically but also spectrophotometrically. Initial evidence that phytochrome from green and etiolated peas also differ appeared in the same year (Shimazaki and Pratt, 1985; Abe *et al.*, 1985).

Physicochemical comparisons of green- and etiolated-plant phytochromes revealed a number of additional differences (Tokuhisa *et al.*, 1985; Abe *et al.*, 1985; Cordonnier *et al.*, 1986b). They exhibited different monomer molecular weights as evaluated by sodium dodecyl sulphate polyacrylamide gel electrophoresis (SDS PAGE), yielded different peptide maps upon digestion with endoproteases, and exhibited differential sensitivity to proteolytic cleavage under non-denaturing conditions. While these physicochemical differences were most easily explained by assuming that green- and etiolated-plant phytochromes were products of different genes, they could not be taken as rigorous proof that this was the case. It does appear, however, that these two phytochromes have similar quaternary structures, as green- and etiolated-oat phytochromes co-elute from a size exclusion column (Cordonnier *et al.*, 1986b; Tokuhisa and Quail, 1989).

ETIOLATED- AND GREEN-PLANT PHYTOCHROMES ARE ENCODED BY DIFFERENT GENES

The first unequivocal evidence that etiolated- and green-plant phytochrome apoproteins derive from different genes was provided by Abe *et al.* (1989), who obtained microsequence data for degradation fragments of immunopurified green-pea (type II) phytochrome. They detected substantial homology with the amino acid sequence of etiolated-pea phytochrome as expected. Nevertheless, the sequences were clearly different, exhibiting only 70% identity when unidentified residues are excluded from the calculation. Abe *et al.* could therefore conclude unambiguously that there are at least two genes in pea encoding different phytochromes.

We have similarly obtained a partial sequence for a proteolytic fragment of green-oat phytochrome. Hydroxyapatite-purified green-oat phytochrome (Pratt, Shimazaki, Stewart and Cordonnier, submitted) was immunopurified by an Affi-Gel 10 column to which was attached a MAb directed to green-oat phytochrome (GO-5; Pratt, Stewart, Shimazaki, Wang and Cordonnier, submitted). The immunopurified green-oat phytochrome exhibited three bands by SDS PAGE: one near 124 kDa, and two at 58 and 34 kDa (Fig. 1a). Only the ~124-kDa polypeptide exhibited Zn^{2+}-induced fluorescence (Fig. 1b) thereby indicating the presence of chromophore (Berkelman and Lagarias, 1986). This immunopurified green-oat phytochrome was eluted from the affinity column with formic acid, which was subsequently eliminated by repetitive dilution with water followed by concentration with a SpeedVac. Tris-Cl (pH 8.5 at 25°C) was added to 60 μg of green-oat phytochrome to a concentration of 100 mM. Protein was chemically reduced by incubation for 30 min at room temperature under nitrogen with tributylphosphine at 0.05%, after which tributylphosphine was eliminated under vacuum with a SpeedVac. Reduced protein was digested for 24 h with Lys-C (1047-825, Boehringer-Mannheim) at a ratio of 2% endoprotease to total protein. Peptides were separated by reverse phase HPLC on a 1 × 250 mm Aquapore RP-300 column in 0.1% trifluoroacetic acid, using a gradient of 0 to 80% isopropanol:acetonitrile (1:1). The sequence was determined with an Applied Biosystems 470A gas-phase sequencer and 120A PTH Analyzer.

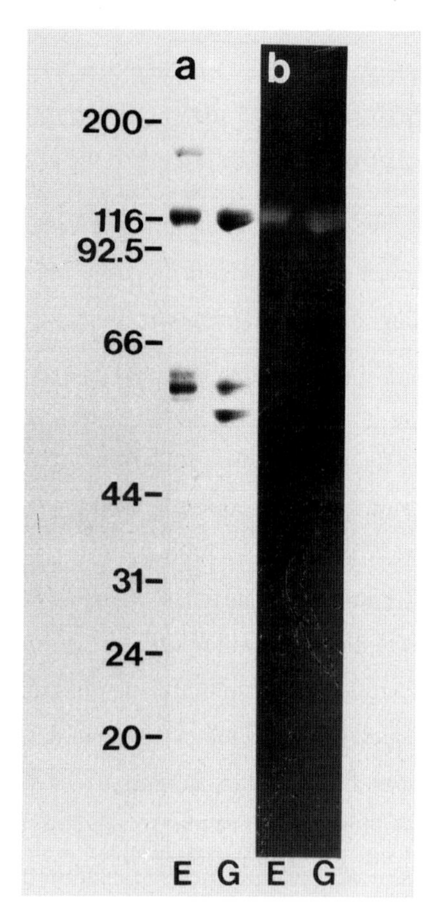

Figure 1. SDS PAGE analysis of immunopurified green-oat phytochrome. Adjacent lanes of a 7.5–15% linear-gradient polyacrylamide gel were loaded with 1 μg of either hydroxyapatite-purified etiolated-oat or immunopurified green-oat phytochrome (lanes E and G, respectively). The gel running buffer contained 1 mM Zn^{2+}, as did the gel itself (Berkelman and Lagarias, 1986). After photographing the gel over an ultraviolet light source (panel b) to visualize fluorescence associated with the phytochrome chromophore, the gel was stained with Coomassie blue (panel a) to visualize total protein. Sizes of molecular mass markers are indicated in kDa.

This sequence exhibits 53% identity to residues 649 through 671 encoded by type 3 etiolated-oat phytochrome cDNA (Hershey *et al.*, 1985), given a three-residue gap as indicated (Fig. 2). Of the remaining nine residues, eight are evolutionarily conserved substitutions. Thus, while clearly distinct from etiolated-oat phytochrome, this 19-residue sequence is nevertheless closely related to it. This green-oat phytochrome sequence also exhibits homology to those recently predicted from putative green-*Arabidopsis* phytochrome cDNAs (Sharrock and Quail, 1989), although the homology in neither case is significantly better than that with etiolated-oat phytochrome (Fig. 2). These data therefore reinforce those of Abe *et al.* (1989), verifying that green- and etiolated-plant phytochrome apoproteins are encoded by different genes in monocotyledons as well as dicotyledons.

```
Etiolated-oat   A A E L T G L R V D D A I G R H I L T L V E
                : | | : | | | : :   : | | |       : : | | :
Green-oat       V A E I T G L P T M E A I G · · · M P L V D
                : | | : | | | : :   | | : |       : : | | :
phyB product    I A E L T G L S V E E A M G · · · K S L V S
                : | | : | | | : :   : | | |       : |   | :
phyC product    A A E V T G L A V E Q A I G · · · K P · V S
```

Figure 2. Comparison of a partial green-oat phytochrome amino sequence with corresponding sequences for etiolated-oat (Hershey *et al.*, 1985) and two putative green-*Arabidopsis* (*phyB*, *phyC*; Sharrock and Quail, 1989) phytochrome sequences. Residues connected by bars are identical; residues connected by two dots are considered to be evolutionarily conserved substitutions. Gaps are indicated by dots.

TWO GREEN-OAT PHYTOCHROMES

MAbs directed to green-oat phytochrome (GO-1, GO-2, GO-4, GO-5, GO-6, GO-7, and GO-8) were prepared with an initial goal of monitoring the temporal and possible tissue-specific expression of the gene that encodes it (Pratt and Cordonnier, 1987; Pratt, Stewart, Shimazaki, Wang and Cordonnier, submitted). As will become evident, however, work towards this goal has been delayed by the discovery that green-oat phytochrome itself is heterogeneous (Cordonnier and Pratt, in press; Wang, Stewart, Cordonnier and Pratt, submitted).

Verification that MAbs GO-1 through GO-8 are directed to green-oat phytochrome

Although all seven MAbs to green-oat phytochrome were initially found to immunoprecipitate photoreversibility from a solution of hydroxyapatite-purified green-oat phytochrome, only three of them (GO-4, GO-5, GO-6) have done so consistently (Pratt, Stewart, Shimazaki, Wang and Cordonnier, submitted). These three MAbs are therefore unambiguously directed to green-oat phytochrome. Each of these MAbs, however, immunoprecipitates no more than about one-third of the photoreversibility from a green-oat phytochrome preparation (Table 1; data for GO-6 not shown). This inability to immunoprecipitate most, if not all, green-oat phytochrome is comparable to the result obtained previously with Oat-9, a MAb directed to etiolated-oat phytochrome. Oat-9 was

Table 1. Immunoprecipitation of hydroxyapatite-purified green-oat phytochrome by MAbs Oat-9, GO-4 and GO-5, either independently or in combination. Aliquots of green-oat phytochrome (2 μg phytochrome) were incubated for 2 h at 4°C with the indicated quantity of MAb in a total volume of 270 μl (150 μl of 0.1 M Na-phosphate, pH. 7.8 + 50 μl of 10 mM Na-phosphate, 140 mM NaCl, 1% bovine serum albumin, pH 7.4 + 70 μl of 0.2 M Na-borate, 75 mM NaCl, pH 8.5). Washed, attenuated *Staphylococcus aureus* cells were added (180 μl of a 10% suspension in 0.2 M Na-borate, 75 mM NaCl, pH 8.5). After incubation for 30 min at 4°C, cells were collected by centrifugation in a microfuge. Pellets were resuspended in 500 μl of 0.1 M Na-phosphate, pH 7.8. Both pellets and supernatants were assayed by photoreversibility at 653 and 728 nm after addition of $CaCO_3$, which minimizes scattering differences between the samples. Percentage of photoreversibility in the pellet is 100 × that measured in the pellet divided by the sum of that measured in both the pellet and the supernatant.

MAb	Amount of MAb (μg)	Photoreversibility in the pellet (%)
Oat-9	10	19
	20	21
GO-4	10	34
	20	39
GO-5	10	38
	20	41
Oat-9 + GO-4	10 each	44
Oat-9 + GO-5	10 each	44
GO-4 + GO-5	10 each	39

one of only two such MAbs found to immunoprecipitate green-oat phytochrome (Shimazaki and Pratt, 1985). It was this inability of Oat-9 to precipitate all green-oat phytochrome that led Shimazaki and Pratt to suggest that green-oat phytochrome might consist of two or more populations. The possibility that the newly developed MAbs might recognize one or more pools of green-oat phytochrome other than that precipitated by Oat-9 was tested by determining whether any two MAbs would exhibit additivity in an immunoprecipitation assay. As can be seen, however, only marginal additivity was observed (Table 1). It thus appears that all three of these MAbs bind to the same population of green-oat phytochrome.

MAb GO-1 was also unequivocally demonstrated to be directed to phytochrome by its ability to detect etiolated-oat phytochrome apoprotein expressed in *Escherichia coli* (Pratt, Stewart, Shimazaki, Wang and Cordonnier, submitted). Because GO-1 detects etiolated-oat phytochrome apoprotein, its epitope could be determined by analysis of overlapping, nested sets of fusion protein as described by Thompson *et al.* (1989). By this strategy, its epitope was found to lie between residues 618 and 686 of the type 3 etiolated-oat phytochrome sequence (Hershey *et al.*, 1985).

Green-oat phytochromes of 125 and 123 kDa Unequivocal documentation that GO-2, GO-7 and GO-8 are directed to phytochrome is inextricably linked to the observation that there are two green-oat phytochromes. Whereas GO-4, GO-5, and GO-6 immunostain polypeptides of 125 and 120 kDa by immunoblot assay, GO-1, GO-2, GO-7 and GO-8 immunostain a polypeptide of 123 kDa (Fig. 3a; Wang, Stewart, Cordonnier and Pratt, submitted). By direct comparison of these results to those obtained with SDS sample buffer extracts of rapidly frozen, lyophilized green oat leaves, it appears that the 120-kDa polypeptide is a degradation product of 125-kDa phytochrome (Wang, Stewart, Cordonnier and Pratt, submitted). These same data also indicate that both the 125-kDa and the 123-kDa polypeptides in the hydroxyapatite-purified samples are the same size as the respective polypeptides *in vivo*.

As already noted, the 125-kDa polypeptide recognized by GO-4, GO-5, and GO-6 is clearly phytochrome. What conclusive evidence is there that the 123-kDa polypeptide is also phytochrome, and that GO-2, GO-7 and GO-8 are thus also directed to this chromoprotein? (1) The 123-kDa polypeptide is also stained on immunoblots, albeit

weakly, by rabbit antibodies directed to and specific for etiolated-oat phytochrome (Wang, Stewart, Cordonnier and Pratt, submitted), including one of the antibody preparations used by Tokuhisa and Quail (1989) to probe green-oat phytochrome.
(2) GO-8 cross reacts weakly with the 125-kDa green-oat phytochrome immunopurified by a column of immobilized GO-4 (Wang, Stewart, Cordonnier and Pratt, submitted).
(3) The 123-kDa polypeptide immunostained by GO-2, GO-7, and GO-8 is also immunostained by GO-1 and Pea-25 (Fig. 3a; Wang, Stewart, Cordonnier and Pratt, submitted), MAbs already shown unequivocally to be directed to phytochrome (Pratt, Stewart,

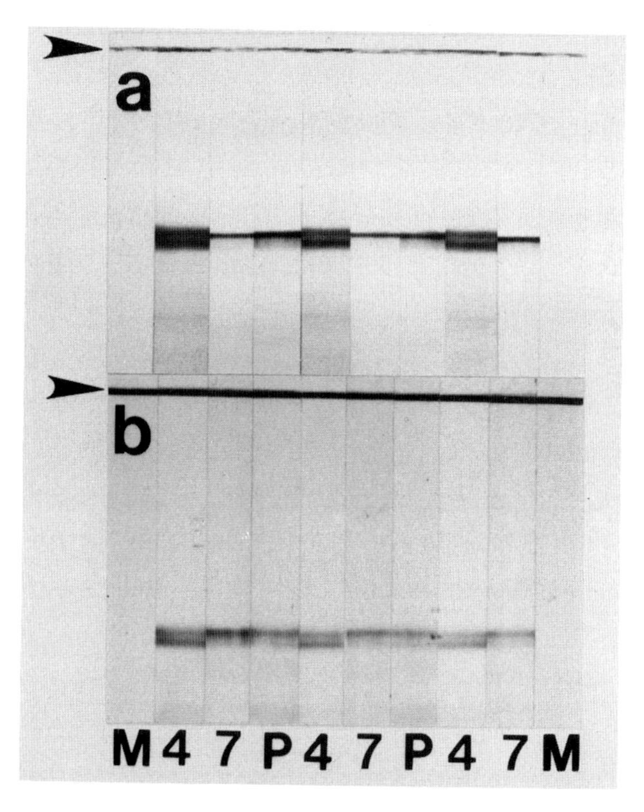

Figure 3. Immunoblot analysis of hydroxyapatite-purified green-oat phytochrome electrophoresed in the absence (panel a) or presence (panel b) of 1 mM Zn^{+2}. Lanes 4.3 cm wide in a 7.5–15% linear-gradient polyacrylamide gel were loaded with 100 ng of green-oat phytochrome. When Zn^{2+} was present, it was included in both the gel solutions and running buffers. A line was drawn across each blot at the origin of the gel (arrowheads) prior to cutting into strips. This line permitted precise reconstitution of the blots after immunostaining. Individual strips were stained with non-immune mouse immunoglobulins (M; 3 μg/ml), GO-4 (4; 1 μg/ml), GO-7 (7; 3 μg/ml) and Pea-25 (P; 0.1 μg/ml). Only the upper portion of each blot is shown.

Shimazaki, Wang and Cordonnier, submitted; Cordonnier *et al.*, 1986a). (4) The 123-kDa polypeptide undergoes a Zn^{2+}-induced mobility shift during SDS PAGE, which is not observed for 125-kDa green-oat phytochrome. This mobility shift is detected not only by the three MAbs in question (GO-2, GO-7, GO-8), but also by GO-1 and Pea-25 (Fig. 3b; Wang, Stewart, Cordonnier and Pratt, submitted). It is therefore possible to conclude with certainty that all seven of the MAbs to green-oat phytochrome are indeed directed to this chromoprotein.

Immunoblot assay of pellets prepared in attempts to immunoprecipitate green-oat phytochrome with GO-1, GO-2, GO-7 and GO-8, as well as Pea-25, indicate that these MAbs fail to precipitate antigen by this assay (unpublished observations). Thus, it is not that any one of these MAbs is immunoprecipitating a polypeptide that is similar in size to phytochrome, but does not bear a photoreversible chromophore. Instead, these MAbs simply do not function well for this application. Unfortunately, none of them are effective for immunopurification of phytochrome as well. Pea-25, for example, will not immunopurify the pea phytochrome to which it is directed, even after the phytochrome has been mildly denatured (unpublished observations). It is therefore impractical with present antibodies to immunopurify 123-kDa green-oat phytochrome.

Relative abundance of etiolated-oat phytochrome in green oat leaves

Of the seven MAbs to green-oat phytochrome, only one (GO-1) cross reacts with etiolated-oat phytochrome. The other six failed to stain immunoblots prepared with phytochrome apoprotein expressed in *E. coli*, even when the SDS polyacrylamide gel was heavily loaded with protein (Wang, Stewart, Cordonnier and Pratt, submitted). Conversely, it is well documented that most MAbs directed to etiolated-oat phytochrome fail to detect any of the phytochrome purified from green oat leaves, either harvested after growth in continuous light (Tokuhisa *et al.*, 1985) or at the end of day (Fig. 4; Shimazaki and Pratt, 1985). It is apparent, therefore, that while at least one type of green-oat phytochrome is present in etiolated oat shoots (Tokuhisa and Quail, 1987), little if any etiolated-oat phytochrome is found in light-grown, green oat leaves.

The data of Shimazaki *et al.* (1983) indicate that the level of etiolated-oat phytochrome drops by at least 1000 fold within the first 24 h after transfer of 5-day-old etiolated oat seedlings to continuous light. ELISA data obtained with Oat-22, a MAb directed to etiolated-oat phytochrome, are consistent with this observation (Fig. 4). Oat-22 begins to detect etiolated-oat phytochrome at a concentration of about 3 ng/ml, while even at 3 μg/ml it fails to respond to a green-oat phytochrome preparation. At 3 μg/ml the small amount of absorbance detected in the ELISA is indistinguishable from that obtained with a comparable amount of non-immune mouse immunoglobulins used as a control. While the green-oat phytochrome preparation with which wells were coated

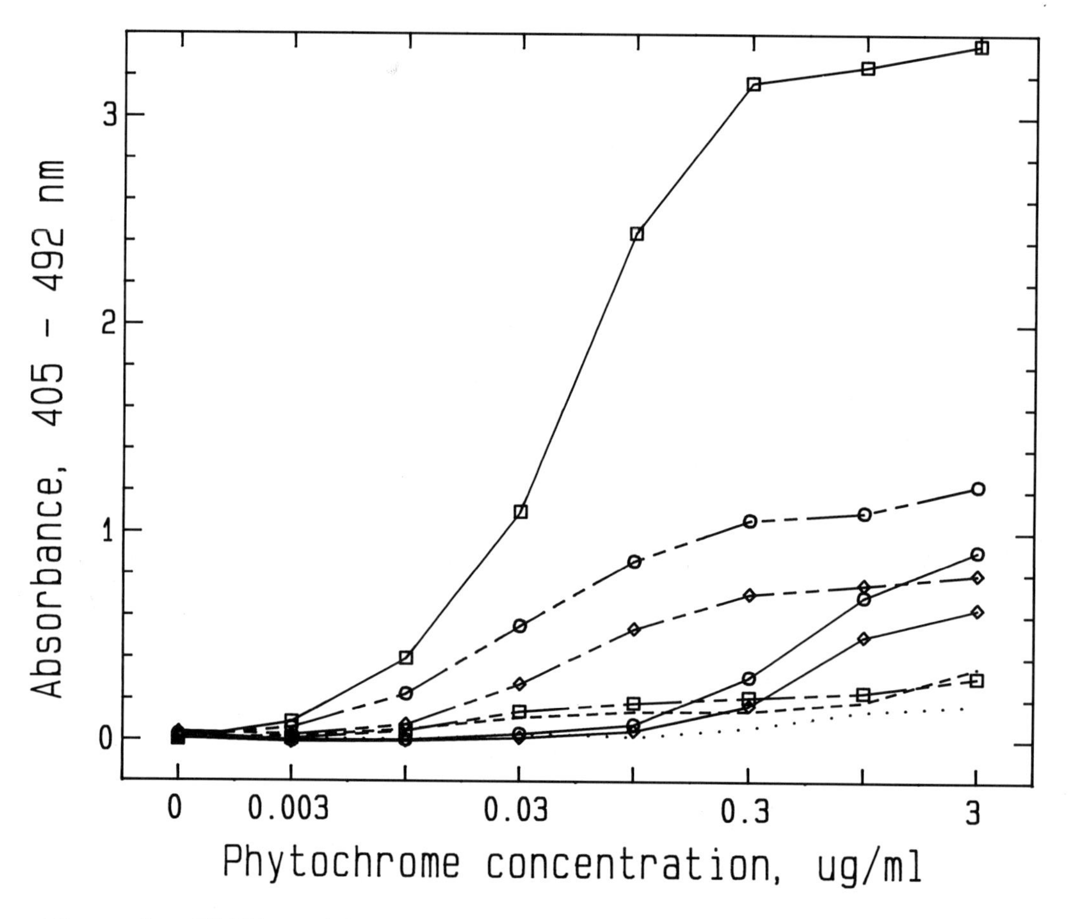

Figure 4. ELISA activity of selected MAbs. Wells were coated with either hydroxyapatite-purified etiolated-oat (————) or hydroxyapatite-purified green-oat (—— – – ——) phytochrome at the indicated concentration. MAbs Oat-22 (□), GO-6 (○) and GO-8 (◇) were tested at 1 μg/ml. Non-immune mouse immunoglobulins at 1 μg/ml were used as a negative control (·····, coated with etiolated-oat phytochrome; – – –, coated with green-oat phytochrome). Alkaline phosphatase-conjugated antibodies to mouse immunoglobulins were used as the label. Details of the assay were as described in Pratt *et al.* (1986). Modified from Pratt, Stewart, Shimazaki, Wang and Cordonnier (submitted to Planta).

is significantly less pure than the etiolated-oat phytochrome preparation, and thus should be expected to coat the ELISA wells less efficiently with phytochrome, it is nevertheless evident that the green-oat phytochrome must contain less than about 1% etiolated-oat phytochrome. Given that the level of green-oat phytochrome in dark-grown tissue is only about 2% that of etiolated-oat phytochrome, there must be in light-grown oats less than about 0.02% (1% × 2%) as much etiolated-oat phytochrome as there is in dark-grown oats. Regardless of these quantitative considerations, it may be concluded with certainty that the two phytochromes found in green oat leaves are both distinct from etiolated-oat phytochrome.

The two green-oat phytochromes not only differ from etiolated-oat phytochrome in monomer size and antigenic properties, but they also differ from each other in both respects. Comparative immunoblot analysis of hydroxyapatite-purified green-oat phytochrome before and after immunoadsorption with a column of immobilized GO-4 illustrates this immunochemical difference. While as expected GO-4 detects its antigen only weakly after adsorption (Fig. 5a, lanes A), GO-7 stains its antigen much more effectively when it has been enriched by removal of 125-kDa green-oat phytochrome by immunoadsorption (Fig. 5b, lanes A). Additionally, the two green-oat phytochromes also differ in their susceptibility to a Zn^{2+}-induced mobility shift during SDS PAGE (Fig. 3; Wang, Stewart, Cordonnier and Pratt, submitted). While these differences do not prove that they are encoded by different genes, this hypothesis appears reasonable in view of the recent data of Sharrock and Quail (1989), who documented the presence of at least three markedly different phytochrome genes in *Arabidopsis*. One of these three, *phyA*, almost certainly encodes etiolated-*Arabidopsis* phytochrome, while the other two (*phyB* and *phyC*) are both highly divergent from *phyA* and from each other. It is possible, therefore, that *phyB* and *phyC* might be responsible for encoding two green-*Arabidopsis* phytochromes. This conclusion is consistent with their relatively low levels of expression, in both darkness and light (Sharrock and Quail, 1989). The two green-oat phytochromes described here might well represent the products of corresponding genes in oat.

RELATIONSHIP TO PREVIOUS OBSERVATIONS

The biological role(s) of these two green-oat phytochromes remain(s) to be established. Nevertheless, their identification does resolve several apparent inconsistencies in the

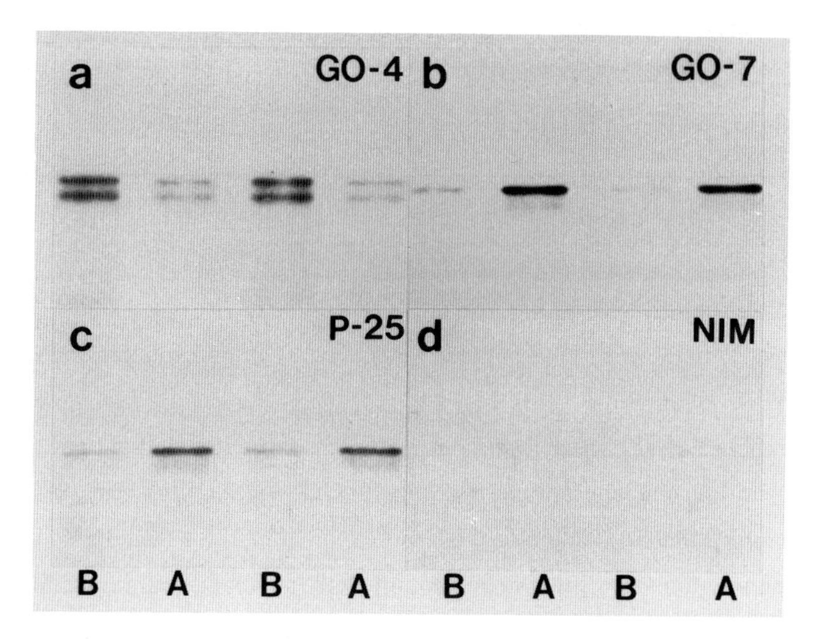

Figure 5. Immunoblot analysis of hydroxyapatite-purified green-oat phytochrome before (lanes B) and after (lanes A) repeated passage through a column of immobilized GO-4. Each lane received the same amount of phytochrome (10 ng) as established by dual wavelength photoreversibility assay. Replica panels were immunostained with (a) GO-4 (1 μg/ml), (b) GO-7 (1 μg/ml), (c) Pea-25 (0.1 μg/ml) or (d) non-immune mouse immunoglobulins (1 μg/ml). While all four lanes in each panel were incubated in substrate solution for the same period of time, this time was selected to maximize observed differences in staining intensity. Only the central portion of each blot is shown.

literature. For example, Cordonnier *et al.* (1986b) noted that the apparent monomer size of green-oat phytochrome as evaluated by SDS PAGE was variable, being slightly greater when immunoblots were prepared in Geneva, Switzerland, as opposed to Athens, Georgia. It is now apparent that the MAb (Pea-25) used by Cordonnier *et al.* for detection of green-oat phytochrome binds preferentially to the 123-kDa polypeptide (Fig. 3, 5; Wang, Stewart, Cordonnier and Pratt, submitted), and that it is this type of green-oat phytochrome whose electrophoretic mobility is sensitive to Zn^{2+}. It is likely, therefore, that the variable mobility results from the variable presence of Zn^{2+}, or some other substance with comparable effect, during electrophoresis. Moreover, it seems that the green-oat phytochrome characterized by Cordonnier *et al.* (1986b) is the 123-kDa pool reported here.

Cordonnier *et al.* (1986b) also reported an observation that at the time was difficult to explain. They found that the green-oat phytochrome immunoprecipitated by Oat-9, a MAb directed to etiolated-oat phytochrome, was not stained by Pea-25 in an immunoblot assay. Immunoblot assay of a precipitate of green-oat phytochrome prepared with Oat-9 and probed with the new MAbs to green-oat phytochrome (Fig. 6) indicates clearly that Oat-9 selectively immunoprecipitates the 125-kDa polypeptide and its 120-kDa degradation product. The protein that Oat-9 precipitates is not etiolated-oat phytochrome as evidenced by the failure of Oat-22 to recognize it (Fig. 6). And, since Pea-25 selectively immunostains the 123-kDa species (Fig. 5), the observation of Cordonnier *et al.* is not surprising in retrospect. As is evident here, however, Pea-25 immunostains, albeit weakly, the 125-kDa polypeptide as well (Fig. 6). Presumably the sensitivity of the immunoblot assays performed here is greater than that realized by Cordonnier *et al.* (1986b). These observations also reinforce the previous conclusion derived from data like those in Table 1 that Oat-9, GO-4, GO-5 and GO-6 all bind to the same type of green-oat phytochrome.

The observations summarized here also explain apparent discrepancies between results obtained in our own laboratories and in that of Quail (Tokuhisa *et al.*, 1985; Tokuhisa and Quail, 1989). Tokuhisa *et al.* initially described two pools of phytochrome in green-oat extracts. One pool had a monomer size comparable to that of etiolated-oat phytochrome and was immunoprecipitated by rabbit antibodies to etiolated-oat phytochrome. They therefore concluded that this pool is like etiolated-oat phytochrome. The other pool, which was not precipitated by the same rabbit antibodies, was assigned a monomer size of 118-kDa. It seems likely that the 125-kDa, green-oat phytochrome described here is the 'etiolated-oat-like' phytochrome described by Tokuhisa *et al.*. It is, however, clearly immunochemically distinct from etiolated-oat phytochrome (Fig. 4), as well as slightly but significantly different in electrophoretic mobility (Wang, Stewart, Cordonnier and Pratt, submitted). The observations of Tokuhisa *et al.* are consistent with the present observation that an immunoprecipitating MAb directed to etiolated-oat phytochrome, Oat-9, selectively immunoprecipitates this pool of green-oat phytochrome (Fig. 6). The 118-kDa pool described by Tokuhisa *et al.* is almost certainly the same as the 123-kDa green-oat phytochrome described here. The relatively large difference in

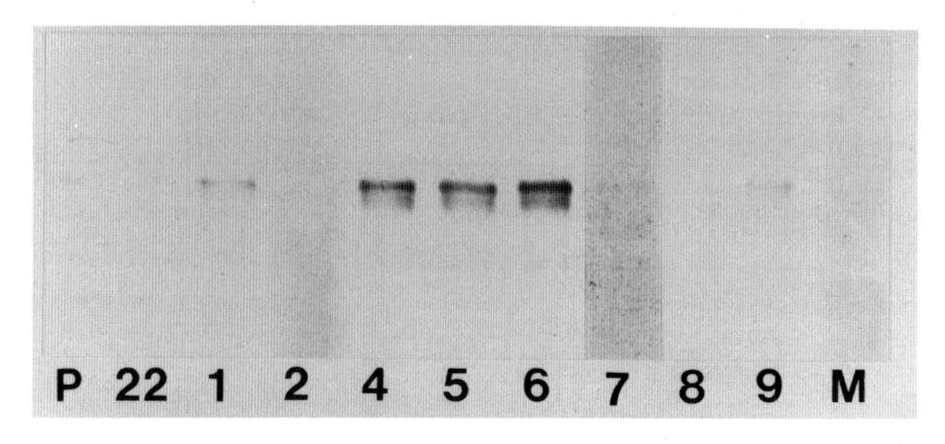

Figure 6. Immunoblot analysis of an immunoprecipitate between Oat-9 and hydroxyapatite-purified green-oat phytochrome. The immunoprecipitate was prepared by incubating 4 μg phytochrome with 30 μg of Oat-9 for 2 h at 4°C in a total volume of 750 μl (550 μl of 0.1 M Na-phosphate, pH 7.8 + 100 μl of 10 mM Na-phosphate, 140 mM NaCl, 1% bovine serum albumin, pH 7.4 + 100 μl of 0.2 M Na-borate, 75 mM NaCl, pH 8.5), after which 150 μl packed volume of agarose-conjugated goat antibodies to mouse immunoglobulins was added. After an additional 50-min incubation at 4°C on a rotating tumbler, the agarose beads were collected by centrifugation and washed once by resuspension in 0.1 M Na-phosphate, pH 7.8, followed by recentrifugation. The washed pellet was suspended in 700 μl of 0.1 M Na-phosphate, pH 7.8. Photoreversibility in the suspended pellet was assayed at 653 and 728 nm for a 400-μl aliquot to which 500 mg of $CaCO_3$ was added. SDS sample buffer was added to the remainder of each pellet. Each lane of a 5–10% linear-gradient polyacrylamide gel was then loaded with 15 ng of this immunoprecipitated green-oat phytochrome. Individual lanes were immunostained with Pea-25 (P, 1 μg/ml), Oat-22 (22, 1 μg/ml), GO-1 (1, 3 μg/ml), GO-2 (2, 3 μg/ml), GO-4 (4, 1 μg/ml), GO-5 (5, 1 μg/ml), GO-6 (6, 1 μg/ml), GO-7 (7, 1 μg/ml), GO-8 (8, 1 μg/ml), Oat-9 (9, 1 μg/ml), or non-immune mouse immunoglobulins (M, 3 μg/ml). Details of the assay were as described in Pratt *et al.* (1986). Only the central portion of the blot is shown.

size assignment might well reflect the differential sensitivity of this pool of phytochrome to potential contaminants during SDS PAGE (Fig. 3; Wang, Stewart, Cordonnier and Pratt, submitted). This conclusion is consistent with the suggestion by Tokuhisa and Quail (1989) that the difference in size assignments most likely reflects some unknown difference in conditions for SDS PAGE.

Are the microsequence data obtained for peptide fragments of green-pea phytochrome (Abe *et al.*, 1989) consistent with their derivation from a mixture of two populations? Each of the four peptides reported by Abe *et al.* exhibits greater identity with predicted *phyB* than with *phyC* gene products, the two putative green-*Arabidopsis* phytochromes (Sharrock and Quail, 1989). Overall, the four peptides have 86% sequence identity with the *phyB* product and only 61% identity with the *phyC* product. Perhaps a better indicator of evolutionary homology is obtained if residues that are fully conserved among the eight phytochrome sequences in the literature (etiolated-oat, -rice, -maize, -pea, -zucchini and -*Arabidopsis*, and the two putative green-*Arabidopsis* phytochromes) are eliminated from consideration. When this is done, there is 76% identity of the four green-pea phytochrome sequences with the *phyB* gene product and only 30% identity with the *phyC* product. For each of the four green-pea phytochrome sequences independently, identity with the *phyC* product is less than half that with the *phyB* product (from 20% to 45%). It therefore seems most likely that the green-pea phytochrome with which Abe *et al.* (1985, 1989) have been working represents only a single type, most closely related to the putative *phyB* gene product. It would be of interest, therefore, to determine whether a second type of green-pea phytochrome exists, comparable to the situation in oat.

FUTURE

As just discussed, several apparent discrepancies in the literature are resolved by the realization that there are two green-oat phytochromes that differ not only from etiolated-oat phytochrome but also from each other. Simultaneously however, this discovery requires re-evaluation of all recent work comparing etiolated-plant (type I) to green-plant (type II) phytochrome. Certainly, it will be necessary to consider the temporal and tissue-specific expression of not just two phytochromes (Tokuhisa and Quail, 1987), but at least three. Moreover, evaluation of data obtained both with mutants and with transgenic plants becomes correspondingly more complex. Not only must the potential interplay among these three (or more) phytochromes be considered, but at least three gene-product-specific probes to evaluate expression of phytochrome genes in such plants must be developed. Thus, while this discovery resolves some issues, it creates new ones. Perhaps now is an appropriate time to pause and to determine not only the total number of different phytochrome genes expressed by an angiosperm, but also whether this number can be generalized among a wide range of taxa. The present situation is somewhat reminiscent of that relating to the apparent molecular weight of phytochrome, which grew significantly over the years, from a low of 42 kDa to its present size of *ca.* 125 kDa. Before investing too much effort in analysis of mutants and transgenic plants it might be wise to resolve completely the system with which we are working.

ACKNOWLEDGEMENTS

This research was supported by the US Department of Energy (DE-AC-09-81SR10925 to LHP). We thank Mrs. Donna Tucker and Mrs. Danielle Neal for their excellent technical assistance.

REFERENCES

Abe H, Yamamoto KT, Nagatani A, Furuya M (1985) Characterization of green tissue-specific phytochrome isolated immunochemically from pea seedlings. Plant Cell Physiol 326:1387-1399

Abe H, Takio K, Titani K, Furuya M (1989) Amino-terminal amino acid sequences of pea phytochrome-II fragments obtained by limited proteolysis. Plant Cell Physiol 30:1089-1098

Berkelman T, Lagarias JC (1986) Visualization of bilin-linked peptides and proteins in polyacrylamide gels. Anal Biochem 156:194-201

Boylan MT, Quail PH (1989) Oat phytochrome is biologically active in transgenic tomatoes. Plant Cell 1:765-773

Butler WL, Lane HC (1965) Dark transformation of phytochrome *in vivo*. II. Plant Physiol 40:13-17

Cordonnier M-M, Pratt LH (to be published) Phytochrome from green *Avena* characterized with monoclonal antibodies directed to it. In: Riklis E (ed) Photobiology. Plenum Press, New York

Cordonnier M-M, Greppin H, Pratt LH (1986a) Identification of a highly conserved domain on phytochrome from angiosperms to algae. Plant Physiol 80:982-987

Cordonnier M-M, Greppin H, Pratt LH (1986b) Phytochrome from green *Avena* shoots characterized with a monoclonal antibody to phytochrome from etiolated *Pisum* shoots. Biochemistry 25:7657-7666

Hershey HP, Barker RF, Idler KB, Lissemore JL, Quail PH (1985) Analysis of cloned cDNA and genomic sequences for phytochrome: amino acid sequences for two gene products expressed in etiolated *Avena*. Nucleic Acids Res 13:8543-8559

Hillman WS (1967) The physiology of phytochrome. Annu Rev Plant Physiol 18:301-324

Kay, SA, Nagatani A, Keith B, Deak M, Furuya M, Chua N-H (1989) Rice phytochrome is biologically active in transgenic tobacco. Plant Cell 1:775-782

Keller, JM, Shanklin J, Vierstra RD, Hershey HP (1989) Expression of a functional

monocotyledonous phytochrome in transgenic tobacco. EMBO J 8:1005-1012
Pratt LH, Cordonnier M-M (1987) Phytochrome from green *Avena*. In: Furuya M (ed) Phytochrome and photoregulation in plants. Academic Press, Tokyo, pp 83-94
Pratt LH, McCurdy DW, Shimazaki Y, Cordonnier M-M (1986) Immunodetection of phytochrome: immunocytochemistry, immunoblotting, and immunoquantitation. In: Linskens HF, Jackson JF (eds) Modern methods of plant analysis, new series, vol 4. Springer-Verlag, Berlin, pp 50-74
Sharrock RA, Quail PH (1989) Novel phytochrome sequences in *Arabidopsis thaliana*: structure, evolution, and differential expression of a plant regulatory gene family. Genes and Development 3:1745-1757
Shimazaki Y, Pratt LH (1985) Immunochemical detection with rabbit polyclonal and mouse monoclonal antibodies of different pools of phytochrome from etiolated and green *Avena* shoots. Planta 164:333-344
Shimazaki Y, Cordonnier M-M, Pratt LH (1983) Phytochrome quantitation in crude extracts of *Avena* by enzyme-linked immunosorbent assay with monoclonal antibodies. Planta 159:534-544
Thompson LK, Pratt LH, Cordonnier M-M, Kadwell S, Darlix J-L, Crossland L (1989) Fusion protein-based epitope mapping of phytochrome: precise identification of an evolutionarily conserved domain. J Biol Chem 264:12426-12431
Tokuhisa JG, Quail PH (1983) Spectral and immunochemical characterization of phytochrome isolated from light-grown *Avena sativa* (Abstr). Plant Physiol 72(Suppl):85
Tokuhisa JG, Quail PH (1987) The levels of two distinct species of phytochrome are regulated differently during germination in *Avena sativa* L. Planta 172:371-377
Tokuhisa JG, Quail PH (1989) Phytochrome in green tissue: partial purification and characterization of the 118-kilodalton phytochrome species from light-grown *Avena sativa* L. Photochem Photobiol 50:143-152
Tokuhisa JG, Daniels SM, Quail PH (1985) Phytochrome in green tissue: Spectral and immunochemical evidence for two distinct molecular species of phytochrome in light-grown *Avena sativa* L. Planta 164:321-332

PHYTOCHROME IN LOWER PLANTS

W. Rüdiger and F. Thümmler
Botanisches Institut der Universität München,
Menzingerstrasse 67,
D-8000 München 19,
Germany.

INTRODUCTION

Most observations and investigations on phytochrome deal with higher - i.e. seed-producing - plants. Reports on phytochrome or phytochrome action in lower plants have only been scattered in the literature (recent reviews: Dring, 1988; Wada and Kadota, 1989). In higher plants, multiple functions can be ascribed to phytochrome (see Shropshire and Mohr, 1983; Furuya, 1987). Recent investigations on phytochrome genes of *Arabidopsis thaliana* (Sharrock and Quail, 1989) and on phytochrome mutants of several dicotyledons (Adamse *et al.*, 1988 a,b) point to 3 or more phytochromes and eventually as many transduction chains in higher plants. This situation may render it difficult to investigate mechanisms of phytochrome action in higher plants.

Many phytochrome functions can be summarized under the general view of adaptation of the photosynthetic apparatus to the actual light conditions. Such a problem has to be handled also by lower plants. The question of the phylogenetic origin of phytochrome arises in this connection. When discussing this question, one should differentiate between the operational criterion ("red/far-red photoreversibility") typical for a "true phytochrome" and photoreceptors which are related in their protein structure to phytochrome but must not necessarily have identical optical properties ("phytochrome-like proteins"). The operational criterion is fulfilled in many ferns, mosses and green algae (Wada and Kadota, 1989). Dring (1988) stated that phytochrome has been positively identified only in green algae. According to Dring, so far reported red/far-red effects in non-green algae must still be treated with caution and the presence of phytochrome in these plants has yet to be confirmed. We started a broad search to phytochrome in lower plants, i.e. in ferns, mosses and algae. This chapter reviews the applied methods and our present state of knowledge.

NATO ASI Series, Vol. H 50
Phytochrome Properties and Biological Action
Edited by B. Thomas and C. B. Johnson

STUDIES WITH MONOCLONAL ANTIBODIES

One usual way of screening for homologous proteins in various organisms is the screening with antibodies. Especially those antibodies which are directed against sequential epitopes, i.e. defined amino acid sequences, are suitable for this purpose. Such antibodies should recognize their epitopes in a Western blot, i.e. after SDS electrophoresis and blotting of unfolded proteins on to nitrocellulose. Antibodies directed against conformational epitopes react in a Western blot only in those (rare) cases where refolding of the peptide chain occurs on nitrocellulose. Conservation of amino acid sequences is more pronounced in phylogeny than conservation of DNA sequences due to degeneration of the genetic code. It is therefore easier in many instances to find cross-reactivity with antibodies than cross-hybridization with gene probes.

We have used several monoclonal antibodies produced against oat or maize phytochrome (Schneider-Poetsch *et al.*, 1988). The position of the epitopes were determined by Western blot analysis of proteolytic fragments of oat phytochrome; mapping of the fragments was achieved by microsequencing (Grimm *et al.*, 1986). The position in the phytochrome sequence of one of the most used antibodies, Z3B1, had to be corrected: it was previously attributed to the chromophore region (Grimm *et al.*, 1986) but is indeed localized in the C-terminal domain as shown by careful re-evaluation of all available data (Schneider-Poetsch, Schendel and Rüdiger, unpublished results).

Some of these antibodies recognize a protein with the apparent size 120-130 kDa in various macroalgae mosses, ferns and a liverwort (see Table 1). We named the algal protein a "phytochrome-like" protein because the present data do not yet allow the unequivocal identification as phytochrome. Extraction of the protein from *Ulva*, for which the detergent Triton X-100 was needed, and partial purification yielded a solution with an absorption maximum at 660 nm but with only partial photoreversibility (López-Figueroa *et al.*, 1989a). This property is similar to extractable "phytochrome" isolated from the moss *Atrichum* (Lindemann *et al.*, 1989). The protein from *Mesotaenium* which was extracted without detergents behaved however like "true phytochrome", i.e. it is fully photoreversible (Kidd and Lagarias, 1990). It should be noted that we find cross-reactivity not only with proteins from *Charophyta* and *Chlorophyta* but also from *Phaeophyta* and *Rhodophyta*. There is no systematic difference

Table 1. Western blot analysis of extracts from lower plants with monoclonal antibodies directed against phytochrome from *Avena* or *Zea*.

Algae Mosses Ferns	Monoclonal antibody (domain)					Ref
	Z2B3 Z4B5 N-terminal	Z3B1	P25	ACC5 AFD4 C-terminal	A2A3 Z4A5	
Phaeophyta						
Cystoseira						
- abies marina	-	+		nt	-	1
- tamariscifolia	-	+		nt	-	1
Rhodophyta						
Corallina elongata	+	+		+	+	1
Gelidium sp.	+	+		-	-	1
Chlorophyta						
Ulva rigida	+	+		+	+	1
Enteromorpha compressa	+	+		-	+	1
Chaetomorpha linum	+	+		nt	+	1
Chlamydomonas		+	+			2,4
Mesotaenium			+			4
Mougeotia			+			4
Charaphytum						
Chara hispida	+	+		-	-	1
Bryophyta						
Physcomitrella patens			+			4
Marchantia polymorpha	nt	+		nt	nt	3
Sphagnum auriculatum	nt	+		nt	nt	3
Funaria hygrometrica	nt	+		nt	nt	3
Pteridophyta						
Asplenium sp.	nt	+		nt	nt	3
Psilotum triquetrum	nt			nt	nt	3
Selaginella martensii	nt	+		nt	nt	3

1 = López-Figuera et al., 1990. 2 = Schendel and Rüdiger, unpublished.
3 = Schneider-Poetsch et al., 1988. 4 = Cordonnier et al, 1986.
+ = Positive reaction. - = No reaction. nt = not tested.

either in the strength of the signal or in the number of recognized epitopes: all applied antibodies are recognized by the protein from *Corallina* (*Rhodophyta*) as well as by that from *Ulva* (*Chlorophyta*). On the other hand, we did not find cross-reactivity with extracts from some other species of ferns, mosses and algae. It is not yet clear whether this

Sense primers (left primers)

```
                L121                        L456
          5'                3'       5'                         3'
Primer    AAATCATTCGACTAXTC   X=C+T  GAATTCTATCCTJGTJCAJTGCAAGAC
Oat       G*C**C**T*****C**          ****C*******G**T**G********
Arab A    *GC**C**T**T**C**          ****C*C**T**T**G**C****G***
Arab B    **************C**          A***C*GG*TTGGA*C**T*C*****A
Arab C    CGT*T***T*****T**          ****C*****ACGC*T**T**TAG*T*
```

```
                    L1318
          5'                              3'
Primer    ATAATGGACCTTGTGAAATGTGACGGXGCXGC   X=A+G+C+T
Oat       **C********A**C********T**T**T**
Arab A    ***********************T**A**A**
Arab B    **C******T*A**************T**A**
Arab C    ********T*****T********T**AG*A**
```

Antisense primers (right primers)

```
                  R740
          5'                        3'
Primer    TCATCTTCATCAAACTTATAAACCAT   X=A+G
Oat       **********G******G***G****
Arab A    **********G**********G****
Arab B    **************************
Arab C    C****CT***GG********C*****
```

```
                    R1676
          5'                              3'
Primer    AGAAAAGCCTTGAATGACGACCTAGGXTGCAT   X=A+G
Oat       **G*****T*****A***A*******G*****
Arab A    **G***********C**T*****T**G*****
Arab B    **********GA**G*****A*G***A*****
Arab C    *T***************G**T*****A*****
```

```
                    R1859
          5'                                 3'
Primer    GGAACAGCTGCAGTCTCAATTAACCXYACCATZTC   X=G+T; Y=C+A; Z=T+C
Oat       *******T***T**T**C****GA*GA*****T**
Arab A    **C**C*TA**************A*GA*****C**
Arab B    *******T************G*G**TA*****C**
Arab C    ***********T**A*****C****GC*****T**
```

Figure 1. Homology of primers used for amplification of phytochrome sequences to known phytochrome sequences. The numbers indicate the position of the first (last) nucleotides of the sense (antisense) primers within the *Avena* phytochrome cDNA.

means altered epitopes in the phytochrome-like protein or lack of production of detectable amounts of this protein.

STUDIES ON THE GENE LEVEL

Another usual way for investigation of phylogenetic relationship is comparison of gene sequences. Cross hybridization with heterologous probes has often been used for this purpose. Hybridization under appropriate stringency with an 0.9kb probe of *Arabidopsis* phytochrome type A gene led to the identification of the different phytochrome types B and C in *Arabidopsis thaliana* (Sharrock and Quail, 1989). Recently a new powerful technology to screen for specific DNA sequences was developed - the polymerase chain reaction (PCR) - using the thermostable Taq DNA polymerase (Saiki *et al.*, 1988). We used this new method to screen several lower plants for phytochrome related sequences.

For this purpose oligonucleotides were synthesized to serve as sense and antisense primers in the PCR (Fig. 1). The primers were designed such that they hybridize to highly conserved regions within the phytochrome genes. Homology between the phytochrome genes is most pronounced around the chromophore binding site (Sharrock and Quail, 1989), therefore all primers were directed to this region of the phytochrome gene. The 5′ region of all investigated phytochrome genes show a very similar structure. A short exon of around 100bp (exon-I) is followed by a large intron up to 2585bp. The following exon-II of around 2kb harbours the chromophore binding site (Kay *et al.*, 1989, Hershey *et al*, 1987; Sato, 1988). Therefore additional care was taken that all primers hybridize within exon-II because we wanted to minimalise the possibility, that non-conserved intron sequences are amplified during PCR. Last not least, the 3′ ends of most primers were degenerated to allow best fitting of the ends where DNA polymerisation starts to obtain highest yields of specific PCR products (Sommer and Tautz, 1989).

The primers were first tested with *Avena* and *Arabidopsis* DNA. With the primer combination L121/R740, two PCR products were obtained, namely a 619bp fragment with *Avena* and a 610bp fragment with *Arabidopsis* genomic DNA. Sequence analysis of the PCR products revealed 100% homology with *Avena* phytochrome (Hershey *et al.*, 1987) and 100% homology with *Arabidopsis* phytochrome type B respectively. Using *Arabidopsis* genomic DNA as template, with the primer combination L1318/R1676 two products were obtained, a 355bp and a 362bp fragment, and with the primer combination L1318/R1859 a 577bp fragment was obtained. After sequencing, the 355bp fragment showed 100% homology to the sequence of *Arabidopsis* phytochrome type C. The 362bp fragment turned out to be included within the 577bp fragment. The sequence could not be related to any of the published phytochrome sequences. On the other hand, one of the calculated open reading frames of the 577bp fragment showed a considerable degree of homology to all known phytochrome sequences on amino acid basis (between 47% and 59%, see Table 2, Fig. 2). We have to assume, that this sequence belongs to another phytochrome gene of *Arabidopsis* which was mentioned but not identified by Sharrock and Quail (1989). We name it provisionally phytochrome type D. We do not know, whether it is expressed in *Arabidopsis* or whether it is merely a pseudo gene.

In summary our primers turned out to be very useful to pick up different types of phytochrome genes and thus we tried using these primers to amplify phytochrome related sequences from a variety of lower plants. So far we only got specific products by using genomic DNA isolated from the moss *Ceratodon purpureus*. The protonema of this moss - like other mosses and ferns - show a phytochrome dependent phototropism (Hartmann and Weber, 1988). The presence of phytochrome in this moss can therefore be assumed. The primer combination L456/R740 and the primer combination L121/R740 gave a 243bp and a 564bp (Cer1) product. With the primer combination L1318/R1676 we obtained a 282bp (Cer2) fragment. The fragments were sequenced and their calculated

Table 2. Percent amino acid identity among phytochromes from various plant species and peptides derived from *Ceratodon* and *Arabidopsis* type D phytochrome. The amino acid sequences of Cer1, Cer2 and ArabD were derived from DNA sequences obtained from PCR products amplified with the following primer combinations: Cer1 with L121/R740, Cer2 with L1318/R1676 and ArabD with L1318/R1895

	Cer1	Cer2	ArabD
Oat	53	66	52
Rice	53	68	53
Zucchini	57	65	55
Pea	57	67	54
Arab A	60	67	55
Arab B	58	71	59
Arab C	59	55	47
Arab D	nt	62	

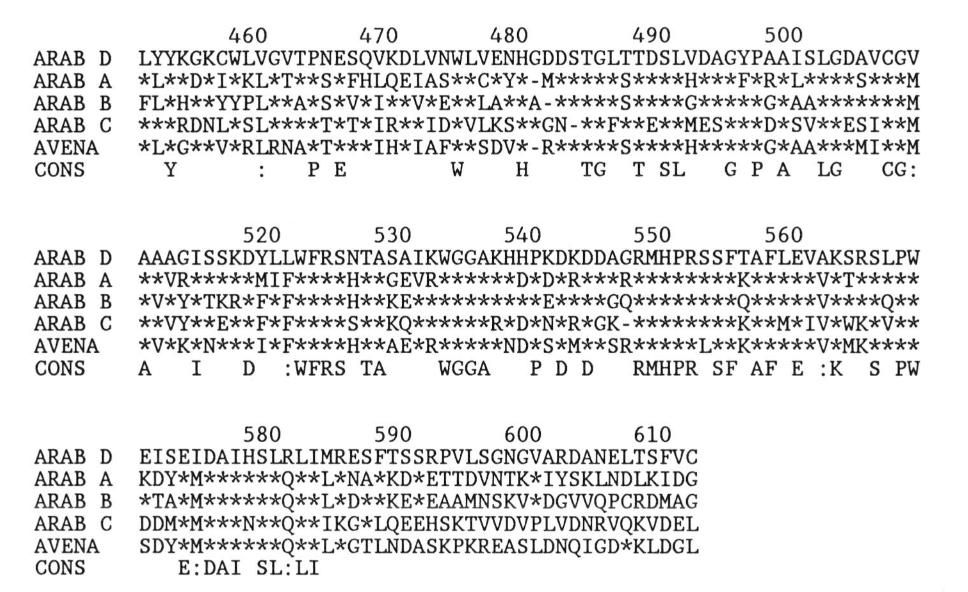

```
                  460       470       480       490       500
ARAB D    LYYKGKCWLVGVTPNESQVKDLVNWLVENHGDDSTGLTTDSLVDAGYPAAISLGDAVCGV
ARAB A    *L**D*I*KL*T**S*FHLQEIAS**C*Y*-M*****S****H***F*R*L****S***M
ARAB B    FL*H**YYPL**A*S*V*I**V*E**LA**A-*****S****G*****G*AA*******M
ARAB C    ***RDNL*SL****T*T*IR**ID*VLKS**GN-**F**E**MES***D*SV**ESI**M
AVENA     *L*G**V*RLRNA*T***IH*IAF**SDV*-R*****S****H*****G*AA***MI**M
CONS        Y      :   P E        W    H    TG  T SL   G P A  LG   CG:

                  520       530       540       550       560
ARAB D    AAAGISSKDYLLWFRSNTASAIKWGGAKHHPKDKDDAGRMHPRSSFTAFLEVAKSRSLPW
ARAB A    **VR*****MIF****H**GEVR******D*D*R***R*********K*****V*T*****
ARAB B    *V*Y*TKR*F*F****H**KE**********E****GQ********Q*****V****Q**
ARAB C    **VY**E**F*F****S**KQ******R*D*N*R*GK-********K**M*IV*WK*V**
AVENA     *V*K*N***I*F****H**AE*R*****ND*S*M**SR*****L**K*****V*MK****
CONS      A   I   D  :WFRS TA    WGGA   P D D   RMHPR SF AF E :K  S PW

                  580       590       600       610
ARAB D    EISEIDAIHSLRLIMRESFTSSRPVLSGNGVARDANELTSFVC
ARAB A    KDY*M******Q**L*NA*KD*ETTDVNTK*IYSKLNDLKIDG
ARAB B    *TA*M******Q**L*D**KE*EAAMNSKV*DGVVQPCRDMAG
ARAB C    DDM*M***N**Q**IKG*LQEEHSKTVVDVPLVDNRVQKVDEL
AVENA     SDY*M******Q**L*GTLNDASKPKREASLDNQIGD*KLDGL
CONS         E:DAI SL:LI
```

Figure 2. Comparison of *Arabidopsis* phytochrome type D amino acid sequence with other phytochrome sequences. The numbers indicate the position of the partial sequence of the *Arabidopsis* type D phytochrome in relation to the *Avena* sequence. The dots in the consensus sequences mark amino acids which are specifically exchanged in the *Arabidopsis* type D phytochrome.

translation products are given in Figure 3 in comparison with sequences of other phytochrome types of higher plants. This is the first (partial) sequence of a moss phytochrome. The degree of homology of the *Ceratodon* phytochrome sequences to all know phytochrome sequences is given in Table 2. From this data it is clear, that *Ceratodon* phytochrome cannot be unequivocally related to one of the known phytochrome types in higher plants (including phytochrome type D). This is somewhat surprising, because phytochrome of lower plants was thought to be more related to type II (green) phytochrome of higher plants: lower plants do not accumulate phytochrome in the dark and are green even when grown in the dark (Wada and Kadota, 1989). We have no data so far, that in *Ceratodon* more than one type of phytochrome is present. Hybridizing *Ceratodon* Southern blots with probes of phytochrome type A,B,C and D did not give clear evidence for the presence of more than one phytochrome type in *Ceratodon*. Interestingly the 243bp probe of the *Ceratodon* gene hybridized in another experiment more strongly to phytochrome type A than with types B, C, D in a Southern blot with *Arabidopsis* genomic DNA. In a Northern experiment we can clearly show that

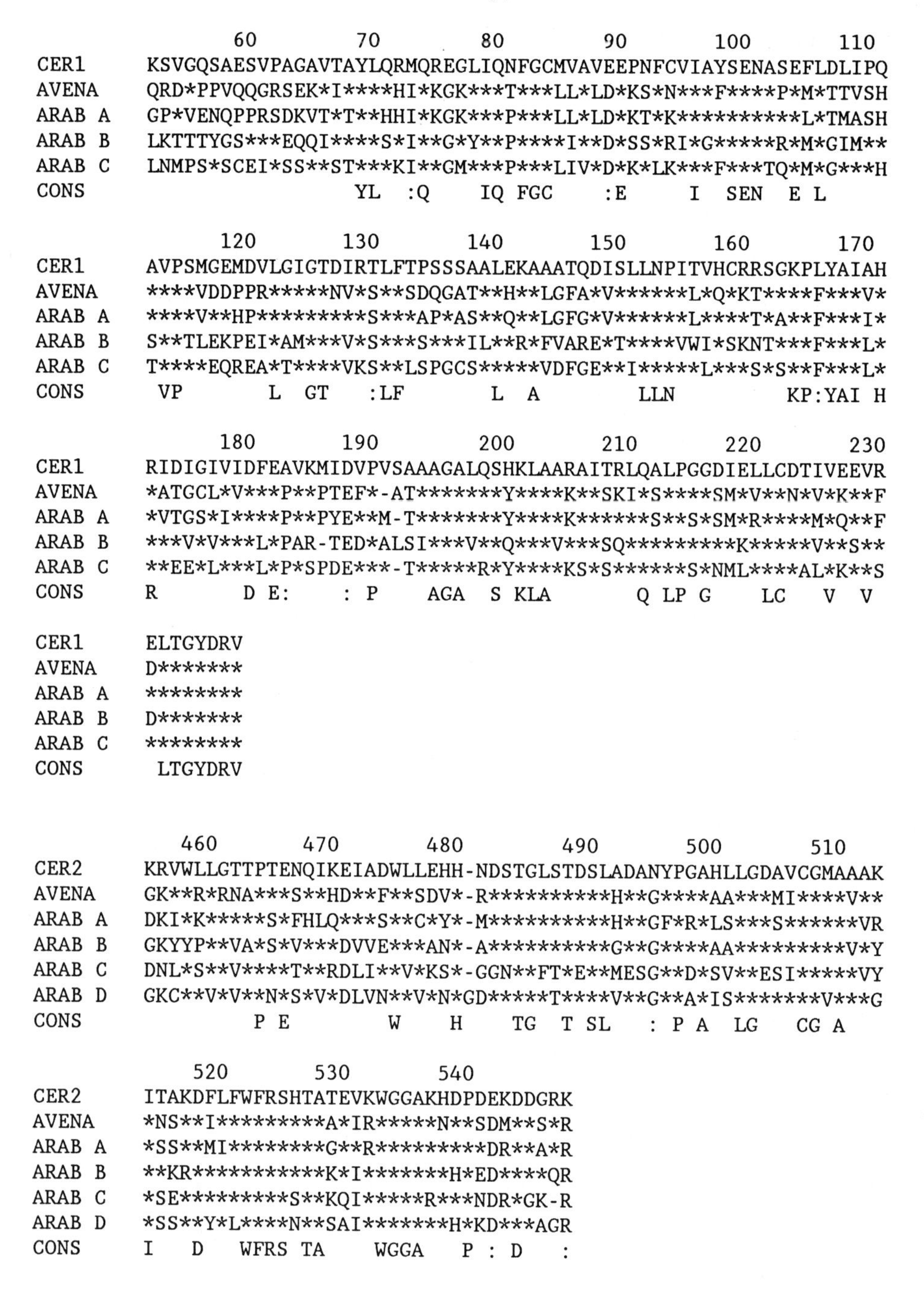

```
              60        70        80        90        100       110
CER1     KSVGQSAESVPAGAVTAYLQRMQREGLIQNFGCMVAVEEPNFCVIAYSENASEFLDLIPQ
AVENA    QRD*PPVQQGRSEK*I****HI*KGK***T***LL*LD*KS*N***F****P*M*TTVSH
ARAB A   GP*VENQPPRSDKVT*T**HHI*KGK***P***LL*LD*KT*K**********L*TMASH
ARAB B   LKTTTYGS***EQQI****S*I**G*Y**P****I**D*SS*RI*G*****R*M*GIM**
ARAB C   LNMPS*SCEI*SS**ST***KI**GM***P***LIV*D*K*LK***F***TQ*M*G***H
CONS                      YL  :Q    IQ FGC    :E     I  SEN  E L

             120       130       140       150       160       170
CER1     AVPSMGEMDVLGIGTDIRTLFTPSSSAALEKAAATQDISLLNPITVHCRRSGKPLYAIAH
AVENA    ****VDDPPR*****NV*S**SDQGAT**H**LGFA*V******L*Q*KT****F***V*
ARAB A   ****V**HP*********S***AP*AS**Q**LGFG*V******L****T*A**F***I*
ARAB B   S**TLEKPEI*AM***V*S***S***IL**R*FVARE*T****VWI*SKNT***F***L*
ARAB C   T****EQREA*T****VKS**LSPGCS*****VDFGE**I*****L***S*S**F***L*
CONS      VP       L  GT   :LF       L  A       LLN          KP:YAI H

             180       190       200       210       220       230
CER1     RIDIGIVIDFEAVKMIDVPVSAAAGALQSHKLAARAITRLQALPGGDIELLCDTIVEEVR
AVENA    *ATGCL*V***P**PTEF*-AT*******Y****K**SKI*S****SM*V**N*V*K**F
ARAB A   *VTGS*I****P**PYE**M-T*******Y****K******S**S*SM*R****M*Q**F
ARAB B   ***V*V***L*PAR-TED*ALSI***V**Q***V***SQ*********K*****V**S**
ARAB C   **EE*L***L*P*SPDE***-T*****R*Y****KS*S******S*NML****AL*K**S
CONS     R       D E:    : P    AGA  S KLA       Q LP G    LC   V  V

CER1     ELTGYDRV
AVENA    D*******
ARAB A   ********
ARAB B   D*******
ARAB C   ********
CONS      LTGYDRV


           460       470       480         490       500       510
CER2     KRVWLLGTTPTENQIKEIADWLLEHH-NDSTGLSTDSLADANYPGAHLLGDAVCGMAAAK
AVENA    GK**R*RNA***S**HD**F**SDV*-R**********H**G****AA***MI****V**
ARAB A   DKI*K*****S*FHLQ***S**C*Y*-M**********H**GF*R*LS***S******VR
ARAB B   GKYYP**VA*S*V***DVVE***AN*-A**********G**G****AA*********V*Y
ARAB C   DNL*S**V****T**RDLI**V*KS*-GGN**FT*E**MESG**D*SV**ESI*****VY
ARAB D   GKC**V*V**N*S*V*DLVN**V*N*GD*****T****V**G**A*IS*******V***G
CONS              P E        W    H    TG  T SL   : P A  LG   CG A

            520       530       540
CER2     ITAKDFLFWFRSHTATEVKWGGAKHDPDEKDDGRK
AVENA    *NS**I*********A*IR*****N**SDM**S*R
ARAB A   *SS**MI********G**R*********DR**A*R
ARAB B   **KR***********K*I*******H*ED****QR
ARAB C   *SE*********S**KQI*****R***NDR*GK-R
ARAB D   *SS**Y*L****N**SAI*******H*KD***AGR
CONS     I   D   WFRS TA    WGGA   P : D   :
```

Figure 3. Comparison of *Ceratodon* phytochrome amino acid sequences with phytochrome sequences of higher plants. The numbers indicate the position of the partial *Ceratodon* sequences in relation to the *Avena* sequence. The dots in the consensus sequences mark amino acids which are specifically exchanged in the *Ceratodon* sequences.

phytochrome transcript is present in considerable amounts in light grown *Ceratodon* protonema.

If we assume that *Ceratodon* phytochrome represents an ancient phytochrome form, none of the known phytochrome types (A,B,C,D) is more closely related to ancient phytochrome. Type I(A) and type II(B,C) phytochrome of higher plants must be therefore specific forms of phytochrome in Angiosperms. We have to admit, that this statement holds only true for parts of the phytochrome molecule probably closely related to its spectral properties. If you compare Figures 2 and 3 it becomes evident, that between all known phytochrome sequences there is a considerable amount of conserved amino acids. In this region around the chromophore binding site they represent about 30% of the amino acids. These "consensus sequences" can be found already in the four phytochrome types of *Arabidopsis* (see Fig. 2) and except a few amino acids they are also present in *Ceratodon* phytochrome (see Fig. 3). These highly conserved amino acids are certainly closely related to the function of phytochrome. In this region of the phytochrome gene, function should mean photoreversibility which is an expected property of "true phytochrome". *Ceratodon* shows clear R/FR reversible physiological reactions (Hartmann and Weber, 1988). Both sequences given in Figure 3 contain the parts of the peptide chain which have been discussed previously to stabilize the Pr and the Pfr chromophore (Rüdiger, 1987).

The site(s) of the phytochrome molecules which is(are) linked to the physiological reaction chain could be located at less conserved parts of the phytochrome molecules, representing the specific roles of the different phytochrome types in the plant cell. Thus it will be of great interest to see, if other parts of the *Ceratodon* phytochrome sequences are more similar to one of the known phytochrome types in higher plants. This investigation is currently in progress. Another statement with respect to the structure of phytochrome genes can be made with the results of our PCR experiments. In none of our experiments we amplified intron sequences. With a new primer combination L721/R1676, we obtained a fragment of around 0.9kb (Thümmler unpublished), the expected size of a PCR product without any large intron sequences. This implicates, that the intron/exon structure of all phytochrome genes so far investigated is highly conserved, at least around the chromophore binding site (see Fig. 4). This could be of great interest in evaluating the phylogeny of the phytochrome gene.

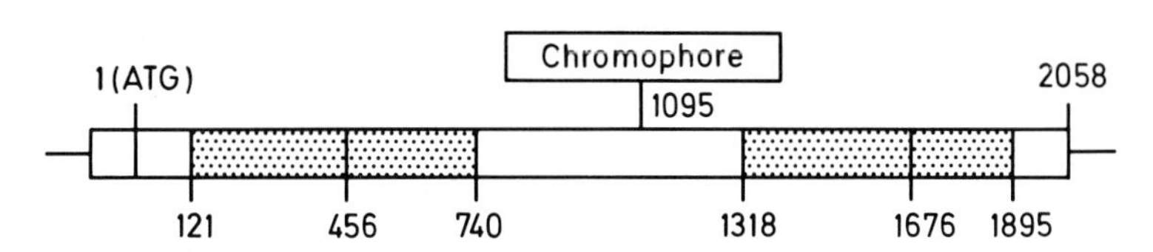

Figure 4. Presumed conserved gene structure of phytochrome genes around the chromophore binding site. Dotted areas show sequenced DNA fragments which were amplified by the polymerase chain reaction from *Ceratodon* and *Arabidopsis* DNA with primers shown in Fig. 1. Numbering is in accordance with the phytochrome gene of *Avena* (Hershey *et al.*, 1987).

NEW FUNCTION OF PHYTOCHROME IN MARINE MACROALGAE

A quite large number of physiological responses in algae mediated by phytochrome have been summarized by Dring (1988) and Wada and Kadota (1989). None of these responses seems to fit for those marine macroalgae in which we had detected a phytochrome-like protein by absorbtion spectroscopy and immunochemistry (see Table 1). We asked therefore whether we could find here a possible function for the photoreceptor.

We chose the green alga *Ulva rigida* for this purpose which gave a strong signal in Western blot and a relatively high phytochrome signal in visible spectroscopy (see Table 1). It had previously been shown that chlorophyll accumulation in this alga is regulated by phytochrome (López-Figueroa and Niell, 1989). The same type of light regulation was found for biliprotein accumulation in the red alga *Corallina elongata* (López-Figueroa *et al.*, 1989b). It is well known that stable accumulation of photosynthetic pigments requires their incorporation into protein complexes. We checked therefore the possibility that the experimental conditions which involved feeding of the algae with exogenous nitrate would lead to increased protein accumulation. The pigment accumulation could then be considered as a secondary effect.

We found that *Ulva* is able to synthesize rapidly large amounts of soluble protein from exogenous nitrate (López-Figueroa and Rüdiger, 1990). As shown in Table 3, this net synthesis of protein is regulated by phytochrome. As pointed out by López-Figueroa and Rüdiger (1990), other photoreceptors absorbing blue or green light might be involved as well. Gel electrophoresis of the extracts showed that single proteins were not predominant after phytochrome-induced accumulation but that all proteins are equally affected. This result led to the suggestion of a general stimulation of protein synthesis

Table 3. Possible control by phytochrome of protein accumulation in the green alga *Ulva rigida*. Extractable protein (mg. g fresh weight^{-1}) after light pulses followed by incubation with 0.5 mM KNO_3 in darkness for 3 h (data from López-Figueroa and Rüdiger, 1990).

Treatment	mg protein
Initial	1.7
dark control	1.9
5 min FR	1.7
5 min R	2.7
5 min R + 5 min FR	1.7
30 min R	3.0
30 min R + 5 min FR	2.1
30 min B	2.6
30 min B + 5 min FR	2.4
30 min G	2.5
30 min G + 5 min FR	2.1

from exogenous nitrate by phytochrome.

Several sites of phytochrome regulation can be discussed to explain the observed effect: it could be e.g. nitrate uptake, nitrate reduction, availability of NADH, stimulation of carbohydrate metabolism (to produce 2-oxo-carboxylic acids) for increased levels of amino acids, general stimulation of protein synthesis (see Fig. 5). In order to provisionally localize phytochrome action in this reaction chain, we determined the steady

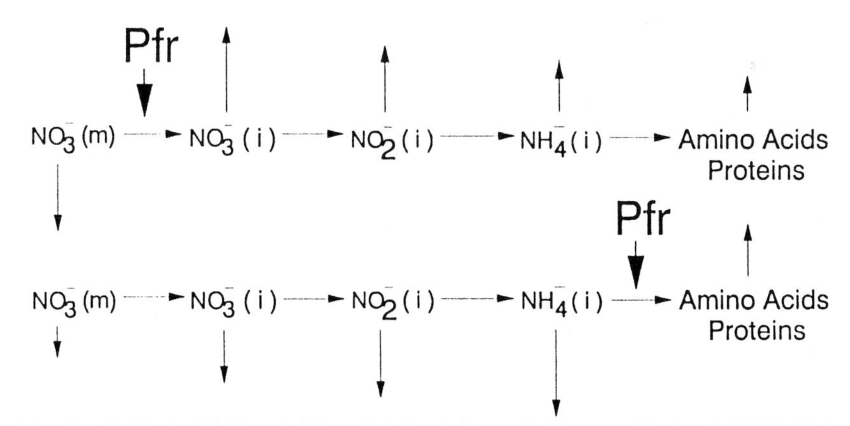

Figure 5. Consequence of phytochrome regulation of nitrate uptake and metabolism at different steps. Upper lane: Pfr stimulates nitrate uptake. The steady state level of nitrate in the medium decreases, that of nitrate in the cell and all subsequent levels go up. Lower lane: Pfr stimulates incorporation into amino acids. Only the steady state level of amino acids and later products increases, that of all previous compounds goes down. The upper possibility was verified (see Table 4).

state levels of the main substrates of this pathway under various light conditions. If one of these steps is specifically up-regulated by phytochrome (i.e. by red irradiation), then the steady-state level of the substrate of this step should decrease and the level of product of this step should increase. The experimental results (Table 4) clearly show that this condition only holds for nitrate uptake: the amount of nitrate in the culture medium after 3 h incubation is reproducibly lower after R or B light pulses than in the dark or after final FR irradiation.

Table 4. Nitrate uptake and reduction by the green alga *Ulva rigida*. Steady-state levels of nitrate in the medium [NO_3^-(m)], intracellular nitrate [NO_3^-(i)], nitrite [NO_2^-(i)] and ammonium [NH_4^+(i)] are given in μmol. g fresh weight^{-1}

Treatment	NO_3^-(m)	NO_3^-(i)	$\frac{NO_3^-(i)}{NO_3^-(m)}$	$\frac{NO_2^-(i)}{NO_3^-(i)}$	$\frac{NH_4^+(m)}{NO_2^-(i)}$
initial situation	75.0	28.5	0.4	0.04	0.8
3h D	28.0	28.0	1.0	0.03	2.3
5min FR, 3h D	25.0	28.0	1.1	0.04	2.5
5min R, 3h D	8.0	39.2	4.9	0.07	2.4
5min R, 5min FR, 3h D	23.0	24.7	0.9	0.05	1.6
30min R, 3h D	10.0	42.4	4.3	0.06	(5.3)
30min R, 5min FR, 3h D	26.0	36.6	1.4	0.04	1.0
5min B, 3h D	9.7	43.5	4.5	0.04	1.3
5min B, 5min FR, 3h D	24.0	29.1	1.2	0.02	1.0
30min B,3h D	9.5	40.0	4.2	0.07	1.1
30min B, 5min FR, 3h D	27.0	28.8	1.1	0.04	1.5

Vice versa, the intracellular amount of nitrate is significantly increased by R or B light pulses; this effect is reversed by subsequent FR irradiation. The concentration of nitrite(i)/nitrate(i) does not change as much as the ratio nitrate(i)/nitrate(m). Also the ratio ammonium(i)/nitrite(i) does not change much except for 30 min R. The changes, even if small, are also for nitrite and ammonium towards larger steady state levels after R or B than in the dark or after FR. This clearly demonstrates that the primary effect of phytochrome regulation is not at the steps of amino acid or protein synthesis but before nitrate reduction. The effect might be more closely related to the phytochrome effect on water permeability in *Mougeotia* (Weisenseel and Smeibidl, 1973) than to red (and blue) light effects on carbohydrate metabolism (Ruyters, 1988; Ruyters *et al.*, 1989).

REFERENCES

Adamse P, Bakker JA, Wesselius JC, Heeringa GH, Kendrick RE, Koornneef M (1988a) Photophysiology of a tomato mutant deficient in labile phytochrome. J Plant Physiol 133:436-440

Adamse P, Koornneef M (1988b) Photomorphogenetic mutants of higher plants. Photochem Photobiol 48:833-841

Cordonnier MM, Greppin H, Pratt LH (1986) Identification of a highly conserved domain on phytochrome from angiosperms to algae. Plant Physiol 80:982-987

Dring MJ (1988) Photocontrol of development in algae. Ann Rev Plant Physiol Plant Mol Biol 39:157-174

Furuya M (1987) Phytochrome and Photoregulation in Plants. Academic Press Inc Tokyo, Orlando, San Diego, New York, Austin, Boston, London, Sydney, Toronto

Grimm R, Lottspeich F, Schneider HjAW, Rüdiger W (1986) Investigation of the peptide chain of 124 kDA phytochrome: localization of proteolytic fragments and epitopes for monoclonal antibodies. Z Naturforsch 41c:993-1000

Hartmann E, Weber M (1988) Storage of the phytochrome-mediated phototropic stimulus of moss protonemal tip cells. Planta 175:39-49

Hershey HP, Barker RF, Kenneth BI, Murray GM, Quail PH (1987) Nucleotide sequence and characterization of a gene encoding the phytochrome polypeptide of *Avena*. Gene 61:339-348

Kay SA, Keith B, Shinozaki K, Chua, NH (1989) The sequence of the rice phytochrome. Nucleic Acids Res 17:2865-2866

Kidd DG, Lagarias JC (1990) Phytochrome from the green alga *Mesotaenium caldariorum*. J Biol Chem 265:7029-7035

Lindemann P, Braslavsky SE, Hartmann E, Schaffner K (1989) Partial purification and initial characterization of phytochrome from the moss *Atrichum undulatum* P.Beauv. grown in the light. Planta 178:436-442

López-Figueroa F, Niell FX (1989) Red light and blue-light photoreceptors controlling chlorophyll a synthesis in the red alga *Porphyra umbilicalis* and in the green alga *Ulva rigida*. Physiol Plant 76:391-397

López-Figueroa F, Rüdiger W (1990) A possible control by phytochrome and other photoreceptors of protein accumulation in the green alga *Ulva rigida*. Photochem Photobiol in press

López-Figueroa F, Lindemann P, Braslavsky SE, Schaffner K, Schneider-Poetsch HjAW, Rüdiger W (1989a) Detection of a phytochrome-like protein in macroalgae. Botanica Acta 102:178-180

López-Figueroa F, Perez R, Niell FX (1989b) Effects of red and far-red light pulses on the chlorophyll and biliprotein accumulation in the red alga *Corallina elongata*. J Photochem Photobiol 4:185-194

López-Figueroa F, Lindemann P, Braslavsky SE, Schaffner K, Schneider-Poetsch HA, Rüdiger W (1990) Detection of some conserved domains in phytochrome-like proteins from algae. J Plant Physiol in press

Rüdiger W (1987) Biochemistry of the phytochrome chromophore. In Phytochrome and Photoregulation in Plants (ed. Furuya M) pp. 127-138 Academic Press Inc Tokyo, Orlando, San Diego, New York, Austin, Boston, London, Sydney, Toronto

Ruyters G (1988) Light-stimulated respiration in the green alga *Dunaliella tertiolecta*: involvement of ultraviolet/blue-light photoreceptor(s) and phytochrome? Planta 174:422-425

Ruyters G, Grotjohann N, Kowallik W, Schneider-Poetsch HjAW (1989) Phytochrome in *Dunaliella*, *Chlorella* and other green algae. Europ Symp Photomorphogenesis in Plants, Freiburg 1989, Abstract No. 20

Saiki RK, Gelfand DH, Stoffel S, Scharf SJ, Higuchi R, Horn GT, Mullis KB, Erlich HA (1988) Primer-directed enzymatic amplification of DNA with a thermostable DNA polymerase. Science 239:487-491

Sato N (1988) Nucleotide sequence and expression of the phytochrome gene in *Pisum sativum*: differential regulation by light of multiple transcripts. Plant Mol Biol 11:697-710

Schneider-Poetsch HjAW, Schwarz H, Grimm R, Rüdiger W (1988) Cross-reactivity of monoclonal antibodies against phytochrome from *Zea* and *Avena*. Planta 173: 61-72

Sharrock RA, Quail PH (1989) Novel phytochrome sequences in *Arabidopsis thaliana*: structure, evolution, and differential expression of a plant regulatory photoreceptor family. Genes & Development 3:1745-1757

Shropshire WJr, Mohr H (1983) Photomorphogenesis. Encyclopedia of Plant Physiology, New Series, Vol 16A/B. Springer, Berlin Heidelberg

Sommer R, Tautz D (1989) Minimal homology requirements for PCR primers. Nucleic Acids Res 17:6749

Wada M, Kadota A (1989) Photomorphogenesis in lower plants. Annu Rev Plant Physiol Plant Mol Biol 40:169-191

Weisenseel MH, Smeibidl E (1973) Phytochrome controls the water permeability in *Mougeotia*. Z Pflanzenphysiol 70:420-431

BIOGENESIS OF PHYTOCHROME APOPROTEIN IN TRANSGENIC ORGANISMS AND ITS ASSEMBLY TO THE CHROMOPHORE

M. Furuya[1], K. Tomizawa[1], N. Ito[1], D. Sommer[1,2], L. Deforce[1,2], K. Konomi[1], D. Farrens[2] and P-S. Song[1,2]
[1]Laboratory of Plant Biological Regulation,
Frontier Research Program, RIKEN Institute,
Wako City, Japan 351-01.

INTRODUCTION

One of the major open questions with regard to phytochrome is what molecular mechanism results in the unique, photoreversible absorbance change between Pr and Pfr (Kendrick & Kronenberg, 1986; Furuya, 1987)? Because of the covalent binding of phytochrome apoprotein to the linear tetrapyrrole chromophore (Lagarias & Rapoport, 1980), it has been very difficult to modify experimentally either the apoprotein or the chromophore.

However, the recent development of molecular biology and gene engineering techniques has now provided us new approaches to studies of the phytochrome molecules. Two different molecular species of the phytochrome apoprotein, in terms of peptide map patterns and amino-acid, sequence were recently found in peas (Abe *et al.*, 1989). Phytochrome existing predominantly in etiolated plant tissues is named as phytochrome I (PI), and that discovered recently from green tissues and relatively stable, irrespective of light conditions, as phytochrome II (PII) (see the review, Furuya, 1989). Either cDNA and/or genomic DNA clones encoding PI apoprotein of several higher plants have been isolated and characterized by now, and became available as very useful tools for the analysis of phytochrome molecules (see reviews by Quail *et al.*, 1987 and Tomizawa *et al.*, 1990). Hence, a great road has now opened to obtain any particular region of PI apoprotein, or ones which are site-specifically mutated, without the chromophore, in transgenic micro-organisms, and assemble them to chemically modified bilin pigments. This article will introduce the latest story centring around these problems.

[2]Department of Chemistry, University of Nebraska, Lincoln, Nebraska 68588, USA.

NATO ASI Series, Vol. H 50
Phytochrome Properties and Biological Action
Edited by B. Thomas and C. B. Johnson

BIOGENESIS OF PHYTOCHROME I APOPROTEIN IN TRANSGENIC ORGANISMS

Bacteria. As the use of an expression system in *E. coli* has appeared promising to provide a large amount of protein as well as to facilitate deletion or replacement of amino acid residues (Maniatis *et al.*, 1986; Marston, 1986), we have tried to establish a system in which phytochrome apoprotein is expressed in a bacterial transformant (Tomizawa *et al.*, in press). The plasmid pPP1001 was constructed to contain a full-length pea PI cDNA with the Shine-Dalgarno consensus sequence by being joined upstream of the cDNA in an expression vector pNUT6. The *E. coli* transformant with PPP1001 produced a typical inclusion body when cultured at 32°C, while bacteriolysis occurred at higher temperatures. The latter might result from a toxic effect on *E. coli* of rapidly accumulating pea PI apoprotein. When the product is synthesized at such a high rate that the cell's degradation system(s) become saturated, typical inclusion bodies are formed in *E. coli* transformants (Prouty *et al.*, 1975). Hence we examined immunocytochemically the intracellular distribution of PI product in an *E. coli* transformant using anti-pea PI monoclonal antibodies (Nagatani *et al.*, 1987), and found that immunochemically stainable material(s) was distributed not only in the inclusion bodies but throughout the cytoplasmic region of the transformant (Tomizawa *et al.*, in press).

In Western blot analysis, 70 and 100 kDa PI polypeptides were immunochemically detected with the fraction of the inclusion bodies (Fig. 1), and a 90 kDa PI polypeptide in the cytosolic fraction. The amino-terminus of 100 kDa sample was blocked as in native pea PI, but minor signals in micro-sequence analysis indicated the same sequence as the reported N-terminal sequence of native pea PI (Sato, 1988). When the plasmid was incubated in a cell-free system of *E. coli*, a small amount of 120 kDa polypeptide was immunochemically detectable. This evidence suggested that the 100 kDa sample lacked 20 kDa of its carboxyl-terminal region, probably resulting from differences of codon usage between pea and *E. coli*.

Secondary structure of the 100 kDa PI apoprotein in the absence of the chromophore was first estimated 20% as alpha-helix, 40% as beta-sheet, 15% as beta-turn and 25% as random coil in terms of CD spectra (Sommer *et al.*, to be published). We are now determining the secondary structure of the apoprotein assembled to the chromophore, to find any difference between PI apoproteins with or without the chromophore attachment.

Yeast. To obtain full-length apoprotein of pea PI, we next tried to express pea PI cDNA in a yeast strain, *Saccharomyces cerevisiae* CG380, using another expression vector, pAA7. A single band of ca. 120 kDa, stainable with an anti PI monoclonal antibody, was detected on a Western blot of lysate from the yeast transformant (Fig. 1). All of our anti-pea PI monoclonal antibodies (Nagatani *et al.*, 1987), which recognize different

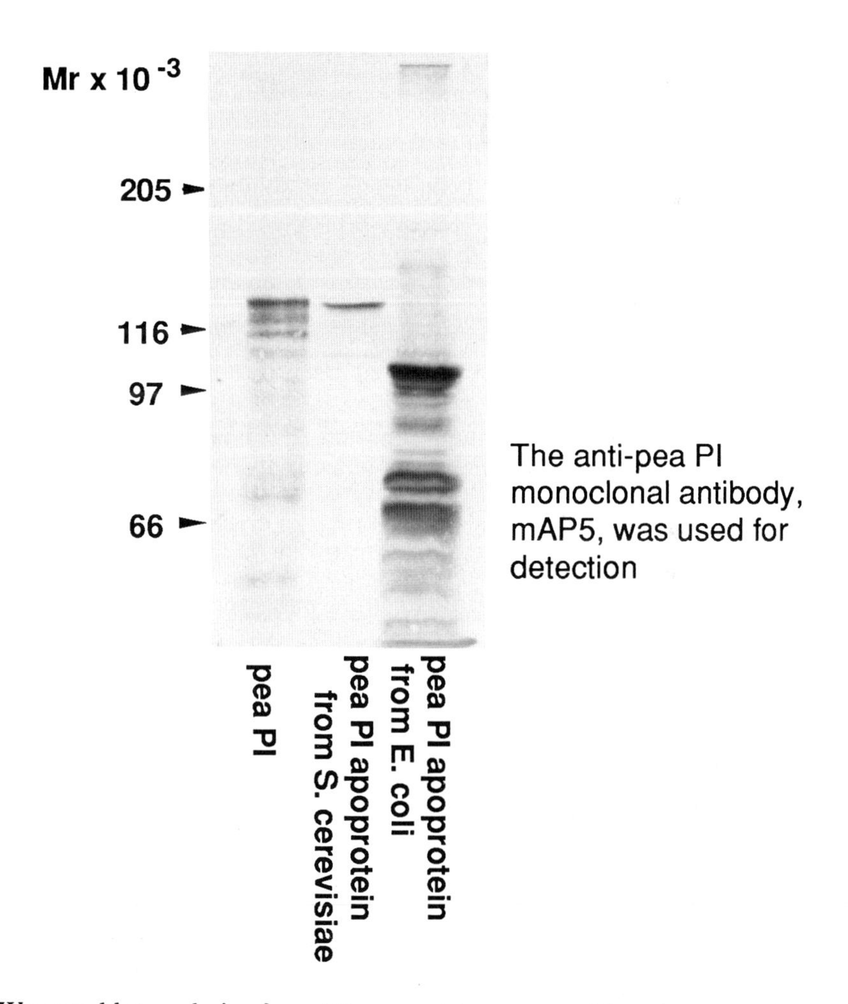

Figure 1. Western blot analysis of pea PI apoprotein expressed in micro-organisms. Either *E. coli* or *Saccharomyces cerevisiae* was transformed with pea PI cDNA-carrying expression vector pPP1001 or pPP1300, respectively. The lysate of the transformant was analyzed by SDS-PAGE and western blotting. Pea PI apoprotein was detected with anti-pea PI monoclonal antibody, mAP5. Partially purified pea PI was used as a control.

epitopes of pea PI, reacted positively with the 120 kDa sample, indicating that the 120 kDa apoprotein prepared from the yeast transformant covers the entire region of pea PI apoprotein. When conventional disruption methods, requiring tedious vortexing with glass beads, were used, the 120 kDa apoprotein was not found in the soluble fraction but went entirely into the pelletable fractions. Remembering that PI in the Pr form is soluble in the cytosol but that as Pfr it becomes pelletable in homogenates of oat (Quail *et al.*, 1973) and pea (Manabe & Furuya, 1975), the surface properties of PI apoprotein look like Pfr rather than Pr. It was not possible to solubilize the PI apoprotein from the pellet unless it had been denaturated by SDS or guanidine-HCl. When protoplasts of the yeast transformant were prepared and disrupted quickly with osmotic cell lysis, as used in *Mesotaenium caldariorum* (Kidd and Lagarias, 1990), only a very small amount of PI apoprotein was isolated in the soluble fraction.

Thus, several trials have been made to solubilize a large amount of the PI apoprotein from the yeast cell and refold it. For example, most of PI apoprotein in the transformant was dissolved in solution of 6 M guanidium hydrochloride. This sample showed a molecular mass of 120 kDa in gel filtration under denaturing conditions, which also enriched the sample. The enriched fraction was then dialysed to remove the denaturant in the presence of reagents for protein refolding and/or the formation of disulphide bridges. The fraction was subsequently applied to a DEAE-sepharose column chromatography, from which the PI apoprotein that eluted out ran in the neighbourhood of 300kDa on the gel filtration column. This indicates that PI apoprotein without the chromophore exists as a dimer (Ito *et al.*, to be published) as does native pea PI with the chromophore (Nakasako *et al.*, 1990). The dimer of PI apoprotein is stable in the presence of 0.5 M NaCl, while an aggregate of PI apoprotein is immediately formed in the absence of the salt. The optimal salt concentration appears essential to stabilize the dimer of PI apoprotein. This dimer preparation, however, has so far been unsuccessful in assembling to the chromophore *in vitro*.

On the other hand, when the transgenic yeast was quickly deep-frozen in liquid nitrogen and disrupted vigorously using a mortar and pestle, PI apoprotein expressed in the transformant could be isolated in the extraction buffer and fractionated by 40% saturated ammonium sulphate. Resolubilized PI apoprotein eluted as a PI dimer on a Sephacryl S-300 column. This apoprotein is ca 1 kDa smaller than native pea PI upon

SDS-PAGE (Fig. 1). When PI apoprotein, without chromophore, was prepared from gabaculin-treated pea seedlings, no significant difference in molecular size on the SDS-PAGE was found between it and native pea PI with chromophore. Hence the above-mentioned difference of approximate 1 kDa between native PI and the yeast product probably does not result from the presence or absence of the chromophore (Deforce *et al.*, to be published). The difference of molecular mass between native PI and the yeast product, however, may possibly be due to somewhat different post-translational modification of PI apoprotein in pea and yeast. In fact, it is known that phytochrome apoprotein is phosphorylated in oat (Wong *et al.*, 1986) and glycosylated in oat, *Sinapis* and pea (Grimm, personal communication). It is interesting to note that the glycosylation machinery of yeast is different from other organisms (Tanner and Lehle, 1987), so that PI apoprotein produced in yeast could be differently glycosylated from native PI in pea tissues (Grimm *et al*, to be published).

Higher plants. In addition to the micro-organisms described above, PI cDNA can also be introduced and over-expressed in higher plants. Full-length PI cDNAs from rice (Kay *et al.*, 1989) and oat (Keller *et al.*, 1989; Boylan and Quail, 1989) have been fused to the cauliflower mosaic virus 35S promoter and expressed in transgenic tobacco or tomato plants, respectively.

Using species-specific, anti-PI monoclonal antibodies in immunoblots (Nagatani *et al.*, 1987), rice and tobacco PI apoproteins were separated and characterized with fractions eluted from a DEAE-Sepharose column of extracts from transgenic and wild tobacco plants (Kay *et al.*, 1989). Tobacco PI is present in both transgenic and wild plants, whereas rice PI is over-expressed only in the transgenic plants.

BIOGENESIS OF PHYTOCHROMOBILIN

The chemical structure of phytochromobilin, the phytochrome chromophore, was first characterized by Siegelman *et al.* (1966) and then by Rüdiger and Correll (1969). The A-ring of phytochromobilin was covalently assembled to a cysteine of the apoprotein (Lagarias and Rapoport, 1980) and the oat mature PI polypeptide is 1128 amino acid-long, with a single chromophore attachment site at Cys-321 (Hershey *et al.*, 1985).

The biogenesis pathway of phytochromobilin has not been thoroughly studied because of technical difficulties. Bonner (1967) reported that ^{14}C-labelled 5-aminolevulinic acid was incorporated into the phytochrome fraction prepared from pea seedlings. However, there has been little subsequent investigation using tracer techniques. The biogenesis pathway of phytochromobilin has been mainly analyzed by using diverse metabolic inhibitors for tetrapyrrole synthesis, and the results from the literature is summarized along with our recent data in Table 1.

Table 1. Hypothesized pathway of biogenesis of phytochromobilin

Step	Inhibitor	References
Succinyl CoA + Glycine	Gabaculin	Elich & Lagarias (1987) Gardner & Gorton (1985) Jones *et al.* (1986) Konomi & Furuya (1986)
	4-amino-5-hexanoic acid	Elich & Lagarias (1988)
5-aminolevulinic acid		
Protophorphyrinogen	S-23142	Konomi *et al.* (to be published)
Protoporphyrin	N-methyl mesoporphyrin IX	Konomi *et al.* (to be published)
Fe-protoporphyrin		
Biliverdin		
Phytochromobilin		

It has been confirmed by the use of gabaculin (Elich and Lagarias, 1987; Gardner and Gorton, 1985; Jones *et al.*, 1986; Konomi and Furuya, 1986) and the specific inhibitor 4-amino-5-hexanoic acid (Elich and Lagarias, 1988), that the assembly of succinyl CoA to glycine is the first step of phytochromobilin synthesis in plants.

Recently, N-(4-chloro-2-fluoro-5-propargyloxy)phenyl-3,4,5,6-tetraphthalimide (abbreviated as S-23142), produced as a herbicide by Sumitomo Chemical Company, was found to inhibit protoporphyrin synthesis in higher plants (Mito *et al.*, 1990). Hence we examined whether S-23142 could prevent the biosynthesis of spectrophotometrically detectable phytochrome using a bioassay with imbibed embryonic axes of pea seeds

(Konomi and Furuya, 1986). Photoreversible absorbance change between Pr and Pfr in the pea embryonic axes was no longer observed in the presence of S-23142 at 0.1 mM or higher, and 50% inhibition was observed ca. 1 μM (Konomi *et al.*, to be published). This indicates that the process from protoporphyrinogen to protoporphyrin may be involved in the biosynthesis of phytochromobilin in pea tissues.

Similarly, the R/FR reversible absorbance change in imbibed pea embryonic axes was lowered in the presence of N-methyl mesoporphyrin IX (Konomi *et al*, to be published), which is an inhibitor of Fe chelatase for heme and inhibits the synthesis of phycocyanobilin but not chlorophyll formation in Cyanidium (Beale and Chen, 1983). These results appear to be consistent with the hypothesis that protoporphyrinogen, protoporphyrin and heme are the intermediates in the biogenesis of phytochromobilin (Brown *et al.*, 1990).

ASSEMBLY OF PHYTOCHROME APOPROTEIN TO THE CHROMOPHORE

In vivo assembly. As already described earlier, when monocot PI cDNA was introduced into dicot plants, monocot PI apoprotein was over-expressed in the resulting transgenic plants. In all cases, monocot PI apoprotein over-expressed in dicots was assembled to dicot chromophore and showed the typical R/FR photoreversible change of absorption spectra.

Furthermore, the over-expressed monocot PI was biologically active in transgenic dicots in terms of photomorphogenesis (Keller *et al.*, 1989; Boylan and Quail, 1989) and capable of regulating the circadian pattern of Cab gene expression (Kay *et al.*, 1989). We know that, assuming that the amount of PII in transgenic plants is too low to influence the detection of over-expresssed PI, the total amount of spectrophotometrically detectable phytochrome *in vivo* is of the same order as that of the immunochemically detectable one, indicating that the over-expressed monocot PI apoprotein would be mostly assembled to the chromophore in transgenic dicots. In another words, it appears difficult to prepare PI apoprotein without the chromophore from extracts of transgenic higher green plants.

In vitro assembly. The reactivity of pea PI apoprotein, expressed in yeast, to different tetrapyrrole chromophores *in vitro* was next examined in spectrophotometric assay. The adduct of pea PI apoprotein was formed by adding C-phycocyanobilin, prepared from

Synechoccus (Glazer and Cohen-Bazire, 1971) at various concentrations at 4°C to the solution of pea PI apoprotein produced in the yeast transformant. The red/far-red photoreversible absorbance change of the adduct was then measured by spectrophotometry (Deforce *et al.*, to be published). The adduct clearly demonstrates a typical difference spectrum of phytochrome with minimum and maximum peaks at 652 and 710.5 nm respectively (Fig. 2), while no such R/FR photoreversible spectral change was observed with the apoprotein or with C-phycocyanobilin alone. The observed blue shift

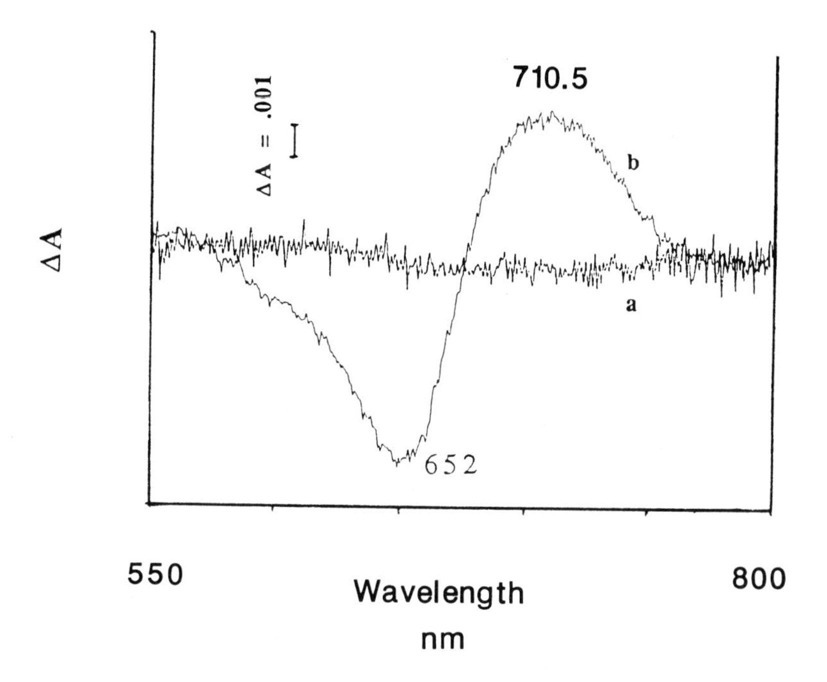

Figure 2. Difference spectrum of 6 μM C-phycocyanobilin (a) and pea phytochrome I apoprotein assembled *in vitro* to C-phycocyanobilin (b) in a 50 mM tris-HCl buffer pH 7.8 at 4°C (Deforce *et al.*, to be published). Both samples were irradiated with 100 W/m^2 red light for 60 seconds and then with 25 W/m^2 far-red light for 120 seconds.

is expected as phytochromobilin is substituted by C-phytocyanobilin in the adduct formation. The Zn^{++}-induced bilin fluorescence (Tokuhisa and Quail, 1989) was clearly found with the adduct, indicating that the chromophore is covalently bound to the apoprotein.

The formation of the thio-ether bond between the apoprotein and the chromophore occurs very rapidly at 4°C in the reaction mixture in the absence of pea homogenate which may contain some enzyme(s) like C-S-lyase for the assembly. It thus appears that assembly *in vitro* is an autocatalytic process as reported with the adduct formation of phycocyanobilin and oat PI apoprotein prepared from tetrapyrrole-deficient seedlings (Elich and Lagarias, 1989). However, as nothing is known at moment about the intracellular site of the chromophore assembly and the concentration *in vivo* of phytochromobilin and PI apoprotein, it remains to be established whether the adduct formation *in vivo* is autocatalytic. Besides C-phycocyanobin, we have tested whether other bilin compounds form the adduct with the yeast product. However, the trial has so far been unsuccessful.

CONCLUDING REMARKS

The production of phytochrome apoprotein in *E.coli* (Tomizawa *et al.*, in press) and *Saccharomyces cerevisiae* (Ito *et al.*, to be published; Deforce *et al.*, to be published) has opened a new approach to structure-function studies of the photoreversible change between Pr and Pfr. This is because we now can not only produce 1) phytochrome apoprotein without chromophore, 2) variously modified apoprotein by gene engineering, and 3) enough amount of apoprotein for determination of its chemical and physical properties, but also carry out the assembly *in vitro* of such designed apoproteins to any candidates of the chromophore.

The cleavage of the covalent linkage between phytochrome apoprotein and phytochromobilin was carried out first by denaturing phytochrome in 5% trichloroacetic acid and refluxing it in methanol for a long time (Siegleman *et al.*, 1966) and later by chromic acid degradation (Rüdiger & Correll, 1969). Both of these methods were so drastic that the resulting apoprotein, in the absence of the chromophore, was severely denatured and no longer useful for structure-function analysis. Hence there have been no publications on the molecular properties of phytochrome apoprotein in the absence of the chromophore during the past two decades. In contrast, pea PI apoprotein produced in yeast is native so that we can measure any molecular properties of the apoprotein in the absence of the chromophore and see the effect of the chromophore attachment on the apoprotein *in vitro* in terms of photoreversible absorbance change (Fig.

Table 2. Yield of pea phytochrome I apoprotein from various materials.

Material	PI gene source	PI Apoprotein (μg/g.fr.wt)	References
Pea shoot (light-grown)	*in vivo*	0.15	Abe *et al.* (1985)
Pea epicotyl (dark-grown)	*in vivo*	4.8	Abe *et al.* (1985)
Yeast transformant	pPP1300	40	Ito *et al.* (to be published)
E.coli transformant	pPPP1001	230	Tomizawa *et al.* (to be published)

Table 3. Samples required for measurement of physical properties with phytochrome.

Method	Sample		References
	PI concentration & Cuvette size	Total amount (μg/each cuvette)	
X-ray scatter	30mg/mlx100μl	3,000	Nakasako *et al.* (1990)
Raman spectra	10mg/mlx50-100μl	500-1,000	Fodor *et al.* (1988)
Fluorescence	1mg/mlx500μl	500	Inoue *et al.* (1985)
Infra-red spectra	10mg/mlx 10μl	100	Sakai *et al.* (1990)
UV-visible spectra	100μg/mlx100μl	100	
GCMS		50-100	
CD spectra	50μg/mlx110μl	25- 50	Sommer & Song (1990)
Surface enhance resonance Raman scattering	1.4 mg/mlx3μl	4	Farrens *et al.* (1989)
Scanning tunnelling microscopy		0.002	Hara *et al.* (1990)

2). This new approach will provide a useful tool to solve several pending problems centring around physical and chemical interactions between phytochrome apoprotein and the chromophore. We are doing further experiments with deletion series of PI cDNA and site-specific mutated PI cDNA in its chromophore-attached region.

Low concentrations of phytochrome *in vivo* (Table 2) in green plants, although it is somewhat higher in etiolated tissues, has presented a technical problem to the study of its molecular properties in a purified form. However, the yield of PI apoprotein from bacterial and yeast transformants is significantly higher even in this preliminary work than that described for higher plants in the literature (Table 2). The minimum amounts of phytochrome sample for single measurements of major physical and chemical properties are summarized in Table 3. When you take a careful look at Tables 2 and 3, you will find how much our labour and time for sample preparation will be saved by the new molecular approaches that we have described in this article, and progress in future along this line will be most promising.

REFERENCES

Abe H, Takio K, Titani K, Furuya M (1989) Amino-terminal amino acid sequences of pea phytochrome II fragments obtained by limited proteolysis. Plant Cell Physiol. 30:1089-1097

Beale SI, Chen NC (1983) N-methyl mesoporphyrin IX inhibits phycocyanin, but not chlorophyll synthesis in *Cyanidium caldarium*. Plant Physiol. 71:263-268

Bonner BA (1967) Incorporation of delta aminolevulinic acid into the chromophore of phytochrome. Plant Physiol. 42: suppl., s-11

Boylan MT, Quail PH (1989) Oat phytochrome is biologically active in transgenic tomatoes. The Plant Cell 1:765-773

Brown SB, Houghton JD, Vernon DI (1990) Biosynthesis of phycobilins. Formation of the chromophore of phytochrome, phycocyanin and phycoerythrin. J. Photochem. Photobiol. B: Biology, 5:3-23

Deforce L, Tomizawa K, Ito N, Farrens D, Song PS, Furuya M (to be published) The expression of pea phytochrome I apoprotein in yeast for bilin-apophytochrome reconstitution

Elich TD, Lagarias JC (1987) Phytochrome chromophore biosynthesis. Both 5-aminolevulinic acid and biliverdin overcome inhibition by gabaculine in etiolated *Avena sativa* L. seedlings. Plant Physiol. 84:304-310

Elich TD, Lagarias JC (1988) 4-amino-5-hexanoic-acid-a potent inhibitor of tetrapyrrole biosynthesis in plants. Plant Physiol. 88:747-751

Elich TD, Lagarias JC (1989) Formation of a photoreversible phycocyanobilin - apophytochrome adduct *in vitro*. J.Biol.Chem. 264:12902-12908

Farrens DL, Holt RE, Rospendowski BN, Song PS, Cotton TM (1989)

Surface-enhanced resonance Raman scattering spectroscopy applied to phytochrome and its model compounds. 2. Phytochrome and phycocyanin chromophores. J. Am. Chem. Soc. 111:9162-9169

Fodor SPA, Lagarias JC, Mathies RA (1988) Rsonance raman spectra of the Pr-form of phytochrome. Photochem. Photobiol. 48:129-136

Furuya M (ed)(1987) Phytochrome and photoregulation in plants. Academic Press, Tokyo

Furuya M (1989) Molecular properties and biogenesis of phytochrome I and II. Adv. Biophys. 25:133-167

Glazer AN, Cohen-Bazire G (1971) Subunit structure of the phycobiliproteins of blue-green Algae. Proc. Nat. Acad. Sci. USA. 68:1398-1401

Hara M, Iwakabe Y, Tochigi K, Sasabe H, Garito AF, Yamada A (1990) Anchoring structure of smectic liquid-crystal layers on MoS_2 observed by scanning tunnelling microscopy. Nature 344:228-230

Hershey HP, Barker RF, Idler KB, Lissemore JL, Quail PH (1985) Analysis of cloned cDNA and genomic sequences for phytochrome: Amino acid sequences for two gene products expressed in etiolated *Avena*. Nucleic Acids Res. 13:8543-8559

Holt SE, Farrens DL, Song PS, Cotton TM (1989) Surface-enhanced resonance raman scattering spectroscopy applied to phytochrome and its model compounds 1. Biliverdin photoisomers. J. Am. Chem. Soc. 111:9156-9162

Inoue Y, Hamaguchi H, Yamamoto KT, Tasumi M, Furuya M (1985) Light induced fluorescence spectral changes in native phytochrome from *Secale cereale* L. at liquid nitrogen temperature. Photochem. Photobiol. 42:423-427

Ito N, Tomizawa K, Furuya M (to be published) Dimerization site of pea phytochrome I apoproteins that were produced in yeast transformant

Jones AM, Allen CD, Gardner G, Quail PH (1986) Synthesis of phytochrome apoprotein and chromophore are not coupled obligatorily. Plant Physiol 81:1014-1016

Kay SA, Nagatani A, Keith B, Deak M, Furuya M, Chua NH (1989) Rice phytochrome is biologically active in transgenic tobacco. The Plant Cell 1:775-782

Keller JM, Shanklin J, Vierstra RD, Hershey HP (1989) Expression of a functional monocotyledonous phytochrome in transgenic tobacco. EMBO J. 8: 1005-1012.

Kendrick RE, Kronenberg GHM (ed) (1986) Photomorphogenesis in plants. Nijhoff, Dordrecht

Kidd DG, Lagarias JC (1990) Phytochrome from the green alga *Mesotaenium caldariorum* purification and preliminary characterization. J. Biol. Chem. 265:7029-7035

Konomi K, Furuya M (1986) Effects of gabaculine on phytochrome synthesis during imbibition in embryonic axes of *Pisum sativum* L. Plant Cell Physiol. 27:1507-1512

Konomi K, Li HS, Furuya M (to be published) Effects of N- (4-chloro-2-fluoro-5-propargyloxyphenyl)-3,4,5,6-tetraphthalimide and N-methyl mesoporphyrin IX on phytochrome synthesis during imbibition in embryonic axes of *Pisum sativum* L.

Lagarias JC, Rapoport H (1980) Chromopeptides from phytochrome. The structure and linkage of the Pr form of the phytochrome chromophore. J. Am. Chem. Soc. 102:4821-4828

Maniatis T, Fritsch EF, Sambrook J (1986) Molecular cloning: A Laboratory Manual. Cold Spring Habor Laboratory, Cold Spring Habor, New York

Marston FAO (1986) The purification of eukaryotic polypeptides synthesized in *Escherichia coli*. Biochem. J. 240: 1-12

Mito N, Sato R, Miyakado M, Oshio H, Tanaka S (1990) Mode of actions of N-phenylimide herbicide, S-23142. Plant Physiol. 93: Suppl. No.37

Nagatani A, Lumsden PJ, Konomi K, Abe H (1987) Application of monoclonal antibodies to phytochrome studies. In "Phytochrome and photoregulation in plants" (ed. M. Furuya), p.95-114. Academic Press, Tokyo

Nakasako M, Wada M, Tokutomi S, Yamamoto KT, Sakai J, Kataoka M, Tokunaga F, Furuya M (1990) Quaternary structure of pea phytochrome I dimer studied with small-angle X-ray scattering and rotary-shadowing electron microscopy. Photochem. Photobiol.

Prouty WF, Karnovsky MJ, Goldberg AL (1975) Degradation of abnormal proteins in Escherichia coli. J. Biol. Chem. 250:1112-1122

Quail PH, Gatz C, Hershey HP, Jones AM, Lissemore JL, Parks BM, Sharrock RA, Barker RF, Idler KB, Murray MG, Koornneef M, Kendrick RE (1987) Molecular biology of phytochrome. In "Phytochrome and photoregulation in plants" (ed. M. Furuya), p.23-37. Academic Press, Tokyo

Rüdiger W, Correll DL (1969) Uber die Struktur des Phytochrom-Chromophors und seine Protein-Bindung. Liebigs Ann.Chem. 723:208-212

Sakai J, Morita EH, Hayashi H, Furuya M, Tasumi M (to be published) Infrared studies of the phototransformation of phytochrome

Sato N (1988) Nucleotide sequence and expression of the phytochrome gene in *Pisum sativum*: Differential regulation by light of multiple transcripts. Plant Mol. Biol. 11:697-710

Siegelman HW, Turner BC, Hendricks SB (1966) The chromophore of phytochrome. Plant Physiol. 41:1289-1292

Sommer D, Farrens D, Song PS, Tomizawa K, Furuya M (to be published) Secondary structure of 100 kDA pea phytochrome type I apoprotein in presence and absence of the chromophore

Sommer D, Song PS (1990) Chromophore topography and secondary structure of 124-kilodalton *Avena* phytochrome probed by Zn^{2+}-induced chromophore modification. Biochemistry 29:1943-1948

Tanner W, Lehle L (1987) Protein glycosylation in yeast. Biochem. Biophys. Acta 906:81-99

Tokuhisa JG, Quail PH (1989) Phytochrome in green-tissue: Partial purification and characterization of the 118-kilodalton phytochrome species from light-grown *Avena sativa* L. Photochem Photobiol 50:143-152

Tomizawa K, Ito N, Komeda Y, Uyeda TPQ, Takio K, Furuya M (1990) Characterization and intracellular distribution of pea phytochrome I polypeptides expressed in *E. coli.* Plant Cell Physiol. (in press)

Tomizawa K, Nagatani A, Furuya M (1990) Phytochrome genes: Studies using the tools of molecular biology and photomorphogentic mutants. Photochem. Photobiol. 52:265-275

Wong YS, Cheng HC, Walsh DA, Lagarias JC (1986) Phosphorylation of *Avena* phytochrome *in vitro* as a probe of light-induced conformational changes. J. Biol. Chem. 261:12089-12097

CONFORMATION AND ITS FUNCTIONAL IMPLICATIONS IN PHYTOCHROME

W. Parker, M. Romanowski and P-S. Song
Institute for Cellular and Molecular Photobiology,
Department of Chemistry,
University of Nebraska,
Lincoln, NE 68588, USA.

INTRODUCTION

An important part of elucidating the biological function of phytochrome is an understanding of the different domains and their function/relationship in the molecule. Phytochrome has two basic domains per monomer, a 69-72 kDa chromophore and a 52-55 kDa non chromophore domain connected by a proteolytically sensitive linker region (Lagarias and Mercurio, 1985). Intact phytochrome has been shown to be a dimer (Lagarias and Mercurio, 1985; Jones and Quail, 1986) and the region responsible for subunit contact is probably localized within 42 kDa of the C-terminus (Jones and Quail, 1986). Recently, small-angle X-ray scattering (Tokutomi *et al.,* 1989), as well as electron microscope studies (Jones and Erickson, 1989) have yielded a model of the phytochrome dimer which entails a distinct separation of the non-chromophore and chromophore domains. This two domain model, as well as a more detailed structural domain model (Romanowski and Song, 1990), is discussed.

We have used Zn^{2+} ions to probe the structure of the phytochrome protein and "apoprotein". Zn^{2+} ions co-ordinate with the phytochrome chromophore in a one to one molar ratio yielding a bleached spectrum as well as a drastically modified conformation in the phytochrome protein (Sommer and Song, 1990; Fig. 6). This chelation, in effect, produces an "apoprotein" which lacks many of the chromophore - protein interactions present in native phytochrome. Information concerning the structure of Zn^{2+}-chelated phytochrome is thus useful since it may directly relate to the structure of the apoprotein. Comparisons of Zn^{2+}-chelated Pfr with native phytochrome can also provide information concerning the native (unchelated) protein.

To further probe the structure of Zn^{2+}-chelated phytochrome we have determined its quaternary structure by size exclusion chromatography. CD results obtained by comparison of native and Zn^{2+}-chelated phytochromes are also presented. In addition,

NATO ASI Series, Vol. H 50
Phytochrome Properties and Biological Action
Edited by B. Thomas and C. B. Johnson

we have analyzed the N-terminal segment (about 6 kDa) of phytochrome in some detail. This is probably the section of the phytochrome polypeptide which has received the most attention. The study of the N-terminus chain is interesting since it appears to have very different conformation and orientation with respect to the rest of the protein depending on chromophore form (Pr or Pfr). This section of the polypeptide appears to undergo a more drastic change in conformation/orientation than any other part of the polypeptide according to proteolytic (Jones *et al.,* 1985; Lagarias and Mercurio, 1985), antibody affinity (Cordonnier *et al.,* 1985), circular dichroic (Vierstra *et al.,* 1987) and antibody/circular dichroic (Chai *et al.,* 1987a) studies.

The 6 kDa N-terminal segment of *Avena* phytochrome is known to form α helix upon conversion from Pr to Pfr (Viestra *et al.,* 1987). This helical formation entails a 3.3 - 5.0% increase (depending on buffer used) in helical content based on CD studies (Sommer and Song, 1990). This change involves about 37 residues in 1 mM EDTA and 20 mM potassium phosphate, pH 7.8. However, approximately 56 residues are involved in the presence of 20 mM Tris, pH 7.8. This difference may result from the stabilization of non-helical conformations by the organic Tris buffer. The shifting of structural equilibrium toward a less ordered or more open structure by Tris may be expected given the solubilizing (mildly detergent like) nature of the molecule. This Tris-induced stabilization is also evident in the overall analysis of phytochrome structure by CD. The percent α-helix drops from 51.2% (Pr) and 54.5% (Pfr) in phosphate to 45.5% (Pr) and 50.5% (Pfr) in Tris (Sommer and Song, 1990). It therefore seems that the Pr to Pfr transition actually stabilizes helical formation in more than 37 residues. This may not be observable by CD in phosphate buffer since helix may exist in equilibrium with a random coil or turn structure in the Pr form. This helix-coil equilibrium (involving more than 37 residues) would then be "locked" into helical conformation upon conversion to Pfr.

The phytochrome chromophore lies in a hydrophobic pocket (Hahn and Song, 1981; Choi *et al.,* 1990) and becomes more exposed in the Pfr form (Hahn *et al.,* 1984; Thuemmler *et al.,* 1985; Farrens et al, 1989). From spectroscopic [comparison of large (118 kDa) vs. intact (124 kDa) Pfr spectra; Vierstra and Quail (1983)] as well as chemical accessibility (Chai *et al.,* 1987a; Hahn *et al.,* 1984) studies, it is apparent that the N-terminal sequence interacts with the phytochrome chromophore in the Pfr form. It is likely that this interaction with the hydrophobic chromophore and presumably the chromophore environment (hydrophobic pocket) induces α helix in the N-terminus chain. The 6 kDa N-terminal segment is more exposed in the Pr form than in the Pfr form, but proteolytic studies (Grimm *et al.,* 1988) as well as hydropathy profiles (Quail *et al.,* 1987; Romanowski and Song, 1990) indicate the N-terminus is surface exposed in both the Pr and Pfr forms. The N-terminus segment (Pfr form), in addition to interaction with a hydrophobic pocket and chromophore, must also interact with the polar solvent. This suggests the helices of the Pfr N-terminal segment should be amphiphilic.

A large number of α helices in globular proteins are amphiphilic. These helices have a hydrophobic face interacting with the nonpolar core of the protein and a hydrophilic face which interacts with the polar solvent. Generally, about 50% of the helices in globular proteins are amphiphilic (Cornette *et al.,* 1987). Perutz *et al.,* (1965) observed that all of the helices in myoglobin have amphiphilic character. This is also the case for other proteins which have a very hydrophobic core. All helices which interact with a hydrophobic core and a polar aqueous solvent should be amphiphilic. Helical wheel analysis of several proteins (Schiffer and Edmundson, 1967) has demonstrated that helices in some globular proteins may actually be located or predicted by the occurrence of helical amphiphilic character in the primary sequence. The presence of helical amphiphilicity has been used to successfully predict helical regions in the Antennapedia homeodomain sequence (Bowie *et al.,*1990) as well as the Arc repressor protein (Bowie and Sauer, 1989). These predictions were made by use of a helical hydrophobic moment analysis.

The helical hydrophobic moment ($<\mu_H>$) of an α helix is a measure of amphiphilicity of that helix (Eisenberg, 1982). $<\mu_H>$ is calculated by assigning a hydrophobic value to each amino acid in a given helix (Table 1). These magnitudes are then assigned a direction 90 degrees from the helical axis and 100 degrees from adjacent residues (3.6 residues/turn = 100 degrees/residue). The resulting vector quantities (Fig. 1) are then summed, yielding $<\mu_H>$. There are two helices in Figure 1, one (helix b) being highly amphiphilic and the other (helix a) being less amphiphilic due to a histidine residue on the hydrophobic side of the helix. As is demonstrated in Figure 1, a higher $<\mu_H>$ is a result of amphiphilic character and a lower $<\mu_H>$ indicates less amphiphilicity. Thus, the helical amphiphilic nature of a given protein should be evidenced by high $<\mu_H>$ values. This principle has been demonstrated previously (Bowie

Table 1. Hydrophobic value assignments of each amino acid according to Eisenberg *et al.* (1984). Values used have a mean of 0.0 and a standard deviation of 1.0. The assignments are based on five scales using different criteria for hydrophobicity.

Residue	Hydrophobic Value
Arginine	-2.53
Lysine	-1.50
Aspartic acid	-0.90
Glutamine	-0.85
Asparagine	-0.78
Glutamic acid	-0.74
Histidine	-0.40
Serine	-0.18
Threonine	-0.05
Proline	0.12
Tyrosine	0.26
Cysteine	0.29
Clycine	0.48
Alanine	0.62
Methionine	0.64
Tryptophan	0.81
Leucine	1.06
Valine	1.08
Phenylalanine	1.19
Isoleucine	1.38

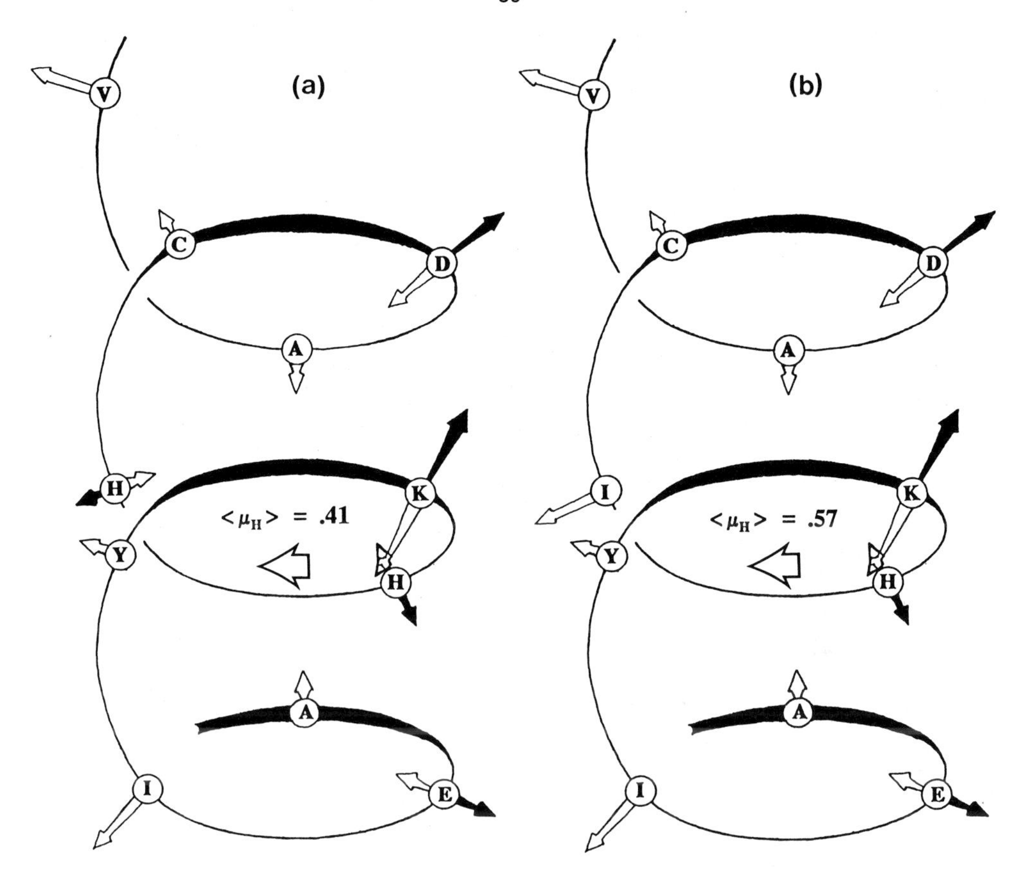

Figure 1. The hydrophobic moments of residues in two amphipathic helices are represented. Solid arrows represent negative hydrophobic moments and outlined arrows represent positive hydrophobic moments. Positive moments are included 180 degrees from negative moments (The two vectors are equivalent.) so that the result of summing the positive vector quantities may clearly be seen. The resulting vector sums divided by the total number of residues ($<\mu_H>$) are also shown. Helix (b) differs from (a) in that there is an isoleucine rather than a histidine at the seventh position from the N-terminus (bottom of helix).

et al., 1990; Parker and Song, 1990), and is evident in myoglobin (Fig. 2c). Other traditionally used methods are also effective in conveying the nature of amphiphilic helices as shown in Figure 2. Notably, the Edmundson wheel also demonstrates amphiphilicity in helices. We have used these methods to analyze the N-terminus of phytochrome.

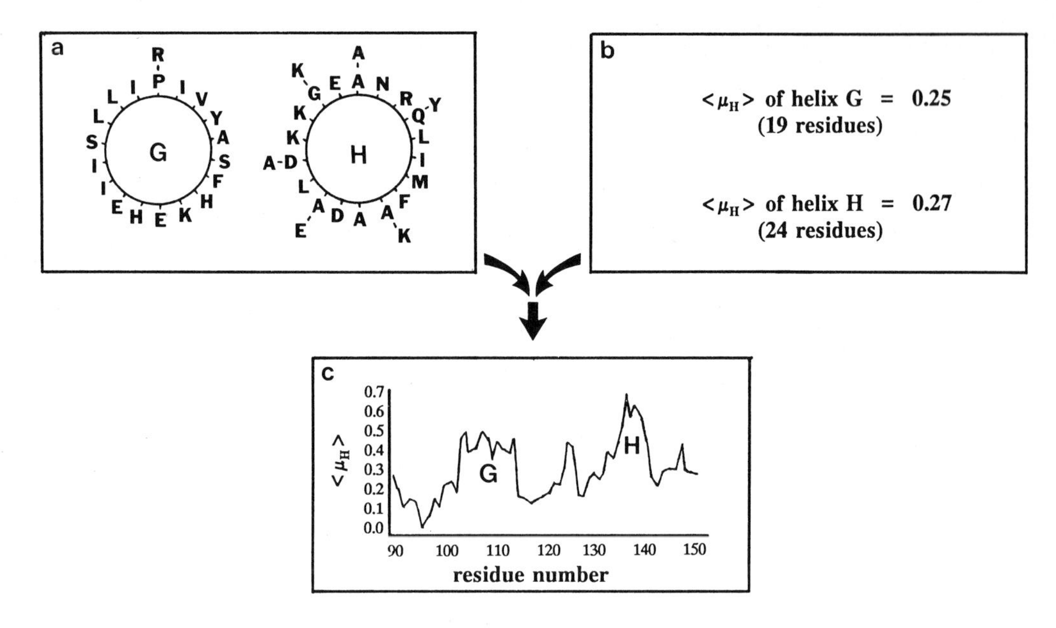

Figure 2. Block diagram illustrating the differences between helical wheel analysis (a), standard use of $<\mu_H>$ values on globular proteins (b), and a graphical $<\mu_H>$ analysis [(c); Parker and Song (1990)]. The last 60 residues (containing two helices) of sperm whale myoglobin are analyzed. The "G" and "H" helices (residues 100-118 and 125-148, respectively) are labeled accordingly.

MATERIALS AND METHODS

Chemicals and phytochrome. Size exclusion standards were purchased from Pharmacia (Piscataway, NJ). Zinc chloride was purchased from Matheson Coleman and Bell Manufacturing Chemist (Norwood, OH). Tris [Tris(hydroxymethyl)aminomethane] was purchased from Sigma, St Louis, MO. NaCl was purchased from Fisher Scientific, Fair Lawn, NJ. All water used was purified by filtering previously deionized water through three Barnstead deionizing filters followed by distillation. Phytochrome was isolated in our laboratory as previously described (Chai *et al.,* 1987b).

Statistical analysis. A correlation function was used to analyze conservation in *Avena* (Hershey *et al.,* 1985), *Pisum* (Sato, 1988), *Cucurbita* (Sharrock *et al.,* 1986), *Oryza* (Kay *et al.,* 1989), *Maize* (Christensen and Quail, 1989), and type A *Arabidopsis* (Sharrock and Quail, 1989). The correlation function used was a modification of the method of Argos *et al.* (1983) which has been previously described (Romanowski and Song, 1990). According to this method, there are six parameters which are compared for each amino acid. These six parameters are (1) helical, (2) sheet and (3) turn conformational parameters derived from Chou-Fasman calculations, as well as (4) hydration potential, (5) hydrophobicity, and (6) polarity. These six parameters are used to derive a correlation function diagram (homology plot) for a given set of sequences, which is a measure of conservation in those sequences. This type of analysis has previously been used to identify homologous structures as well as possibly important subsequences (Argos *et al.*, 1983; Kubota *et al.*, 1982).

Secondary structure predictions were made according to the method of Chou and Fasman (1978). A mean of consensus secondary structure was derived as previously reported (Romanowski and Song, 1990). From the homology plots as well as the secondary structure results, a set of structural domains was proposed (Romanowski and Song, 1990).

Zn^{2+} Chelation of phytochrome. $ZnCl_2$ solutions were prepared by serial dilution in deionized distilled water. An aliquot containing a two fold molar excess of Zinc chloride was added to phytochrome solutions in 20 mM Tris buffer, pH 7.8. Although Zn^{2+} chelates phytochrome in a one to one molar ratio (Sommer and Song, 1990), a two fold molar excess was used to ensure a complete reaction without excessively long incubation times. Bleaching of the spectral band was observed with a Hewlett Packard 8451A diode array spectrophotometer. Continuous 664 nm irradiation was used to convert Pr to Pfr as the reaction proceeded.

HPLC. A Shimadzu HPLC system consisting of an SCL-6A system controller, LC-6A pump, SPD-6AV detector, and a C-R3A recorder was used. A 10 mm ID TSK-3000 column was employed for all size exclusion work. The mobile phase was 40 mM NaCl and 10 mM Tris, pH 7.8. Elution of proteins was monitored at 280 nm with a flow rate of 0.7 ml/min. Phytochrome solutions for HPLC injection were transferred to mobile phase buffer and concentrated to approximately 1.5 mg/ml (Pfr) or 0.7 (Zn^{2+}-chelated Pfr) using a Centracon 30 Concentrator from Amicon. The injection volume was 20 μl for all solutions.

Circular dichroism. Circular dichroic spectra were taken on a JASCO J-600 Spectropolarimeter standardized with ammonium d-10-camphorsulfonate (Takakuwa *et al.,* 1985). Conditions of measurement were as follows: band width = 2 nm, scan speed = 20 nm/min, time constant = 2, step resolution = 0.2 nm/min. Phytochrome concentrations of 0.6 to 1.2 μM and a 2 mm pathlength cell were used. All spectra were taken in 20 mM Tris, pH 7.8. Structural analysis was made from the spectra according to Chang *et al.,* (1978).

$<\mu_H>$ *Calculations.* $<\mu_H>$ calculations were performed using the method of Eisenberg *et al.,* (1982) according to the following equation:

$$<\mu_H> = \{[\sum_{n=1}^{N} H_n Sin(\delta n)]^2 + [\sum_{n=1}^{N} H_n Cos(\delta n)]^2\}^{1/2} / N$$

where n is a specific residue in a peptide segment of N residues, δ is the distance between residues as viewed down the helical axis (100 degrees for an α helix) and H_n is the hydrophobic value assigned to residue n. Normalized hydrophobic values (mean of 0 and standard deviation of 1) were assigned to each amino acid according to Eisenberg *et al.*, (1984; Table 1). The segment length (N) or window was defined as 11 amino acids and a "moving window" algorithm was used for all sequences as described by Rose (1978). This algorithm takes the first 11 amino acids (residues 1-11), performs calculations, moves one amino acid toward the C-terminus (residues 2-12), performs calculations, and continues in this fashion until the C-terminus is reached. This generates a series of $<\mu_H>$ values for a sequence in which the first and last 5 residues have no assigned value. All other residues are the central amino acid in different 11 residue segments and have the assigned values of the $<\mu_H>$ of their corresponding segments.

RESULTS

Structural analysis. Figure 3 shows the correlation plot of six "type A" phytochromes. This plot contains several local minima with generally high correlation over the entire sequence. Prominent minima occur around residues 60, 359, 603, and the region 845 to the C-terminus, which is generally somewhat less conserved than the rest of the protein. No significant differences were found between monocot and dicot phytochromes (Romanowski and Song, 1990).

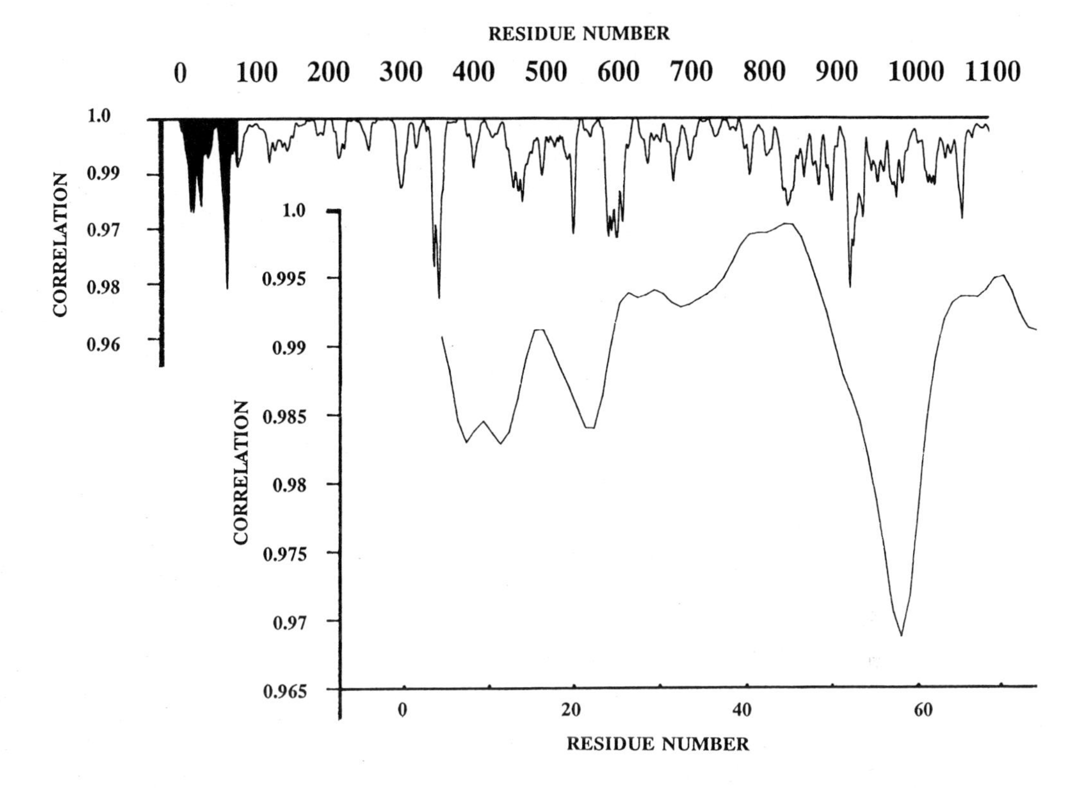

Figure 3. The correlation plot of six phytochrome sequences (Romanowski and Song, 1990), including an enlarged view of the N-terminal diagram. Numbering is not according to any one given sequence, but is based on an alignment of four different phytochromes (Furuya, 1989). The upper diagram (large numbering) represents the entire sequence, while the lower diagram represents the N-terminal region, shown in black in the upper diagram.

The phytochrome sequence may be divided into structural domains according to correlation and structural prediction results. Uninterrupted regions of high correlation may be considered to be structurally significant, or intact structural (and perhaps functional) domains. Some of these conserved domains exhibit specific structural patterns which are typical of known structural domains. In general, three basic types of domains have been observed in the structures thus far determined by X-ray. These are α helical, ß sheet, and α\ß domains (Richardson and Richardson, 1989).

The results of the division of phytochrome into structural domains is shown in Figure 4. There are 11 domains (A - K). The four C-terminal domains (H - K) are short and poorly conserved, and may be lumped into one large "C-terminus" domain (Fig. 5).

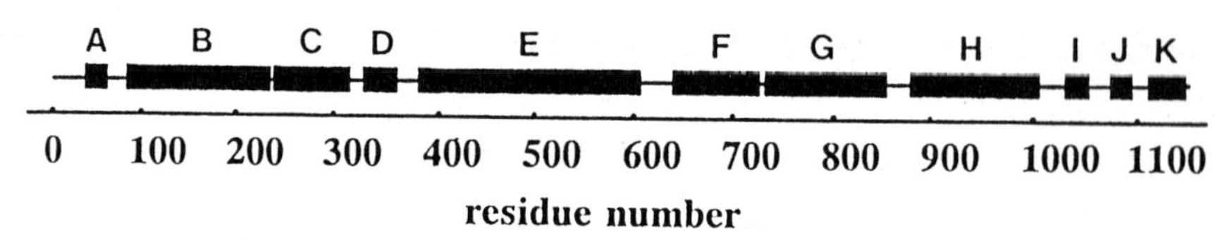

Figure 4. The structural domain regions (A through K) of "type A" phytochrome according to Romanowski and Song (1990). The numbering is the same as in Figure 3.

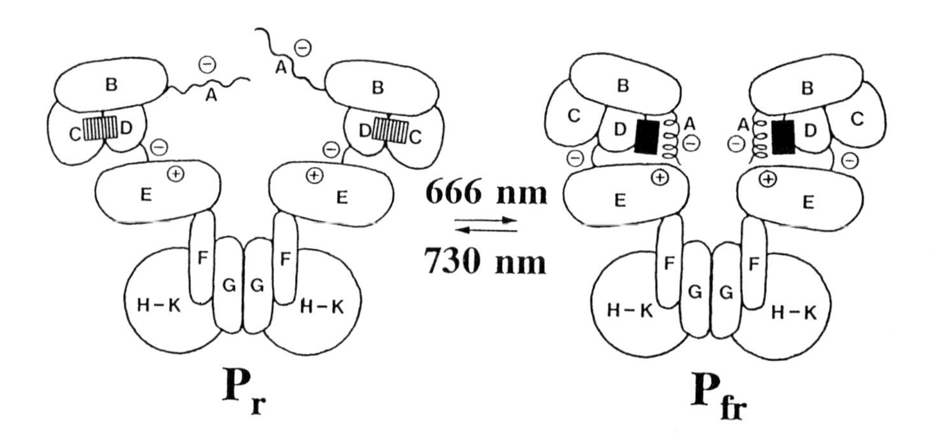

Figure 5. Structural model of phytochrome as proposed by Romanowski and Song (1990). This model includes 11 distinct structural subunits. The chromophore domain contains regions A through E and the nonchromophore domain contains regions G through K (region F is the linker region). The solid triangle represents the chromophore and is not drawn to scale with the peptide portion of the molecule. The encircled plus and minus signs represent regions of concentrated positive or negative charge in the polypeptide chain. Negative regions are present in the N-terminus as well as the PEST region between the D and E domains. In this model, the possible "displacement" of the negatively charged PEST sequence by the N-terminus is evident. A possible movement of the A through E domains is also pictured in this diagram. The chromophore is associated with the C domain in the Pr form and is associated with the A, B, and possibly E domains in the Pfr form. Greater exposure of the Pfr chromophore is evident in the diagram.

Domain C and domain F are all α helical domains, while domains B and E are α\ß domains. Domain G is the only all ß domain located in the phytochrome molecule.

Zn^{2+} Chelation. Figure 5 shows the differences in the CD spectra of Pfr and Zn^{2+}-chelated Pfr in *Avena* phytochrome. The difference between the Pfr and Pr spectra is also shown. An apparent change from 50.5% to 28.0% α helix occurs as determined by circular dichroism (Fig. 6, Sommer and Song, 1990). The loss of helix upon Zn^{2+} binding is evident in Figure 5. The dramatic increase in ellipticity at 208 and 222 nm in the CD spectra of the Zn^{2+}-chelated species (Fig. 6) represents a 45% loss of the helix present in the Pfr form according to calculations made by the method of Chang *et al.*, (1978).

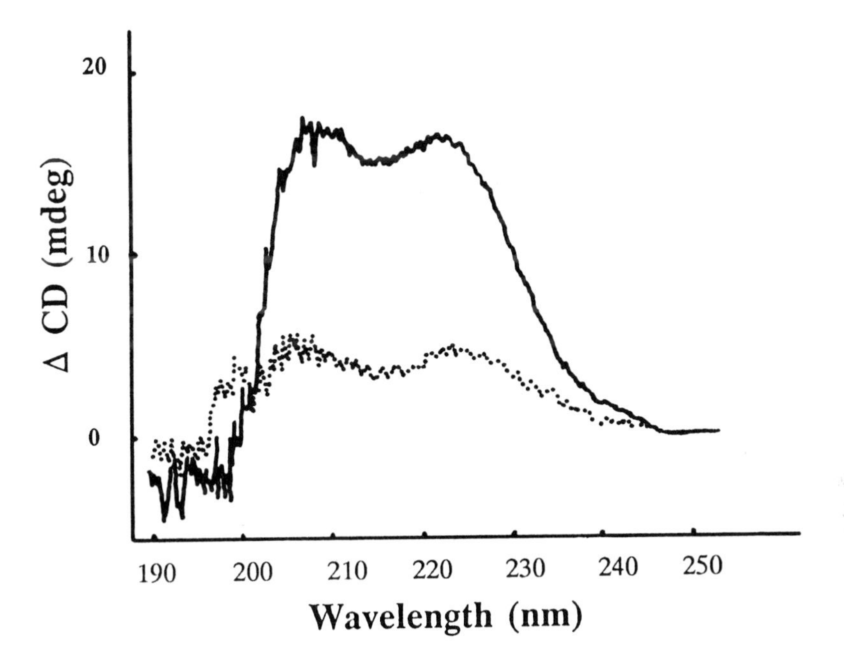

Figure 6. The CD difference spectra of Pr, Pfr, and Zn^{2+}-chelated Pfr. Pr-Pfr (-------); Zn^{2+}-chelated Pfr - Pfr (·······). Positive peaks indicating loss of helix are located around 208 and 222nm.

There is no predicted ß sheet structure in Pr or Pfr in phosphate or Tris buffers (Sommer and Song, 1990) according to the Chang *et al.* (1978) analysis. However, 39.5% ß sheet is calculated in the Zn^{2+}-chelated species. ß turn is calculated at 17.5% in Pfr, 21.0% in Pr and 7.5% in Zn^{2+}-chelated Pfr. Random coil is calculated to be 32.0% in Pfr, 33.5% in Pr and 25.0% in Zn^{2+}-chelated Pfr.

Phytochrome typically elutes from HPLC size exclusion columns as two peaks (Choi *et al.*, 1990; Fig. 7). The first peak, which elutes near the void volume, corresponds to a concentration and salt dependent oligomeric form (Choi *et al.*, 1990). The second peak corresponds to the apparent dimer which is the predominant form at lower concentrations (Jones and Quail, 1986; Lagarias and Mercurio, 1985). Zn^{2+}-chelated phytochrome apparently maintains the same quaternary structure as phytochrome since it shows the same characteristic elution profile (Fig. 7). The apparent Pfr dimer of MW 224,000 is also present in the Zn^{2+}-chelated form (apparent MW = 234,000). The oligomeric form of phytochrome present in Pfr (apparent MW = 372,000) also elutes in the Zn^{2+}-chelated species with an apparent MW of 467,000. The shifts in MW of the Zn^{2+}-chelated species may reflect an increase in the size of the Zn^{2+}-chelated molecule, but the method used may be sensitive to changes in hydrophobicity of the protein. For this reason, further experiments are needed to conclusively determine the relative sizes of the Zn^{2+}-chelated and native phytochromes. Additionally, the Zn^{2+}-chelated species contains a higher dimer to oligomer ratio than does the native (unchelated) species. This may or may not partially result from a change in the monomer-dimer equilibrium constant, since concentrations are not the same for the two samples.

N-Terminus conformation analysis. Figure 2c shows the correlation between $<\mu_H>$ values for myoglobin and two helices in that protein. Although this correlation can be seen in sperm whale myoglobin (Fig. 2c), these diagrams are generally more clear in composite analysis utilizing consensus sequences (Bowie *et al.*, 1990). The "noise" which is present in the single sequence analysis is probably due to hydrophobic residues occurring on the surface of the protein (Bowie *et al.*, 1990). The correlation is also interrupted by the proximal and distal histidines, which occur on the hydrophobic side of amphiphilic helices. Plots of $<\mu_H>$ versus residue number for the N-termini of *Avena, Cucurbita, Pisum, Maize, Arabidopsis*, and *Oryza* phytochromes are shown in Figures 8, 9, and 10. In the *Avena, Cucurbita*, and *Pisum* phytochromes there are five maxima or peaks in the $<\mu_H>$ diagrams of the N-terminus. These five maxima correspond to five possible helices, labeled (a) through (e) in the composite analysis in Figure 10. The hydrophobic arcs of these possible helices (for the *Avena* sequence) are shown on Edmundson wheels in Figure 11. In the *Pisum, Maize*, and *Oryza* phytochromes there is not a clear

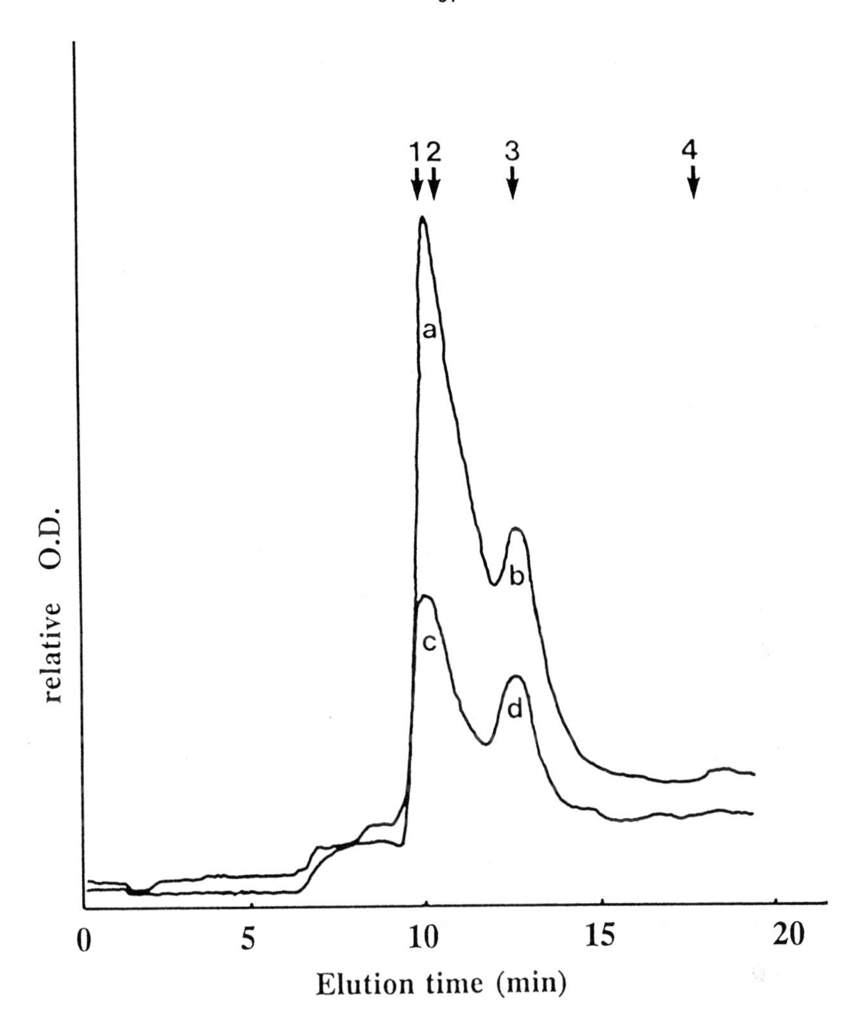

Figure 7. Elution profiles of Pfr (top profile, peaks a and b) and Zn^{2+}-chelated Pfr (lower profile, peaks c and d). Peaks a and c, eluting near the void volume, represent the apparent oligomer form of phytochrome, while peaks b and d represent the apparent dimer. The elution of standards is indicated at the top as follows: (1) Thyroglobulin, MW 669,000 (2) Ferritin, MW 440,000 (3) Catalase, MW 232,000 (4) Aldolase, MW 158,000.

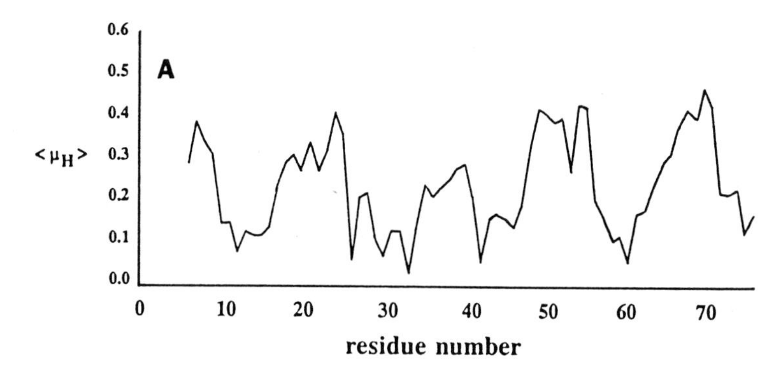

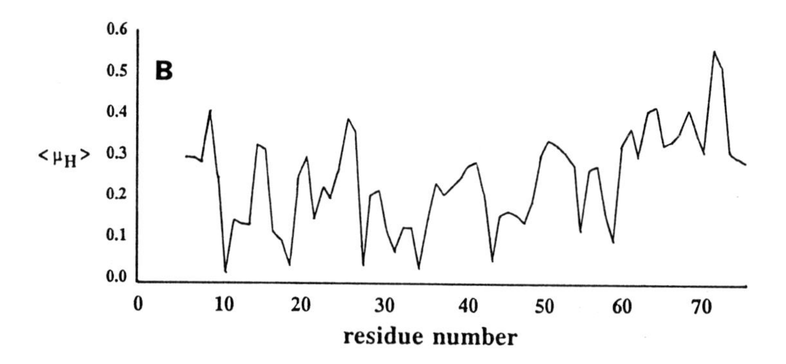

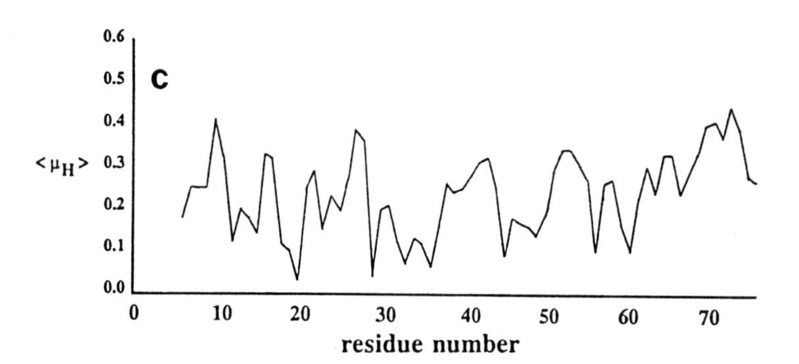

Figure 8. The $<\mu_H>$ diagrams for the N-termini of three monocot phytochromes: (A) *Avena* (B) *Maize* (C) *Oryza*. The numbering is not according to a consensus and is plotted as it was originally published for each sequence, except that the N-terminal methionine is not included.

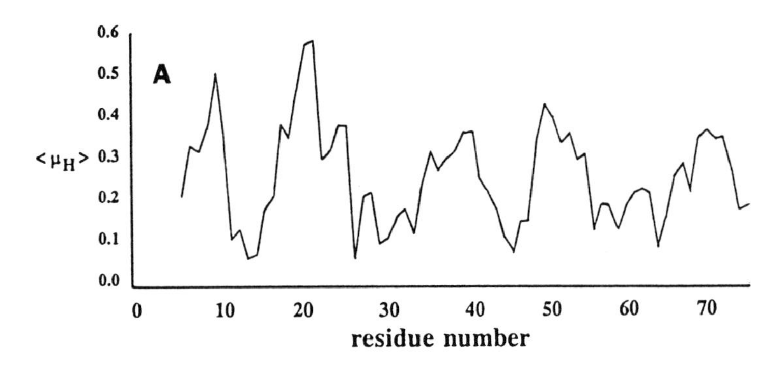

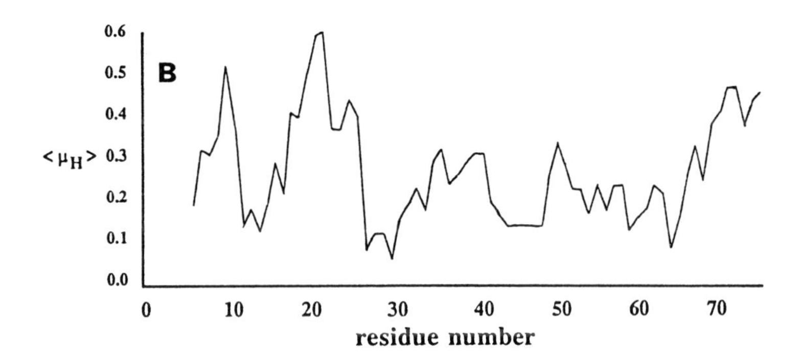

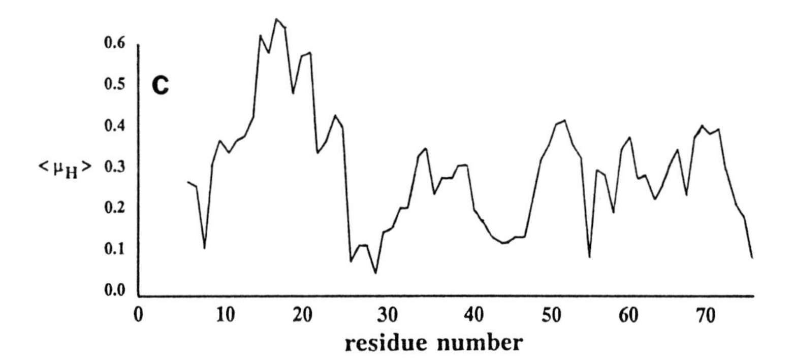

Figure 9. The $<\mu_H>$ diagrams for the N-termini of three "type A" dicot phytochromes: (A) *Cucurbita* (B) *Pisum* (C) *Arabidopsis A*. Again, each sequence is numbered sequentially, with no adjustments made for alignment purposes. The N-terminal methionine is excluded.

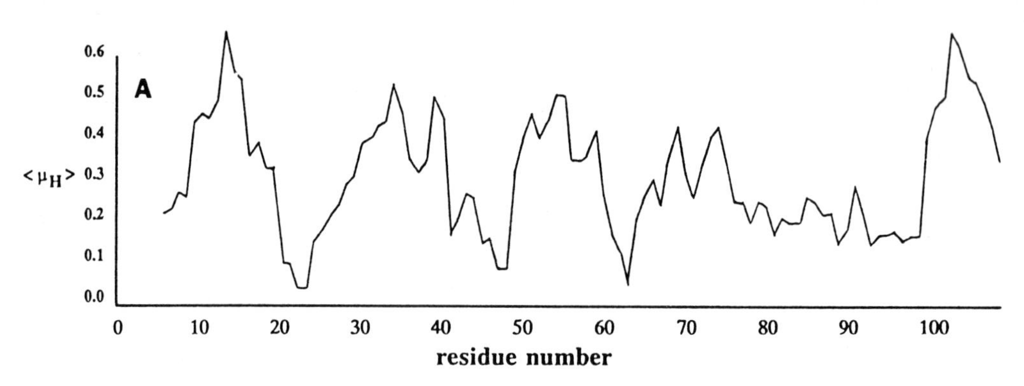

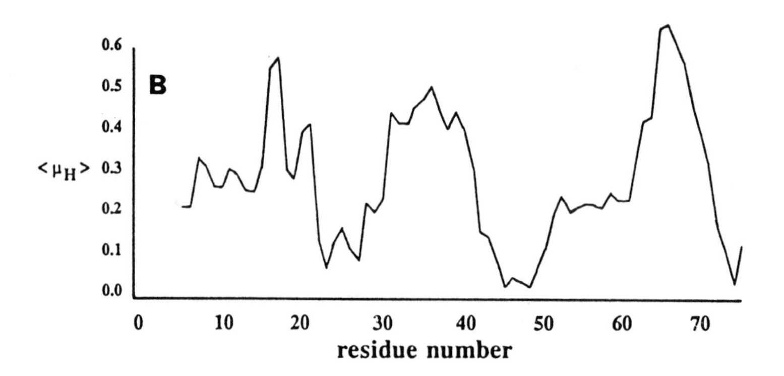

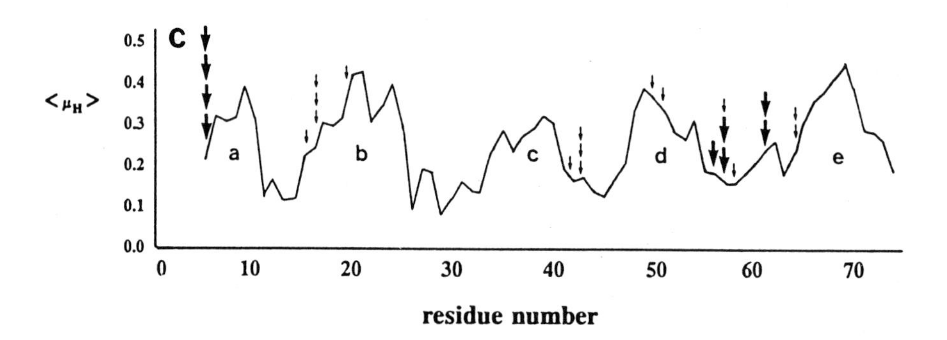

Figure 10. The $<\mu_H>$ diagrams for the N-termini of two Arabidopsis sequences are shown: (A) Arabidopsis B (B) Arabidopsis C. Part (C) of the figure shows the consensus $<\mu_H>$ diagram alignment (An average at each point was taken.) of *Avena*, *Oryza*, *Pisum*, and *Cucurbita* phytochromes. Alignment according to Furuya (1989) was used. The *Maize* diagram (Fig.7B) was not used since it is very similar to the *Oryza* diagram, thus not adding and more "smoothing" advantage to the composite analysis. The predicted (Rose, 1978; small arrows) as well as expected (proline residues, large arrows) turns are shown in the composite diagram. All turns from the four sequences used are shown in the composite diagram.

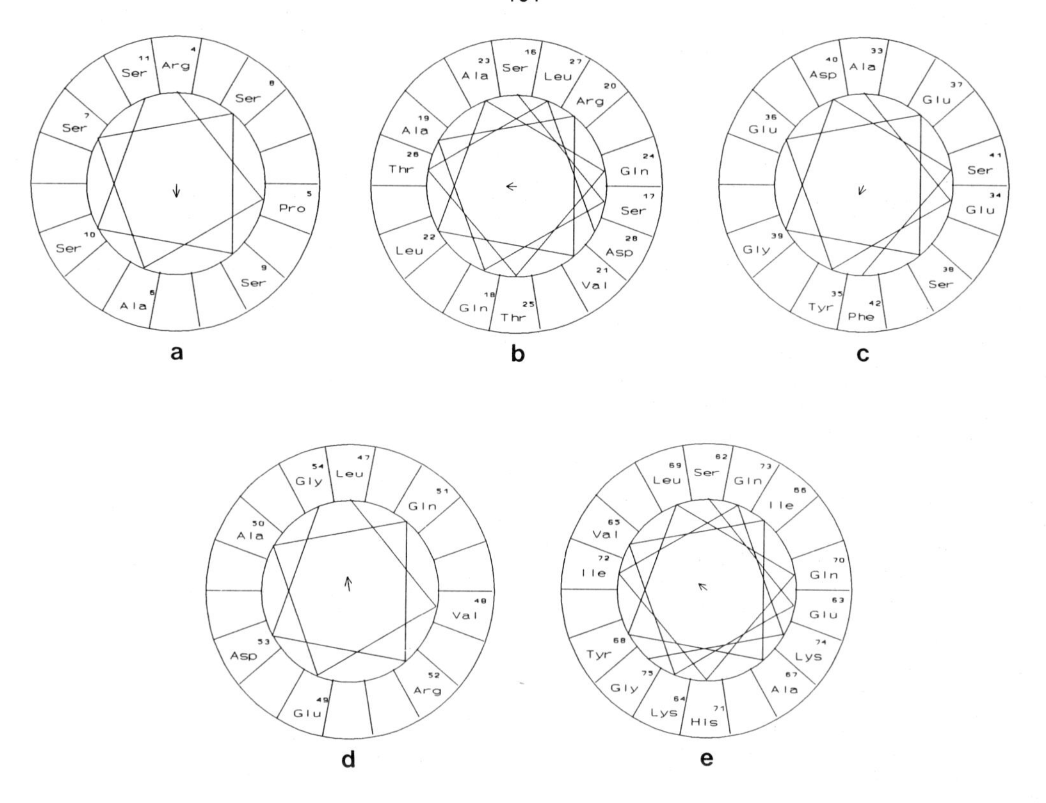

Figure 11. The helical wheels of all five possible helices in the N-terminus of Avena phytochrome are shown. Hydrophobic "arcs" or hydrophobic clusters are clear in the b, c, d, and e helices. The concentration of hydrophobic residues is not as clear in the (a) helix, but the $<\mu_H>$ value for the region is relatively high. The (c) helix, which occurs in the most conserved region of the N-terminus (Fig. 3) has two closely associated aromatic residues (Tyr 35 and Phe 42) which may interact with the chromophore in the Pfr form of phytochrome.

separation of helices (d) and (e) in the $<\mu_H>$ analysis. Separation of the (d) and (e) helices is still possible, since there is a glycine-proline pair (residues 57 and 58 of *Oryza*) or proline-proline pair (residues 56 and 57 of *Maize*) which would interrupt any continuous helix. There is also a large hydrophobic minimum in the *Pisum* sequence occurring at residue 63. It has been demonstrated that these minima generally correspond to turns or bends adjacent to regular structures such as helix or sheet (Rose, 1978). This separation of helices (d) and (e) is more evident in the composite diagram (Fig. 10).

Turns predicted according to the method of Rose (1978) are indicated with small arrows in the composite diagram (Fig. 10) while turns or bends imposed by the presence of proline residues are shown with large arrows. For the *Maize* and *Oryza* sequences (Fig. 8, the $<\mu_H>$ diagrams are almost superimposable), the (a) and (b) helices are not clear. There are three small peaks from residues 1-30, but we have observed that small peaks such as these often do not correspond to helical structure. (Observations have been made for over a hundred helices in a variety of proteins). Also, the predicted turns (Fig. 10) made from hydropathy profiles (Rose, 1978) discourage the assignment of three helices in the (a) and (b) helix region of the *Maize* and *Oryza* sequences.

The N-terminal $<\mu_H>$ profiles for three *Arabidopsis* sequences are shown in Figures 9 and 10. There are significant deviations from other phytochromes in the *Arabidopsis* $<\mu_H>$ profiles. The loss of a peak corresponding to helix (d) is noticeable in *Arabidopsis* B and C. The apparent "fusion" of the (a) and (b) helices is noticeable in both the *Arabidopsis* A and to a lesser extent the *Arabidopsis* C sequence. [The fused peak is still divided by predicted (Rose, 1978) turns]. Finally, in the *Arabidopsis* B sequence an additional segment has been added to the extreme N-terminus. This extra segment apparently extends helix (a) as well as incorporating an additional helix (Fig. 10).

Chou-Fasman analysis of the N-terminal segments invariably predicted the region between the (b) and (c) helices to be helical for all phytochromes including the three *Arabidopsis* phytochromes. This region, which has no indications of turn formation (Fig. 10) is possibly helical without any influence of a hydrophobic pocket or chromophore since it contains helix forming residues and is apparently not amphiphilic. The 11 residue segment (residues 26-36, 27-37 or 29-39, depending on the phytochrome) may be a nucleation site for helix formation. The five apparent helical regions [(a)-(e)] in Figure 10 are not generally predicted to be helical by a Chou-Fasman analysis. The only exceptions to this are the prediction of a part (six residues) of helix (d) in *Avena* and *Maize* and part (five residues) of helix (e) in *Oryza*. A larger part of helix (c) is predicted as helical in the *Arabidopsis* C sequence.

DISCUSSION

Structural domain diagram. The structural domain diagram (Fig. 4) is based on conformational (Chou and Fasman, 1978) and homology (Argos *et al.,* 1983; Romanowski and Song, 1990) analysis. It is surprising that the domains of this model match several lines of previous experimental work rather closely. The α helical domain C (residues 220 to 300 using the numbering of an aligned sequence of oat, rice, zucchini, and pea phytochromes; Furuya, 1989) and the chromophore binding domain D (residues 312 to 348 using the same numbering system) correspond to an observed 16 kDa proteolytic fragment of phytochrome (Jones and Quail, 1989). This fragment stabilizes spectrally intact Pr but not Pfr. The inclusion of the two α\b domains (B, residues 67 to 218; E, residues 366 to 590; numbering as described previously; Furuya, 1989) into the 16 kDa fragment yields a peptide which approximately corresponds to the 60 kDa fragment of phytochrome. This fragment is spectrally intact in both the Pr and Pfr forms. A model of phytochrome photoisomerization involving structural domains may thus be developed. In this model, the Pr chromophore would interact with an α helical domain (C). Upon photoisomerization, the chromophore and/or protein would then be repositioned such that some interaction with one or both α\ß domains (B and E) stabilizes the chromophore (Fig. 5).

There are several other domains of interest. The only ß sheet domain (G, residues 688 to 832, previous numbering system) has been found to be a possible dimer contact region by a combined "complementary interaction analysis", "super-secondary structure preferences analysis", and "interior surface analysis" (Romanowski and Song, 1990; Fig. 5). One of the most common types of antiparallel ß sheet structure is often involved in subunit-subunit contact (Richardson and Richardson, 1989). The N-terminus (including domain A) and last three C-terminal domains (including domains I through K) are somewhat less conserved than the rest of the protein and are perhaps the most proteolytically sensitive regions of the molecule. [See peptide mapping by Grimm *et al.*, (1988), Jones and Quail (1989), and Lagarias and Mercurio (1985)] The division of many of the proposed domains (Fig. 4) by previous proteolytic mapping studies is interesting, and adds to the significance of the work.

Zn^{2+} Perturbed conformation. Loss of spectroscopically crucial chromophore-protein interaction and/or insertion of the divalent cation in the hydrophobic chromophore pocket is apparently responsible for the large change in secondary structure occurring upon Zn^{2+} chelation. This suggests that the chromophore stabilizes α-helical structure

in phytochrome. This is not surprising, since chromophores in an internal hydrophobic pocket are known to stabilize α-helical folding in other proteins. This structural motif has been found in cytochrome c (Stellwagen and Rysavy, 1972), hemoglobin (Hrkal and Vodrazka, 1967), and in myoglobin (Breslow *et al.,* 1965).

Perhaps one of the most interesting results obtained from the structural analysis (method of Chang *et al.,* 1978) of the CD spectra is the calculated presence of 39.5% ß sheet structure in the Zn^{2+}-chelated species. The projection of almost 40% ß sheet is certainly an indication of considerable ß sheet formation given the reasonable accuracy and reproducibility of the spectral measurements and subsequent calculations. This highly ordered form of protein structure would not be expected in a protein which is "denatured" or conformationally non-native. ß sheet is not formed unless the primary and to some extent the secondary structure very specifically accommodate or allow ß sheet. Much more "precise" primary and secondary structure is required for ß sheet than for α helix formation. This idea is based on polymer and copolymer studies, as well as *de novo* design of proteins containing different secondary structures. α helical proteins have been successfully designed (Hecht *et al.,* 1889; Ho and DeGrado, 1987) on the first attempt. The design of a ß sheet containing protein has been more elusive, requiring nine attempts before success (Richardson *et al.,* 1984; Richardson and Richardson, 1987). More direct evidence of the "preciseness" required for ß sheet structure is the fact that many amino acid polymers and copolymers form α helix in a variety of solutions, but true ß sheet has not been found in these simple model systems.

The appearance of almost 40% ß sheet indicates that the Zn^{2+}-chelated phytochrome is not merely "denatured" in the chromophore region, but has some amount of ordered structure which differs from that of native phytochrome. It must also be noted that the percent calculated random coil and ß turn are lower in Zn^{2+}-chelated phytochrome than in either native Pr or native Pfr. The possibility exists, based on this data, that the apoprotein possesses a distinct structure rich in ß sheet. It may be that this conformation allows or accommodates insertion of the chromophore. That insertion of the chromophore would then facilitate the α helical folding (concomitant formation of the hydrophobic chromophore pocket) present in the native protein. This line of thinking is not without precedent, as lactate dehydrogenase (Abad-Zapatero *et al.,* 1987), fatty acid binding protein (Sacchettini *et al.,* 1989) and demetallized concanavalin A (Shoham *et al.,*

1979) are structurally distinct apoproteins which presumably allow for insertion of prosthetic groups. However, apoprotein structures in these proteins do not differ significantly in secondary structure from the holoprotein. In the case of myoglobin, the apoprotein structure resembles an "uncompact" or somewhat unwound form of the holoprotein, with an open pocket for chromophore insertion. The open pocket exists in apo-riboflavin binding protein (Kumosinski *et al.*, 1982), but there is no appreciable change in secondary structure as detected by CD. Apo-lactate dehydrogenase, on the other hand, is almost identical to the holoprotein in secondary structure, the only difference being in the relative position of one loop structure with respect to the rest of protein (Richardson and Richardson, 1989; Abad-Zapatero *et al.*, 1987). Fatty acid binding protein (Sacchettini *et al.*, 1989) and concanavalin A (Shoham *et al.*, 1979) have even less difference in apoprotein vs. holoprotein structure, differing mostly in side chain positions. Thus, a large disparity between apophytochrome and phytochrome was not originally expected. In order to experimentally determine if a ß sheet-rich structure (apoprotein) precedes a helix rich structure in the holoprotein, structural studies must be performed on the apoprotein and hopefully a reconstituted holoprotein. If this idea is borne out experimentally, it would be a unique example of largely different yet distinct secondary structures in apoprotein versus holoprotein.

The formation of ß sheet at the expense of α helix upon Zn^{2+} binding is clear from the CD studies. The lack of CD predicted ß sheet in the native protein is interesting, as proteins containing only α helix (no ß sheet) are not generally much larger than 200 residues (compared with 1128 residues in *Avena* phytochrome). In addition, Chou-Fasman calculations indicate that a significant percentage of ß sheet is present (Romanowski and Song, 1990). Also, helical analysis using $<\mu_H>$ values (Parker and Song, 1990) fails to reveal significant amounts of helical structure in regions which Chou-Fasman methods predict largely ß sheet structure. The presence of ß sheet in the native molecule is possible, but chromophore-protein interactions may stabilize helix in the native molecule, thus eliminating a significant amount of ß sheet structure. This idea must await crystallographic data before being resolved completely.

The present work concerning Zn^{2+}-chelated phytochrome indicates there is specific secondary as well as quaternary structure in the modified protein. It seems likely that this ß sheet-rich dimer may resemble the apoprotein in structure. The insertion of a divalent cation as well as cyclization of the chromophore (conformational and configurational changes) could well account for loss of protein-chromophore interactions, thus producing, in effect, an apoprotein. If this is not the case, a more unattractive (or at least unsettling) alternative may be considered to account for the very specific structure of Zn^{2+}-chelated Pfr. The suggestion that metal chelated phytochrome (chelation would occur after Pfr formation) may serve a specific function, thus having specific structure, is speculative. Such a function, if it exists, may pertain to the degradation of Pfr.

The topographic relationship between the chromophore and non-chromophore domains merits analysis. Connected by a hinge region, the two domains are clearly separated. However, there may be some effect of the chromophore domain on non-chromophore domain structure and/or hinge region structure. There is a change in proteolytic accessibility of at least one site which may be near the hinge region (Glu 354; Grimm *et al.,* 1988). This site is located in a PEST region which has been postulated to play an important role in degradation of proteins *in vivo* (Rogers *et al.,* 1986). This particular region of phytochrome was located by a computer search performed on a number of proteins with varying half-lives (Rogers *et al.,* 1986). It was found that proteins with short half-lives contained sequences rich in proline, glutamate, serine, and threonine (PEST sequences). Exposure of this PEST region may be solely responsible for the degradation of Pfr. It is possible that Zn^{2+}-chelated Pfr further exposes this region for its rapid proteolytic degradation (Sommer and Song, 1990).

The structural association between the different domains may be important in the dimeric model of phytochrome action, which is implicated in some physiological responses (VanDerWoude, 1985). In this model, a receptor would be able to distinguish between a Pr-Pfr dimer and a Pfr-Pfr dimer. Movement or conformational change of a chromophore domain dependent on the conformation of the other monomer in a dimer pair may facilitate this type of response. Movement of the chromophore domain with respect to the axis of symmetry in the molecule has been previously suggested by experiment (Sugimoto *et al.,* 1987). Hinge region and/or nonchromophore domain movement would necessarily be involved in one monomer's effect on the other monomer. Thus, there should be some separation of the two major domains by a flexible "hinge". Given this model, it seems reasonable that the chromophore domain may be modified (in this case by Zn^{2+} chelation) while the dimer contact region maintains native structure.

N-Terminus segment. The five helical peaks in the $<\mu_H>$ plot of *Avena* phytochrome can be more precisely divided into five predicted helices using predicted (Rose, 1970) and expected (proline residues) turns. These five predicted helices shown in Figure 11 (Edmundson wheel) and Figure 10 ($<\mu_H>$ diagram) contain about 53 total residues (in *Avena*). Helix (a) contains 8 residues (Arg 4 - Ser 11), helix (b) contains 13 residues (Ser 16 - Asp 28), helix (c) contains 10 residues (Ala 33 - Phe 42), helix (d) contains 8 residues (Tyr 47 - Gly 54), and helix (e) contains 14 residues (Ser 62 - Gly 75). It is possible that these five regions fold into an α-helical conformation upon conversion of Pr to Pfr in Tris buffer, thus accounting for the 5% increase in α helix (Fig. 5). The regions corresponding to predicted helices (c) and (d) are likely candidates for α-helical formation in phosphate buffer since the regions are partially predicted to be helical by Chou-Fasman methods. This would leave only helices (a), (b), and (e) to form helix upon conversion of Pr to Pfr in the presence of phosphate. That would entail approximately 36 residues and may account for the 3.3% increase in helix observed in phosphate buffer.

The lack of Chou-Fasman helical prediction for the N-terminus segment of phytochrome is in agreement with the idea that the 6 kDa N-terminal chain is not helical in its free or "unbound" form. In the Pr molecule the N-terminus chain is considered to be "dangling" or not interacting with the rest of the protein based on UV-Vis spectroscopic [comparison of large (118 kDa) vs. intact (124 kDa) phytochrome; Vierstra and Quail (1983)], enzyme accessibility (Jones *et al.,* 1985) and antibody binding evidence (Cordonnier *et al.,* 1985). In the Pr state, the *Avena* N-terminal segment is non-helical, as indicated by CD studies (Chai *et al.,* 1987a; Vierstra *et al.,* 1987). The N-terminus chain is rich in helix-breaking residues, especially serine and threonine. This, in addition to Chou-Fasman predictions, indicates that the N-terminal segment will not become helical unless the formation of helix is in some way stabilized. This may be done by the interaction of the amphiphilic helices with a chromophore/hydrophobic pocket, as the current model suggests. This model of the 6 kDa N-terminus is very analogous to structural changes which occur in one loop in the triose phosphate isomerase polypeptide. With no substrate bound, there is an "unanchored" or mobile loop structure that folds down (the loop structure becomes rigid or defined) over the substrate after binding (Alber *et al.,* 1981). Our current model of phytochrome suggests the same basic motif.

That is, after conversion to Pfr, the N-terminal segment becomes structurally defined ("locked down") by interaction with the newly exposed chromophore/chromophore pocket (Fig. 5).

Apparently, the 6 kDa N-terminal segment is the only region in phytochrome which undergoes a change in secondary structure detectable by CD upon conversion of Pr to Pfr. It therefore seems likely that the N-terminus chain plays an important functional role in the cell. Amphiphilic helical character is common in biologically active peptides such as hormones and/or neurotransmitters. These effectors require helical amphiphilic character for activity (Blanc *et al.,* 1983). In fact, Kaiser and Kezdy (1984) have determined that synthetic helical amphiphilic proteins bind to receptors of specific amphiphilic hormones even though the primary sequence is not conserved. Thus, the synthetic peptides have significant biological activity based extensively on helical amphiphilic nature. This may explain the tolerance for substitution in the N-terminal segment. The N-terminus chain, in general, is less conserved than almost any given 70 residue segment in the phytochrome protein. This is because there are two relatively non-conserved sections of the N-terminus as shown in Figure 3. These non-conserved regions are split by a relatively conserved region from residues 20 to 50. It is possible, based on large structural differences between Pr and Pfr and conservation analysis (Fig. 3), that this region [residues 20 to 50, approximately helical regions (b) and (c) in Fig.10] is important in the signal transduction of phytochrome. The lack of conservation in the N-terminus of phytochrome has been pointed out previously (Quail *et al.,* 1987). It was, however, noted that the middle of the N-terminus is somewhat conserved (essentially the same residues we have found conserved) and that the substitutions in the 6 kDa N-terminal domain are rather conservative. Quail *et al.,* (1987) also argued that the N-terminal domain may be functionally important, in agreement with the present model.

ACKNOWLEDGEMENTS

This work was supported by a grant from U.S.P.H.S. NIH (No. GM36956) and by a grant from the Center of Biotechnology-University of Nebraska. We thank Debbie Sommer for technical assistance.

REFERENCES

Abad-Zapatero C, Griffith JP, Sussman JL, Rossman MG (1987) Refined crystal structure of dogfish M_4 apo-lactate dehydrogenase. J Mol Biol 198:445-467

Alber T, Banner D, Bloomer A, Petsko G, Phillips D, Rivers P. Wilson I (1981) On the three-dimensional structure and catalytic mechanism of triose phosphate isomerase. Phil. Trans R Soc Lond [Biol] 293:159-171

Argos P, Hanei M, Wilson JM, Kelley WN (1983) A possible nucleotide-binding domain in the tertiary fold of phosphoribosyltransferases. J Biol Chem 258:6450-6457

Blanc JP, Taylor JW, Miller RJ, Kaiser ET (1983) Examination of the requirement for an amphiphilic helical structure in ß-endorphin through the design, synthesis, and study of model peptides. J Biol Chem 258:8277-8284

Bowie JU, Reidhaar-Olson JF, Lim WA, Sauer RT (1990) Deciphering the message of protein sequences: tolerance to amino acid substitutions. Science 247:1306-1310

Bowie JU, Sauer RT (1989) Identifying determinants of folding and activity for a protein of unknown structure. Proc Natl Acad Sci USA 86:2152-2156

Breslow E, Beychok S, Hardman K, Gurd FRN (1965) Reversible conformational changes of myoglobin and apomyoglobin. J Biol Chem 240:304-309

Chai Y-G, Song P-S, Cordonnier M-M, and Pratt LH (1987a) A photoreversible circular dichroism spectral change in oat phytochrome is suppressed by a monoclonal antibody that binds near its N-terminus and by chromophore modification. Biochemistry 26:4947-4952

Chai Y-G, Singh BR, Song P-S, Lee J, Robinson W (1987b) Purification and spectroscopic properties of 124-kDa oat phytochrome. Anal Biochem 163:322-330

Chang CT, Wu C-SC Yang JT (1978) Circular dichroic analysis of protein conformation: inclusion of ß turns. Anal Biochem 91:13-31

Choi JK, Kim I-S, Kwon TI, Parker W, Song P-S. (1990) Spectral perturbations and oligomer/monomer formation in 124-kilodalton *Avena* phytochrome. Biochemistry in press

Chou PY, Fasman GD (1978) Empirical predictions of protein conformation. Ann Rev Biochem 47:251-276

Christensen AH, Quail PH (1989) Structure and expression of a Maize phytochrome-encoding gene. Gene 85:381-390

Cordonnier MM, Greppin H, Pratt LH (1985) Monoclonal antibodies with differing affinities to the red-absorbing and far-red-absorbing forms of phytochrome. Biochemistry 24:3246-3253

Cornette JL, Cease KB, Margalit JH, Spouge JL, Berzofsky JA, DeLisa C (1987) Hydrophobicity scales and computational techniques for detecting amphipathic structures in proteins. J Mol Biol 195:659-685

Eisenberg D, Schwartz E, Komaromy M, Wall R (1984) Analysis of membrane and surface protein sequences with the hydrophobic moment plot. J Mol Biol 179:125-142

Eisenberg D, Weiss RM, Terwilliger TC (1982) The helical moment: a measure of the amphiphilicity of a helix. Nature 299:371-374

Farrens D, Song P-S, Rudiger W, Eilfeld P (1989) Site selected chromophore oxidation of phytochrome with tetranitromethane. J Plant Physiol 134:269-275

Furuya M (1989) Molecular properties and biogenesis of phytochrome I and II. Adv Biophys 25:133-167

Grimm R, Eckerskorn C, Lottspeich F, Zenger C, Rudiger W (1988) Sequence analysis of proteolytic fragments of 124-kilodalton phytochrome from etiolated *Avena sativa* L. Planta 174:396-401

Hahn TR, Song P-S (1981) The hydrophobic properties of phytochrome as probed by 8-Anilinaphthalene Sulfonate fluorescence. Biochemistry 20:2602-2609

Hahn TR, Song P-S, Quail PH, Vierstra RD (1984) Tetranitromethane oxidation of phytochrome chromophore as a function of spectral form and molecular weight. Plant Physiol 74:755-758

Hecht MH, Richardson DC, Richardson JS, Ogden R (1989) Design, expression, and preliminary characterization of Felix a model protein. J Cell Biochem 13A:86

Hershey HP, Baker RF, Idler KB, Lissemore JL, Quail PH (1985) Analysis of cloned cDNA and genomic sequences for phytochrome: amino acid sequences for two gene products expressed in etiolated *Avena*. Nucleic Acids Res 13:8543-8559

Ho SP, DeGrado WF (1987) Design of a 4-helix bundle protein: synthesis of peptides which self associate into a helical protein. J Amer Chem Soc 109:6751-6758.

Hrkal Z, Vodrazka Z (1967) A study of the conformation of human globin in solution by optical methods. Biochem Biophys Acta 133:527-534

Jones AM, Erickson HP (1989) Domain structure of phytochrome from *Avena sativa* visualized by electron microscopy. Photochem Photobiol 49:479-483

Jones AM, Quail PH (1986) Quaternary structure of 124-kilodalton phytochrome from *Avena Sativa* L. Biochemistry 25:2987-2995

Jones AM, Quail PH (1989) Phytochrome structure: peptide fragments from the amino-terminal domain involved in protein-chromophore interactions. Planta 178:147-156

Jones AM, Vierstra RD, Daniels SM, Quail, P (1985) The role of separate molecular domains in the structure of phytochrome from etiolated *Avena sativa* L. Planta 164:501-506

Kaiser ET, Kezdy FJ (1984) Amphiphilic secondary structure: design of peptide hormones. Science 223:249-255

Kay SA, Keith B, Shinozaki K, Chua NH (1989) The sequence of the rice phytochrome gene. Nucleic Acids Res 17:2865-2866

Kumosinski TF, Pessen H, Farrell HMJr (1982) Structure and mechanism of action of riboflavin-binding protein: small-angle X-ray scattering, sedimentation, and circular dichroism studies on the holo and apoproteins. Arch Biochem Biophys 214:714-725

Lagarias JC, Mercurio FM (1985) Structure and function studies on phytochrome. J Biol Chem 260:2415-2423

Parker W, Song P-S (1990) Location of helical regions by a helical hydrophobic moment analysis: application to phytochrome. Biol Chem, in Press

Perutz MF, Kendrew JC, Watson HC (1965) Structure and function of haemoglobin. J Mol Biol 13:669-678

Quail PH, Gatz C, Hershey HP, Jones AM, Lissmore JL, Parks BM, Sharrock RA, Barker RF, Idler K, Murray MG, Koornneef M, Kendrick R (1987) Molecular biology of phytochrome. In: Furuya M (ed) Phytochrome and Photoregulation in Plants. Academic Press, Tokyo New York, p 23-37

Richardson JS, Richardson DC (1987) Some design principles: betabillin In: Oxender DL, Fox CF (eds) Protein Engineering. Alan R. Liss, New York p 149

Richardson JS, Richardson DC (1989) Principles and patterns of protein conformation. In: Fasman GD (ed) Prediction of Protein Structure and the Principles of Protein Conformation. Plenum Press, New York London p 86

Richardson JS, Richardson DC, Erickson BW (1984) *De-Novo* design and synthesis of a protein. Biophys J 45:25a

Romanowski M, Song P-S (1990) Structural domains of phytochrome deduced from homologies in amino acid sequences. Biophys J submitted

Rose GD (1978) Prediction of chain turns in globular proteins on a hydrophobic basis. Nature 272:586-590

Rogers S, Wells R, Rechsteiner M (1986) Amino acid sequences common to rapidly degraded proteins: the PEST hypothesis. Science 234:364-368

Sacchettini JC, Gordon JI, Banaszak LJ (1989) Refined apoprotein structure of rat intestinal fatty acid binding protein produced in *Escherichia coli*. Proc Natl Acad Sci USA 86:7736-7740.

Sato N (1988) Nucleotide sequence and expression of the phytochrome gene in *Pisum sativum*: differential regulation by light of multiple transcripts. Plant Mol Biol 11:697-710

Schiffer M, Edmundson AB (1967) Use of helical wheels to represent the structures of proteins and to identify segments with helical potential. Biophys J 7:121-135

Sharrock RA, Lissemore JL, Quail PH (1986) Nucleotide and amino acid sequence of a *Cucurbita* phytochrome cDNA clone: identification of conserved features by comparison with *Avena* phytochrome. Gene 47:287-295

Sharrock RA, Quail PH (1989) Novel phytochrome sequence in *Arabidopsis thaliana*: structure, evolution and differential expression of a plant regulatory photoreceptor family. Gene and Dev 3:1745-1757

Shoham M, Yonath A, Sussman JL, Moult J, Traub W, Kalb AJ (1979) Crystal structure of demetallized concanavalin A: the metal-binding region. J Mol Biol 131:137-155

Sommer D, Song P-S (1990) Chromophore topography and secondary structure of 124 kilodalton *Avena* phytochrome probed by Zn^{2+}-induced chromophore modification. Biochemistry 29:1943-1948

Stellwagen E, Rysavy R (1972) The conformation of horse heart apocytochrome c. J Biol Chem 247:8074-8077

Sugimoto T, Ito E, Suzuki H (1987) Interpretation of dichroic orientation of phytochrome. Photochem Photobiol 46:517-523

Takakuwa T, Konno T, Meguro, H (1985) A standard substance for calibration of circular dichroism: Ammonium d-10-camphorsulfonate. Anal Sci 1:215-218

Tokutomi ST, Nakasako M, Sakai J, Kataoka M, Yamamoto KT, Wada M, Tokunaga F, Furuya, M (1989) A model for the dimeric molecular structure of phytochrome base small-angle X-ray scattering. FEBS Lett 247:139-142

Thummler F, Eifeld P, Rudiger W, Moon DK, Song P-S (1985) On the chemical reactivity of the phytochrome chromophore in the Pr and Pfr form. Z Naturforsch 40c:215-218

VanDerWoude WJ (1985) A dimeric mechanism for the action of phytochrome: evidence from photothermal interactions in lettuce seed germination. Photochem Photobiol 42:655-661

Vierstra RD, Quail PH (1983) Photochemistry of 124 kilodalton *Avena* phytochrome *in vitro*. Plant Physiol 72:264-267

Vierstra RD, Quail PH, Hahn T-R, Song P-S (1987) Comparison of the protein conformations between different forms (Pr and Pfr) of native (124 kDa) and degraded (118/114 kDa) phytochromes from *Avena sativa*. Photochem Photobiol 45:429-432

THE USE OF TRANSGENIC PLANTS TO STUDY PHYTOCHROME DOMAINS INVOLVED IN STRUCTURE AND FUNCTION

J.R. Cherry, D. Hondred, J.M. Keller[1], H.P. Hershey[1], and R.D. Vierstra
Department of Horticulture,
University of Wisconsin-Madison,
Madison, WI 53706,
USA.

INTRODUCTION

During the 30 years since its initial isolation, a great body of information has accumulated concerning the structure of phytochrome, the physiological responses it controls, and the genes whose expression it affects, yet little is known about the molecular mechanisms of phytochrome action. The recent advent of technologies allowing the expression of heterologous phytochrome genes in transgenic plants provide an important new method for research into the mechanisms of phytochrome action (Keller *et al.*, 1989; Boylan and Quail, 1989; Kay *et al.*, 1989). In the first report of this approach, Keller *et al.* (1989) described the expression of a functional oat phytochrome in tobacco. Transgenic plants expressing the oat protein have a radically altered phenotype characterized by decreased stem elongation, increased leaf chlorophyll content, reduced apical dominance, and delayed leaf senescence. Exploiting this "light-exaggerated" phenotype as an assay, it is now possible to identify and examine domains involved in phytochrome structure and function by *in vitro* mutagenesis.

Here we review our construction and characterization of a transgenic tobacco line homozygous for a chimaeric gene that constitutively expresses wild-type oat phytochrome. We also present the preliminary characterization of tobacco plants synthesizing high levels of an oat phytochrome deletion mutant lacking N-terminal amino acids 7-69.

---

[1]Agricultural Products Department, E.I. du Pont Nemours & Co., P.O. Box 80402, Wilmington, DE 19880-0402, USA.

NATO ASI Series, Vol. H 50
Phytochrome Properties and Biological Action
Edited by B. Thomas and C. B. Johnson

GENERATION OF TRANSGENIC TOBACCO EXPRESSING OAT PHYTOCHROME

In 1989, we successfully expressed oat phytochrome in tobacco using a chimaeric oat gene under the control of the cauliflower mosaic virus (CaMV) 35S promoter (Keller *et al.*, 1989). The chimaeric gene was created using oat cDNA and genomic clones pGP8.2-2, pGP2.4-1, and pAP3.1 (Hershey *et al.*, 1985,1987) in such a way that all introns within the structural gene were eliminated. This synthetic gene contained coding sequence for the N-terminus of type 3 phytochrome linked to the C-terminus and polyadenylation signal of type 4 phytochrome. Both genes are normally expressed predominantly in etiolated seedlings in oat. A DNA fragment containing this structural gene was ligated downstream of the CaMV 35S promoter and inserted into a plant transformation vector to create plasmid pCV35phyt. This plasmid was then mobilized into *Agrobacterium tumefaciens* by bacterial conjugation and the resulting *Agrobacterium* was used to infect leaf discs of *Nicotiana tabacum* cv. Xanthi using standard transformation techniques (Horsch *et al.*, 1985).

Transformed plants were screened for oat phytochrome expression by immunoblot analysis of their progeny using two anti-phytochrome antibody preparations (Fig. 1). Dark-grown seedlings contained an abundant immunoreactive species that co-migrated with purified 124kDa oat phytochrome and could be distinguished from two smaller phytochromes native to etiolated tobacco (118 and 124kDa). In light-grown seedlings, tobacco phytochromes were undetectable due to light-dependent degradation (as discussed later) while the oat protein was present at a reduced level. When a monoclonal antibody specific to oat phytochrome was used for immunoblot analysis, only the 124kDa species present in transformed tobacco was recognized, confirming that this protein was expressed from the oat gene (Fig. 1B).

Plants homozygous for the synthetic oat phytochrome gene were obtained by self-fertilization of initial transformants (such as 9A) and selection of progeny that failed to segregate for expression of the protein in subsequent generations as assessed by immunoblot analysis. One homozygous plant line, designated 9A4, was selected for further characterization based on its high level of oat phytochrome expression and strong phenotypic response. Phytochrome content in 9A4 plants was measured by red-minus-far-red difference spectroscopy of partially purified plant extracts. 9A4 plants were found to contain 5- to 9-times more phytochrome per gram fresh weight than corresponding wt plants at all stages of development. Phytochrome levels in wt plants were ~35-fold lower in light-grown tissue than in etiolated tissue, while phytochrome in 9A4 plants decreased by only ~20-fold. The persistence of oat phytochrome in light-grown tissues results from either a decrease in the light-dependent degradation or from an increased rate of synthesis, or both, as discussed below.

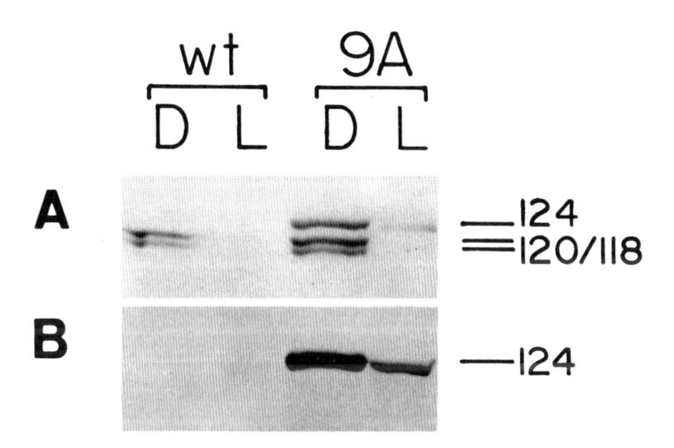

Figure 1. Immunoblot detection of oat phytochrome in transgenic tobacco. Seeds from wild-type (wt) and self-fertilized transformant 9A were grown for 10 days in the dark (D) or 10 days in continuous light (L). Extracts from seedlings were concentrated by ammonium sulphate precipitation and subjected to SDS-PAGE. Immunoblot analysis was performed with a mixture of monoclonal antibodies against pea phytochrome that recognizes either both oat and tobacco phytochrome (A) or a monoclonal antibody specific to oat phytochrome (B). Sample volumes were adjusted to contain equal amounts of protein extracted from wt and transformed tissue. Approximately 10-fold more protein was used in the L lanes than the D lanes to enable visualization of phytochrome from light-grown plants. The molecular mass of the various phytochrome polypeptides in kDa is shown on the right. [Taken with permission from Keller *et al.*, 1986]

PHENOTYPES RESULTING FROM PHYTOCHROME OVEREXPRESSION

As noted by Keller *et al.* (1989) and subsequently by Boylan and Quail (1989), expression of monocot phytochrome in dicotyledonous plants induces a number of phenotypic alterations including an inhibition of stem elongation, delayed leaf senescence, and

Figure 2. Phenotypic changes associated with expression of oat phytochrome in homozygous transgenic tobacco. Transgenic and control plants demonstrating the effect of oat phytochrome expression on plant height, leaf pigmentation, and leaf senescence. Plants were grown in greenhouses under natural lighting and photographed one week after the detection of flower primordia (65 d after planting). WT = untransformed tobacco plants. 9B2 = tobacco plants containing the oat phytochrome gene but not expressing detectable oat phytochrome. 9A4 = tobacco plants homozygous for the oat phytochrome gene which express high levels of oat phytochrome.

reduced apical dominance. These effects have been observed in over 15 independent transformants expressing levels of oat phytochrome from 2- to 9-fold that of untransformed plants; in all cases the phenotypic alterations co-segregated with oat phytochrome expression as assayed by immunoblot analysis (J. Cherry, unpublished). Most striking of the phenotypic alterations was the inhibition of stem elongation (Fig. 2). 9A4 plants were consistently 4- to 5-fold shorter than wt plants over their entire life cycle. A partial loss of apical dominance was also readily apparent as the plants matured. After flowering, 9A4 axial meristems developed nearly fully-expanded mature leaves, while those on wt plants of a similar age remained quiescent. The darker, greener leaves of 9A4 plants resulted from a 34% increase in chlorophyll content as measured spectrophotometrically. Finally, 9A4 leaves were significantly altered in both the pattern and timing of leaf senescence. Leaves of these plants remained green until well after analogous wt leaves had yellowed. Analysis of leaves of the same developmental age revealed that 9A4 plants were able to retain leaf chlorophyll and protein substantially longer than wt plants. The pattern of senescence also differed, with 9A4 leaves often showing an unusual spotted pattern during necrosis. While phytochrome overexpression did affect a number of plant phenotypes, it did not alter a number of phenotypes potentially under phytochrome control, such as the number of leaves to the flower primordia, and the timing of flowering. The morphology of etiolated 9A4 seedlings was also unaffected, supporting the view that Pfr is the biologically active form of phytochrome in dark-grown plants.

These results demonstrate that a monocot photoreceptor can function in a dicot plant, indicating that phytochrome is functionally conserved among distantly related species. They also obfuscate suggested differences between "green" and "etiolated" phytochrome functions (Tokuhisa and Quail, 1989; Pratt and Cordonnier, 1987), since it is clear that etiolated-type phytochrome is active in green tissue.

GA PARTIALLY SUPPRESSES THE PHENOTYPE OF PHYTOCHROME OVER-EXPRESSION

Because of the pleiotropic nature of phytochrome controlled responses, it has long been suggested that regulation may be mediated by changes in the levels of specific plant hormones (Kende and Lang, 1964). We noted that phytochrome-overexpressing plants share some characteristics in common with various genetic dwarfs that are deficient in gibberellin (GA) biosynthesis and with plants treated with gibberellin antagonists (Koorneef *et al.*, 1990; Steffens *et al.*, 1985; Wang *et al.*, 1985). Since the phenotypes of GA biosynthesis mutants are rescued by the foliar application of GA (Koorneef *et al.*, 1990), we attempted to suppress the phenotype of 9A4 plants by foliar applications of GA_3. Conversely, we also tried to create 9A4-like plants by spraying wt plants with the GA biosynthesis inhibitor ancymidol.

We found that weekly treatments with 50μM GA partially rescued 9A4 plants from their "light-exaggerated" phenotype. Plants so treated reached heights approximately 3/4 that of comparable untreated wt plants and over 4-times taller than untreated 9A4 plants. For comparison, wt plants sprayed with GA grew only 20 to 50 % taller than unsprayed plants. Chlorophyll content was also affected in treated 9A4, decreasing to near wild-type levels. Rather than suppressing the increased axial meristem activity characteristic of 9A4 plants, GA actually stimulated axial growth in both phytochrome overexpressors and wt controls. In the converse experiment, wt plants treated with ancymidol became shorter, but remained 3-4 times taller than untreated 9A4 plants, and their leaves became darker. 9A4 plants treated with ancymidol were similarly affected, although to a much lesser extent.

These results suggest that phytochrome control may be partially exerted through alterations in plant sensitivity to GA or modulation of GA levels. Since 9A4 plants were responsive to exogenous GA, phytochrome overexpression does not substanially decrease plant sensitivity to GA, indicating that Pfr may inhibit GA biosynthesis. Additionally, other plant hormone levels may be affected, although measurement of cytokinin levels in 9A4 and wt plants revealed no significant differences (J Cherry, RD Vierstra, and M Brenner, unpublished). Measurement of endogenous GA and auxin levels in phytochrome overexpressors are currently in progress.

BIOCHEMICAL CHARACTERIZATION OF PHYTOCHROME IN WT AND 9A4 TOBACCO

Oat phytochrome synthesized in 9A4 tobacco exhibited physico-chemical properties identical to that expressed in oat. During SDS-PAGE it comigrated with phytochrome purified from oat as a 124kDa species and could easily be distinguished from the native tobacco phytochrome species of 118 and 120kDa (Fig. 2). Phytochrome partially purified from etiolated oat, tobacco, and 9A4 plants were found to be spectrally similar with characteristic absorption maxima at 665 nm for Pr and 730 nm for Pfr. Although phytochrome isolated from 9A4 plants contained a mix of tobacco and oat phytochromes, the high proportion of oat (~5 times that of tobacco) would allow detection of spectrally altered chromoprotein, if it existed.

Because phytochrome from all species characterized to date exists *in vivo* as a dimer, we were interested in determining whether native tobacco phytochrome exists as a dimer, and if so, whether tobacco and oat phytochromes in 9A4 plants formed oat-tobacco heterodimers. Native tobacco phytochrome was found to elute from a calibrated size exclusion column with an apparent molecular mass of ~290kDa, indicating that native tobacco phytochrome exists as a dimer. Chromatography of phytochrome from 9A4 plants indicated that oat phytochrome expressed in tobacco also dimerized, forming a species that eluted approximately ~10kDa earlier than tobacco dimers. This partial separation of the heterologous chromoproteins suggested that oat and tobacco phytochromes did not form heterodimers. The failure of monoclonal antibodies specific for the oat chromoprotein to immunoprecipitate significant amounts of tobacco phytochrome from 9A4 extracts further supports this conclusion (Cherry *et al.*, submitted).

Tobacco, like other dicots characterized (Quail *et al.*, 1973; Pratt, 1978), degraded phytochrome in a light-dependent manner. (Ub-P; Shanklin *et al.*, 1987, Jabben *et al.*, 1989a,b). Using spectral assays and immunoblot analysis, it was found that tobacco phytochrome is rapidly degraded ($t_{1/2}$ 1 hr) *in vivo* after conversion of Pr to Pfr via red light irradiation (Fig. 3A). Phytochrome degradation could be interrupted by conversion of the remaining Pfr to Pr with a short far-red light irradiation (Fig. 3A Inset).

Tobacco also accumulated ubiquitin-phytochrome (Ub-P) conjugates during Pfr degradation. Previously, we showed that soon after photoconversion of Pr to Pfr in etiolated oat seedlings, phytochrome becomes multiply conjugated with the small (76 amino acids) protein ubiquitin (Shanklin *et al.*, 1987). Ubiquitin functions in protein breakdown by becoming covalently ligated to proteins targeted for degradation. The pathway was first described in rabbit reticulocytes (Hershko, 1988) and since has been characterized in yeast (Finley and Varshavsky, 1985) and in higher plants in our laboratory (Vierstra, 1989). Ubiquitin ligation requires ATP and is performed by a three enzyme pathway consisting of ubiquitin activating enzyme (E1), a family of ubiquitin carrier proteins (E2s), and a family of ubiquitin-protein ligases (E3s). Ubiquitin attachment occurs through a novel peptide bond between the C-terminal glycyl carboxyl group of ubiquitin and internal lysyl γ-amino groups on the target protein. Conjugation represents the committed step in the pathway with selectivity appearing to reside in the target protein specificity of the E2 and E3 families (Hershko, 1988). Once conjugated with a multitude of ubiquitins, the modified protein is recognized by a very high molecular mass (1,500kDa) ATP-dependent protease that degrades the target protein with the release of free, functional ubiquitin (Hough and Rechsteiner, 1986). In this way, ubiquitin serves as a reusable recognition signal for proteolysis.

Ub-P formation in tobacco was assayed by irradiating plants with red light for various times, partially purifying phytochrome from these plants, and immunoprecipitating with anti-phytochrome antibodies. Immunoprecipitates were then subjected to SDS-PAGE, transferred to nitrocellulose, and immunoblotted with anti-ubiquitin antibodies. Accumulation and subsequent degradation of Ub-P was assayed by first inducing Ub-P formation with red light irradiation followed by an irradiation with far-red light and an incubation in darkness (Fig. 3B). The immunoblots indicted that, like other species examined (Jabben *et al.*, 1989a), tobacco Ub-P accumulation occurred only after phytochrome was converted to Pfr and ceased after photoconversion of Pfr back to Pr. As many as 7 ubiquitins were covalently attached to the chromoproteins as judged by relative molecular masses of Ub-P on SDS-PAGE. Furthermore, Ub-P formed during red light irradiation were demonstrated to rapidly disappear ($t_{½}$~ 5-10 min) when continued Ub-P formation was halted by far-red light irradiation. These data further support the model of Shanklin *et al.* (1987) suggesting that phytochrome degradation occurs via the ubiquitin-dependent proteolytic pathway.

Analysis of phytochrome from etiolated 9A4 seedlings demonstrated that the oat chromoprotein is also rapidly degraded in tobacco. Red light irradiation induced the loss of the 124kDa protein along with the 118- and 120kDa tobacco species (Fig. 4A). Degradation of oat Pfr, as assayed both spectrally and immunochemically, occurred at a significantly slower rate than tobacco Pfr. The half-life of the oat species, obtained by quantitation of the 124kDa species on immunoblots and by mathematically "peeling" the kinetics of degradation measured spectrally (Fig. 4A), was ~4 hours.

Oat Pfr degradation in tobacco was also correlated with the accumulation and loss of Ub-P in a manner similar to that described for tobacco Pfr studies. In this case, oat Ub-P was quantitated on immunoblots following immunoprecipitated using a monoclonal antibody specific for the oat protein (Fig. 4B). As with tobacco phytochrome, oat Ub-P accumulated to high levels after Pfr formation with multiple ubiquitins attached per phytochrome monomer. Whereas the accumulation of oat Ub-P had kinetics similar to tobacco Ub-P, the disappearance of oat Ub-P after far-red light treatment was much

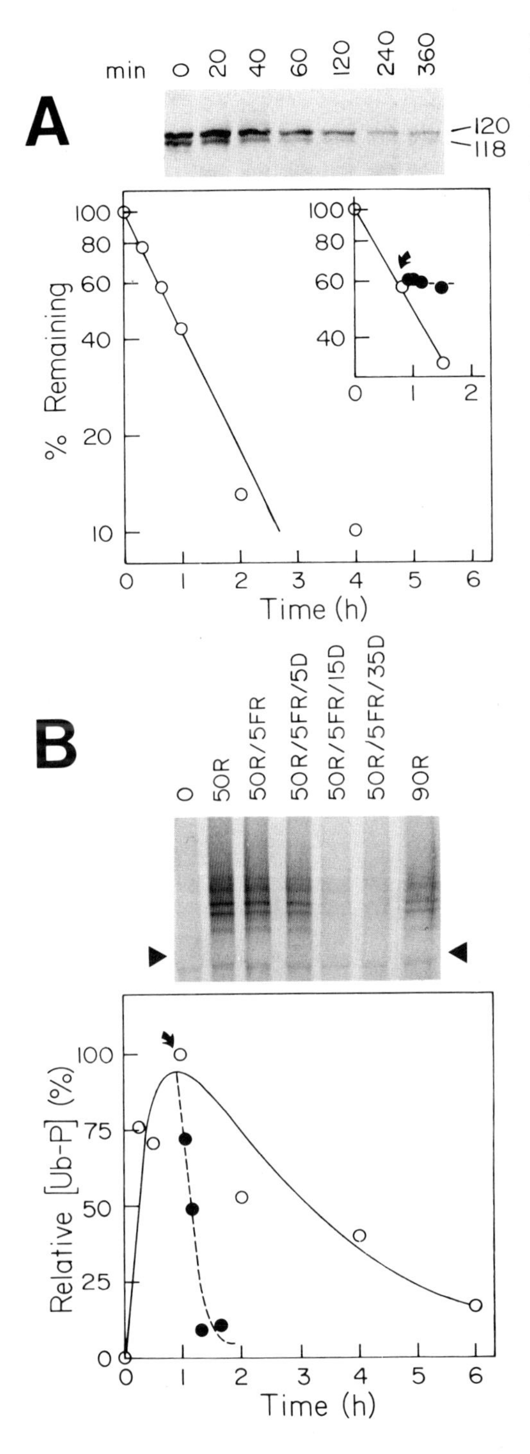

Figure 3. Phytochrome degradation and Ub-P accumulation in 5 day-old etiolated tobacco seedlings (WT). Seedlings were irradiated continuously with red light for various times (0-360, O), or irradiated first with red light for 50min (50R), then with a 5min irradiation with far-red light (5FR, ●; see curved arrows) followed by an incubation in darkness for 5 to 35min (5-35D, ●). At the indicated times tissue was assayed for phytochrome and Ub-P content. (A) Kinetics of phytochrome degradation with the various light treatments as measured by SDS-PAGE and immunoblot analysis with anti-zucchini squash phytochrome antibodies (Upper Panel) or by red-minus-far-red difference spectroscopy [Δ(ΔA), Lower Panel]. Inset shows the effect of a 5min far-red light irradiation (curved arrow) on the kinetics of degradation. The migration positions of the 120- and 118-kDa tobacco phytochromes during SDS-PAGE are indicated on the right of the immunoblot. (B) Accumulation of tobacco Ub-P with the various light treatments. Ub-P were partially purified by immunoprecipitation with a mixture of five monoclonal antibodies against pea phytochrome and subjected to SDS-PAGE and immunoblot analysis with anti-ubiquitin antibodies (Upper Panel). Arrowheads indicate the position of unmodified tobacco phytochrome. The relative levels of tobacco Ub-P during the time courses were determined by reflective densitometry of the immunoblots (Lower Panel).

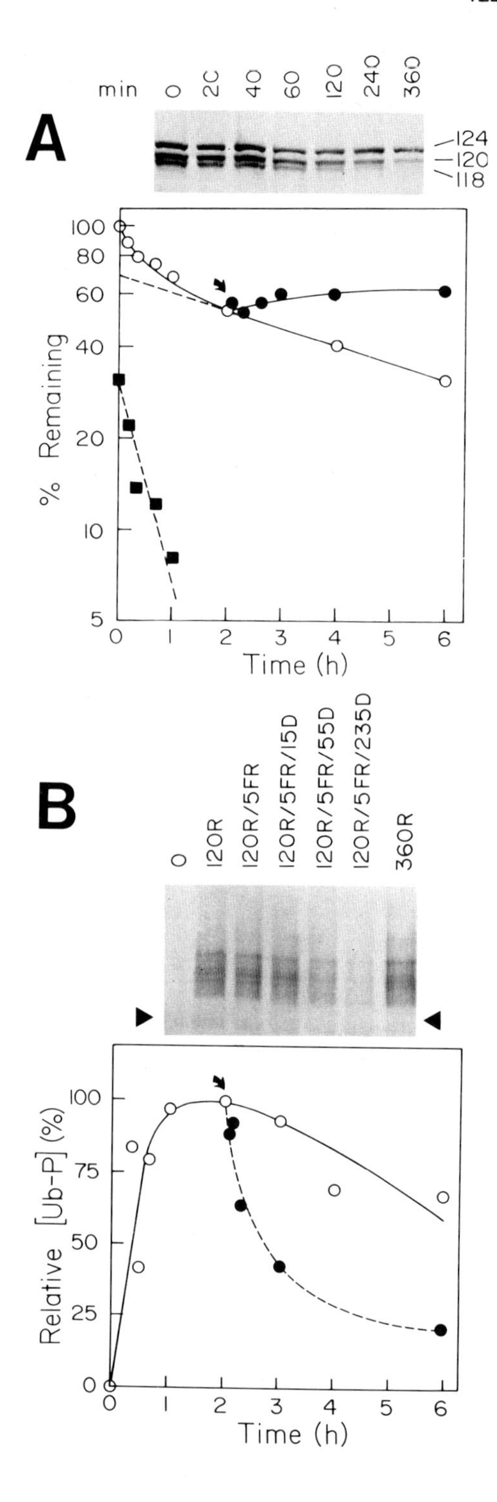

Figure 4. Phytochrome degradation and Ub-P accumulation in 5 day-old etiolated tobacco seedlings expressing oat phytochrome (9A4). Seedlings were irradiated continuously with red light for various times (0-360, ○), or irradiated first with red light for 120min (120R), then with a 5min irradiation with far-red light (5FR, ●; see curved arrows) followed by an incubation in darkness for 15 to 235min (15-235D, ●). At the indicated times tissue was assayed for phytochrome and Ub-P content. (A) Kinetics of phytochrome degradation with the various light treatments as measured by SDS-PAGE and immunoblot analysis with anti-zucchini squash phytochrome antibodies (Upper Panel) or by red-minus-far-red difference spectroscopy [$\Delta(\Delta A)$, Lower Panel]. Dashed lines denote the mathematical peeling of total phytochrome degradation to reveal the predicted degradation kinetics of tobacco phytochrome (■). The migration positions of the 120- and 118-kDa tobacco phytochromes during SDS-PAGE are indicated on the right of the immunoblot. (B) Accumulation of oat Ub-P with the various light treatments. Oat Ub-P were partially purified by immunoprecipitation with a monoclonal antibody specific to oat phytochrome and subjected to SDS-PAGE and immunoblot analysis with anti-ubiquitin antibodies (Upper Panel). Arrowheads indicate the position of unmodified tobacco phytochrome. The relative levels of oat Ub-P during the time courses were determined by reflective densitometry of the immunoblots (Lower Panel).

slower, consistent with the longer half-life of oat Pfr in tobacco. With a half-life of ~30-40 min, oat Ub-P are apparently degraded with kinetics more similar to that seen in oat than that for tobacco Ub-P in tobacco.

These studies suggest that the phytochrome domain(s) responsible for Pfr-dependent Ub-P formation and Pfr-dependent protein degradation are qualitatively conserved between monocots and dicots. That the rates of tobacco and oat Pfr degradation correlate well with the respective rates of Ub-P disappearance suggests that the two phenomena are linked (i.e., oat Pfr and oat Ub-P have longer half-lives than their wt counterparts). The relatively slow apparent rate of oat phytochrome degradation in 9A4 plants may be the result of the high rate of oat protein synthesis. Alternately, oat Pfr may be less efficiently recognized than tobacco Pfr by the ubiquitin-dependent proteolytic pathway of tobacco.

EXPRESSION OF OAT PHYTOCHROME DELETION MUTANTS IN TOBACCO

In an effort to map phytochrome domains involved in structure and function, a variety of deletions in the oat phytochrome structural gene have been constructed and transformed into tobacco. Initial deletions were designed to produce proteins which correspond to previously characterized proteolytic products of the full-length oat phytochrome such as the 118kDa "large" and 64kDa "small" species (Vierstra and Quail, 1986) to determine whether either of these proteins have biological activity.

Of these deletions, only one has been characterized sufficiently to discuss. This deletion mutant, called NA, was designed to express a protein roughly corresponding to the 118kDa "large" phytochrome. The NA gene was constructed by replacement of DNA sequences coding for the oat phytochrome N-terminus with a synthetic double-stranded oligonucleotide which causes an in-frame fusion of codons 5 and 70. In deleting residues 7-69, we intentionally left the N-terminal six residues in place in an effort to preserve sequences potentially important for N-terminal acetylation.

Immunoblot analysis of NA transformant extracts revealed the presence of a 118kDa protein whose size agreed well with the predicted molecular mass of the NA deletion mutant. This protein is recognized by anti-oat antibodies that do not recognize the 118- and 120kDa tobacco phytochromes (Fig. 4), indicating that it is derived from the NA gene. Further analysis of the NA protein indicated that this deletion mutant was capable of chromophore attachment, photoreversibility, and dimcrization. The presence of chromophore was detected by the zinc fluorescence assay following SDS-PAGE of NA extracts (Jones *et al.*, 1989). Difference spectra of partially purified NA phytochrome detected absorption maxima shifted to lower wavelengths in both the Pr and Pfr forms relative to full-length oat or tobacco phytochromes (Fig. 4B). Its spectra was much like that reported for "large" phytochrome (Vierstra and Quail, 1982, 1983; Jones *et al.*, 1985).

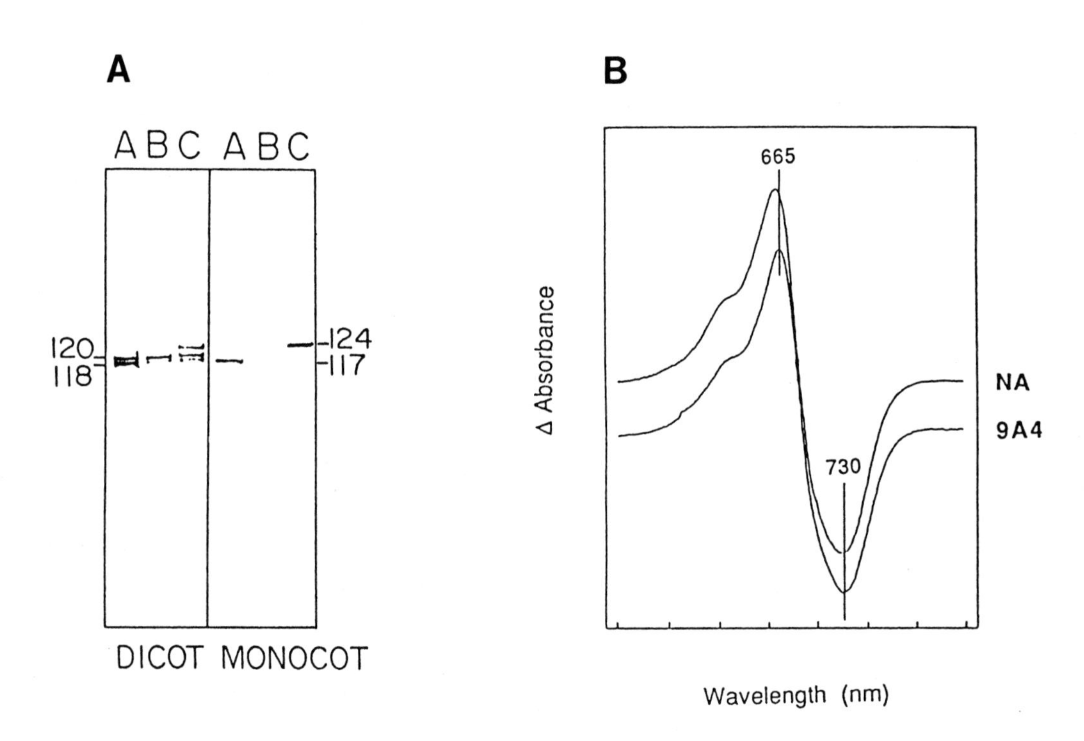

Figure 5. Oat phytochrome deletion NA expressed in transgenic tobacco. (A) Immunoblot analysis of crude extracts from deletion NA (A). wild-type tobacco (B), and full-length oat phytochrome expressor 9A4 (C). Extracts were prepared from 5 day-old seedlings were separated by SDS-PAGE and immunoblotted with anti-phytochrome antibodies raised against purified zucchini squash (DICOT) or oat (MONOCOT) phytochromes. Tobacco contains two etiolated phytochrome species with Mr's of 118 and 120kDa which are not detected by the monocot antibody, while both antibody preparations detect full-length (124kDa) and truncated (117kDa) oat phytochromes. (B) Red-minus-far-red difference spectra of full-length and truncated oat phytochrome expressed in tobacco. Phytochrome was partially purified from etiolated 9A4 or NA tobacco and concentrated to by ammonium sulphate precipitation. Difference spectra were recorded using a Shimadzu UV3000 dual wavelength spectrophotometer. Numbers indicate absoption maxima (nm) of the full-length oat chromoprotein.

The shift is most likely larger than it appears in Figure 4 because the spectra recorded is that of a mix of tobacco and NA phytochromes partially purified from a plant heterozygous for the NA gene. Size exclusion chromatography of native NA extracts indicated that deletion NA has an apparent molecular mass of 280kDa, indicating that residues 7-69 of oat phytochrome are not involved in phytochrome dimerization (data not shown). Finally, partially purified phytochrome from NA plants undergoes dark reversion of Pfr to Pr at a rate ~5 times that of phytochrome purified from either wt or 9A4 plants (data not shown).

While deletion mutant NA is biochemically quite similar to full-length oat phytochrome in that it binds chromophore, is photoreversible, and dimerizes, plants expressing NA are phenotypically identical to untransformed control plants. It thus appears that NA is not biologically active, despite being expressed at levels comparable to full-length oat phytochrome in 9A4 plants. From this, we conclude that residues 7-69 are crucial to proper phytochrome function *in vivo*. It also provides the first evidence that the residues missing from "large" degraded phytochrome are indeed required for phytochrome activity.

CONCLUSIONS

The advent of transgenic phytochrome expression systems opens a new frontier in photomorphogenesis research. By expressing phytochrome mutants with defined deletions or site-directed mutations it should be possible to identify and define domains involved in chromophore attachment, spectral integrity, dimerization, and form-specific degradation. Mutants deficient in any of these characteristic phytochrome properties can then be scored for biological function based on their ability to alter the host plant phenotype. In this way, transgenic plants may provide a much-needed link between the biochemical properties of the photoreceptor and its biological activity. In addition, the ability of full-length phytochrome overexpression to accentuate phytochrome-mediated responses may allow identification of responses previously not known to be under phytochrome control.

ACKNOWLEDGEMENTS

We thank Drs John Shanklin and Merten Jabben for advice and assistance, Dr Lee Pratt for providing polyclonal anti-zucchini squash phytochrome antibodies, Dr Masaki Furuya for providing anti-pea phytochrome monoclonal antibodies, and Drs Peter Quail and Susan Daniels for providing the anti-oat phytochrome monoclonal antibody 10.7E11D. This work was supported by grants from the U.S. Department of Energy grant (DE-F602-88ER13968) and the Research Division of the UW-College Agriculture and Life Sciences (Hatch 2858) to RDV and E.I. du Pont de Nemours & Co.

REFERENCES

Boylan MT, Quail PH (1989) Oat phytochrome is biologically active in transgenic tomatoes. Plant Cell 1:765-773

Cherry JR, Hershey HP, Vierstra RD (1990) Physiological and biochemical characterization of tobacco expressing functional oat phytochrome. Plant Cell (submitted)

De Greef JA, Fredericq JH (1983) In: Shropshire W, Mohr H (eds)Encyclopedia of plant physiology: Photomorphogenesis and hormones 16A:401-427

Finley D, Varshavsky A (1985) The ubiquitin system: functions and mechanisms. Trends Biochem Sci 10:343-347

Hershey HP, Barker RF, Idler KB, Lissemore JL, Quail PH (1985) Analysis of cloned cDNA and genomic sequences for phytochrome: complete amino acid sequence for two gene products expressed in etiolated *Avena*. Nucleic Acids Res 13:8543-8559

Hershko A (1988) Ubiquitin: roles in protein modification and breakdown. Cell 34:11-12

Horsch R, Frey J, Hoffman N, Walroth M, Eicholtz D, Rogers S, Fraley R (1985) A simple and general method for transferring genes into plants. Science 227:1229-1231

Hough R, Pratt G, Rechsteiner M (1986) Ubiquitin-lysozyme conjugates. J Biol Chem 261: 2400-2408

Jabben M, Shanklin J, Vierstra RD (1989a) Red light-induced accumulation of ubiquitin-phytochrome conjugates in both monocots and dicots. Plant Physiol 90:380-384

Jabben M, Shanklin J, Vierstra RD (1989b) Ubiquitin-phytochrome conjugates: Pool dynamics during *in vivo* phytochrome degradation. J Biol Chem 264:4998-5005

Jones AM, Vierstra RD, Daniels SM, Quail PH (1985) The role of separate molecular domains in the structure of phytochrome from etiolated *Avena sativa* L. Planta 164:501-506

Jones AM, Quail PH (1989) Phytochrome structure: peptide fragments from the amino-terminal domain involved in protein-chromophore interactions. Planta 178:147-156

Kay SA, Nagatani A, Keith B, Deak M, Furuya M, Chua N-H (1989) Rice phytochrome is biologically active in transgenic tobacco. Plant Cell 1:775-782

Keller JM, Shanklin J, Vierstra RD, Hershey HP (1989) Expression of a functional monocotyledonous phytochrome in transgenic tobacco. EMBO J 8:1005-1012

Kende H, Lang A (1964) Gibberellins and light inhibition of stem growth in peas. Plant Physiol 39:435-440

Koornneff M, Bosma TDG, Hanhart CJ, van der Veen JH, Zeevaart JAD (1990) Isolation and characterization of GA-deficient mutants in tomato. Theor Appl Genetics (submitted)

Pratt LH (1978) Molecular properties of phytochrome. Photochem Photobiol 27:81-105

Quail PH, Schafer E, Marme D (1973) Turnover of phytochrome in pumpkin cotyledons. Plant Physiol 52:128-131

Shanklin J, Jabben M, Vierstra RD (1987) Red light-induced formation of ubiquitin-phytochrome conjugates: Identification of possible intermediates of phytochrome degradation. Proc Natl Acad Sci USA 84:359-363

Steffens GL, Byun JK, Wang SY (1985) Controlling plant growth via the gibberellin biosynthesis system I. Growth parameter alterations in apple seedlings. Physiol Plant 63:163-168

Tohuhisa JG, Daniels SM, Quail PH (1985) Phytochrome in green tissue: Spectral and immunochemical evidence for two distinct molecular species of phytochrome in light grown *Avena sativa* Planta 164:321-332

Vierstra RD, Quail PH (1982) Native phytochrome: Inhibition of proteolysis yields a homogeneous monomer of 124 kilodaltons from *Avena* Proc Natl Acad Sci USA 79:5272-5275

Vierstra RD, Quail PH (1983) Purification and initial characterization of 124-kilodalton phytochrome from *Avena* Biochemistry 22:2498-2505

Vierstra RD, Quail PH (1986) Phytochrome: the protein. In:Kendrick RE, Kronenberg GHM (eds) Photomorphogenesis in Plants Martinus Nijhoff Publishers, The Netherlands, p 35-59

Vierstra, R.D., (1989) Protein Degradation. In:(Marcus, A. ed) Biochemistry of Plants: A Comprehensive Treatise Academic Press, NY. 15:521-536

Wang SY, Byun JK, Steffens GL (1985) Controlling plant growth via the gibberellin biosynthesis system-II. Biochemidal and physiological alterations in apple seedlings. Physiol Plant 72:169-175

Section B.

PHYTOCHROME ACTION - MOLECULAR BIOLOGY

IN VITRO PROTEIN-DNA INTERACTIONS IN THE RICE PHYTOCHROME PROMOTER

S. A. Kay
Laboratory of Plant Molecular Biology,
The Rockefeller University,
1230 York Avenue,
New York NY 10021,
USA.

INTRODUCTION

Phytochrome exerts many of its effects on plant morphogenesis by changing patterns of gene expression (Nagy *et al.,* 1988). Although the transcription of many genes has been shown to be under phytochrome control, little is known about the chain of events that transmits the light signal to the genome. One valid approach towards unravelling the signal transduction pathway is to identify the *cis*-acting DNA sequences that mediate the phytochrome response and the protein factors that specifically interact with them. An understanding of the regulation of these factors provides an entry point from which one can eventually work 'backwards' towards the photoreceptor, bearing in mind that multiple and redundant light switches may exist for any one gene (Gilmartin *et al.,* 1990).

Most of the studies on phytochrome-regulated gene transcription have been performed on members of gene families encoding the chlorophyll a/b binding proteins (cab) and the small subunit of ribulose bisphosphate carboxylase (rbcS, Gilmartin *et al.,* 1990; see also Gilmartin *et al.,* this volume.). Several *cis*-acting light-responsive elements (LREs) have been identified that confer light-induced expression upon heterologous promoters. The most detailed studies have been performed upon an element designated Box II in the pea *rbcS-3A* promoter (Gilmartin *et al.,* 1990). This element is required for white light- and phytochrome induction, and a tetramer of this sequence confers light responsiveness upon a heterologous promoter (Gilmartin & Chua, 1990a&b ; Lam and Chua, 1990). A protein factor, termed GT-1, has been identified that specifically binds Box II and related sequences, and *in vitro* binding of GT-1 to Box II is closely correlated

NATO ASI Series, Vol. H 50
Phytochrome Properties and Biological Action
Edited by B. Thomas and C. B. Johnson

to light-induced transcription *in vivo* (Green *et al.*, 1987, 1988a; Kuhlemeier *et al.*, 1988; Lam and Chua, 1990).

To complement our studies on phytochrome-induced transcription, we have chosen to study genes whose transcription is reduced by light. Our long term strategy is to identify *cis*-elements and the cognate protein factors for these "photophobic" genes, and to compare these to the light-inducible components. By comparing the transcriptional mechanisms in this way, it should be possible to identify branch points in the transduction chain between phytochrome and the responding target genes. To initiate these studies, we have cloned genes for phytochrome and protochlorophyllide reductase from rice (Kay *et al.*, 1989a&b). Type I phytochrome is encoded by a single copy *phyA* gene in rice, and exhibits a dramatic decline in transcription following phytochrome photoconversion (Kay *et al.*, 1989a), as has also been shown in oat (Lissemore and Quail, 1988). Given the small genome of rice, coupled with recent success in generating transgenic rice plants (Shimamoto *et al.*, 1989), the rice *phyA* gene provides an excellent system for studying genes that are down-regulated by phytochrome.

CONSERVED SEQUENCE ELEMENTS IN MONOCOT *phyA* PROMOTERS

Genomic sequences encoding the "type I" or "etiolated" form of phytochrome have been reported from oat (Hershey *et al.*, 1987), rice (Kay *et al.*, 1989a&b) and maize (Christensen and Quail, 1989). One obvious approach towards elucidating the autoregulation of these genes is to identify conserved elements present in the promoters. Comparisons of the 5′ upstream regions of the rice, oat and maize *phyA* genes are presented in Figure 1. Several elements have been identified that show sequence homology between two or more of the promoters (Kay *et al.*, 1989a; Christensen and Quail, 1989). Three copies of the GT-1 core sequences (GGTA/TA/TA/T) are present in the rice gene, and the distal two are conserved in oat. However, no such core sequences are present in the maize *phyA* gene. Interestingly, the only sequence that is well conserved between all three *phyA* genes is the GC-rich sequence designated Box I (Christensen and Quail, 1989; Kay *et al.*, 1989a). This sequence is conserved 14/14 at the 5′ end between rice and oat, and the conservation extends for 8 more nucleotides at the 3′ end between oat and maize. Lesser conservation is observed for boxes II and III.

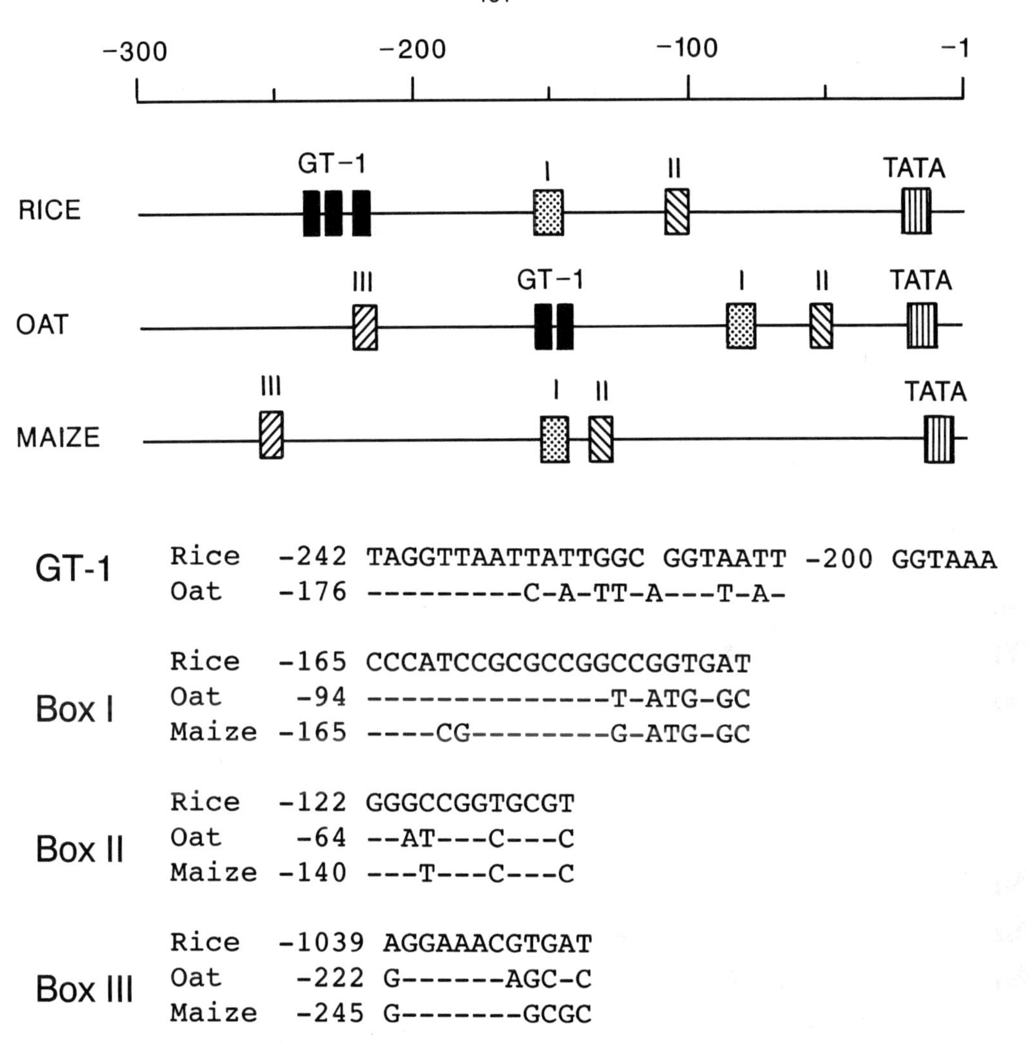

Figure 1. Several sequence motifs are conserved amongst monocot *phyA* promoters. Top: A schematic showing the relative positions of conserved sequence motifs. Bottom: Sequences of the conserved motifs. The GT-1 and Box I homologies were initially identified by comparing the rice and oat sequences (Kay *et al.*, 1989a). The Box number designations and Box II and III homologies were proposed by Christensen and Quail (1989).

IN VITRO FOOTPRINTING IDENTIFIES THREE GT-1 BINDING SITES IN THE RICE *phyA* PROMOTER

To test the significance of evolutionary conserved sequence elements, a number of approaches are plausible. It is possible to delete the sequence, in the context of a minimal promoter, and test for loss of function. Alternatively, one can adopt a positive strategy, whereby the sequence element is tested for protein binding *in vitro*, and independent function as a multimerized element *in vivo*. Correlation of the specific sequence requirements for *in vitro* binding and *in vivo* expression lend functional significance to that particular protein-DNA interaction. Our initial approach has been to identify specific *in vitro* binding of rice nuclear proteins to *phyA* DNA probes. We have previously shown in gel retardation experiments that a probe extending from -441 to -204 of the rice *phyA* gene is specifically bound by GT-1 (Kay *et al.,* 1989a). This probe contains two GT-1 core binding sites that are conserved between rice and oat.

To characterize these interactions further, we have performed DNAse I footprinting experiments, an example of which is shown in Figure 2. Crude rice nuclear extracts were incubated with a rice *phyA* probe extending from -441 to -70, followed by partial cleavage with DNAse I. Two sets of experiments were performed with probe DNA radiolabelled on either the top or bottom strand only. In both cases, addition of extract gives two clearly protected regions, one larger than the other. If a fifty fold molar excess of Box II tetramer competitor is added to the binding reaction, no protection is evident. These data demonstrate that a factor similar to GT-1 is present in rice nuclear extracts and interacts with the *phyA* upstream region. The footprinted regions can be mapped at the nucleotide level using the G cleavage patterns, and are shown in Figure 3. The larger footprinted region encompasses 31 nucleotides on both strands, from -246 to -216 on the top strand and -247 to -217 on the bottom strand. Within this sequence are two GT-1 core binding sites, GGTTAA and GGTAAT. These sites are homologous to the GT-1 binding sites present within box II and box II* of the pea *rbcS-3A* LRE (Green *et al.,* 1987, 1988a). As shown in Figure 1, these sites are conserved between rice and oat, although the corresponding oat sequence contains two box II cores, GGTTAA. We therefore propose that the larger footprint site is bound by two adjacent molecules of GT-1. The smaller footprinted region covers 16 nucleotides on both strands, extending from -205 to -190 on the top strand. At the center of this protected region is the sequence GGTAAA, which previous mutational analysis indicates is a GT-1 core binding site (Green *et al.,* 1988a). It is therefore likely that a third molecule of GT-1 is bound at this site *in vitro.*

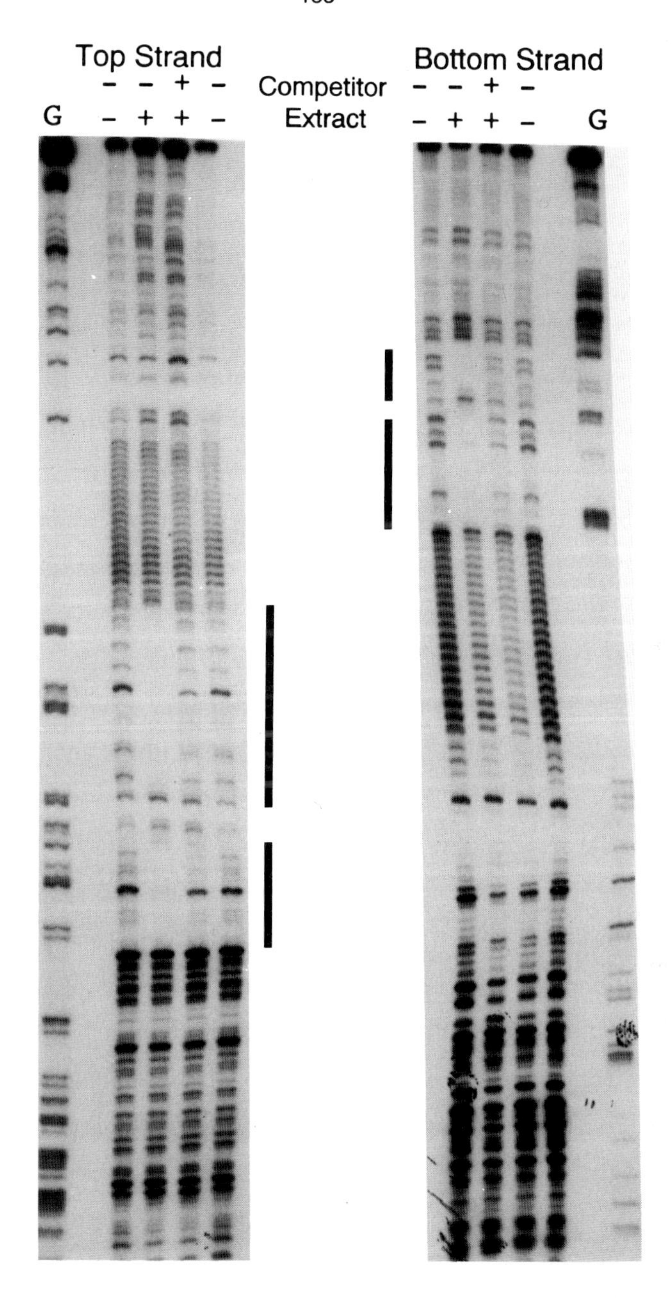

Figure 2. *In vitro* footprinting of the rice *phyA* promoter. A DNA probe extending from -441 to -70 was labelled at the 3′ end on either the top or bottom strands. The probe was then incubated with crude rice nuclear extract, prepared from etiolated rice seedlings. In a parallel incubation, a fifty fold excess of a tetramer of the *rbcS-3A* box II sequence was added to the binding reaction. The protected regions are shown by the vertical black bars. Nuclear extracts, DNAse I footprinting and the G-reactions were performed as described (Green *et al.*, 1988b).

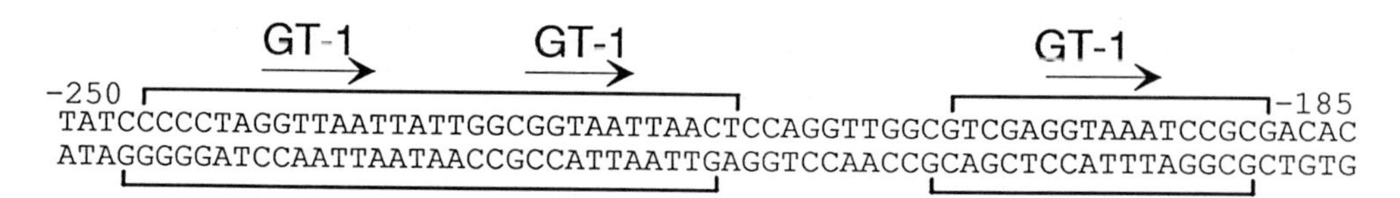

Figure 3. *In vitro* footprinting reveals three GT-1 binding sites in the rice *phyA* promoter. The positions of the footprinted regions were determined from the relative positions of DNAse I cleavage products with the relevant G-reaction.

It is interesting to find GT-1 binding to the rice phytochrome promoter, as this factor is an integral part of the light-inducible switch in the pea *rbcS-3A* promoter. Could GT-1 therefore be a common factor in the pathways of both photophobic and photophilic transcriptional responses to phytochrome? In the *rbcS-3A* LRE GT-1 binding sites are present as boxes II and III, although box II can functionally replace box III (Gilmartin and Chua, 1990b). In this arrangement, the two GT-1 binding sites are situated head to head, with 33 nucleotides between critical G residues. In the rice and oat *phyA* promoters the two GT-1 binding sites are situated head to tail and spaced at 15 and 16 nucleotides, respectively. As it has been established for the *rbcS-3A* that the conformation and physical spacing of GT-1 binding sites affects activity *in vivo* (Gilmartin and Chua 1990a,b), the particular conformation present in the rice and oat *phyA* promoters may contribute to photophobic regulation by phytochrome. The proximity of the two conserved GT-1 binding sites may lead to protein-protein interactions, which await further analysis. It is possible that other factors also bind to this site *in vivo*. Although the maize *phyA* promoter does not contain obvious GT-1 binding sites, it is interesting to note that the transcriptional shut-off following phytochrome activation is less dramatic than in rice and oat (Christensen & Quail, 1989; Kay *et al.,* 1989; Lissemore and Quail, 1988). To delineate the role of the *in vitro* GT-1 interactions *in vivo*, site-specific mutants and multimerized sequences can be tested *in vivo*, both in transient expression assays by particle bombardment (Bruce *et al.,* 1989), and ultimately in transgenic rice (Shimamoto *et al.,* 1989).

IDENTIFICATION OF ϕGC-1, A FACTOR THAT SPECIFICALLY BINDS THE GC-RICH BOX I IN MONOCOT *phyA* PROMOTERS

We have noted several distinct regions in different footprinting experiments that showed weak protection (S. Kay, unpublished observations). In order to investigate this further, we have fractionated rice nuclear extracts on several media. Resolved peak protein fractions were bulked, concentrated and then used in gel retardation experiments. An example of this is shown in Figure 4. Rice nuclear extracts were fractionated on heparin-sepharose and biorex 70, and the major binding peak was collected. A radiolabelled rice *phyA* DNA binding probe, extending from -205 to -50 (containing the conserved box I sequence), was used in gel retardation experiments (Fig. 4). When such a fraction is incubated with the probe, two DNA protein complexes are detected, one large smeary band of high mobility, and a weaker band of low mobility. If an equivalent fraction from tobacco nuclear extracts is used with the same probe, only the higher mobility band is detected. If cold competitor, consisting of a tetramer of the GC-rich box I, is added at 5- or 50-fold molar excess, the upper band is specifically competed. Thus a factor is present at low abundance in rice nuclear extracts that recognizes the box I sequence. We have designated this factor ϕGC-1.

To delineate the sequence specificity of ϕGC-1, we performed DNAse I footprinting experiments. However, due to the low abundance of this factor, we were unable to detect a footprint interaction directly. We therefore had to resolve the bound complex first by gel retardation, before performing DNAse treatment. The resulting footprint is shown in Figure 5. The bound complex reveals a DNAse I footprint that essentially spans the 14 nucleotide region of box I that is highly conserved between rice, oat and maize (see Fig. 1). However, due to the technical difficulty of this experiment, we cannot exclude that the protected region may extend beyond these nucleotides. The high conservation of the ϕGC-1 binding site amongst monocot *phyA* promoters suggests a functional role for this factor *in vivo*. We are currently investigating the sequence-specificity of ϕGC-1 by performing methylation interference studies to identify G-residues that are critical for major groove contacts. This will allow us to define site-specific mutations that interfere with ϕGC-1 binding. In addition, we are testing equivalent fractions of extracts prepared from plants given different illumination regimes, to test if the activity of this factor is altered by light.

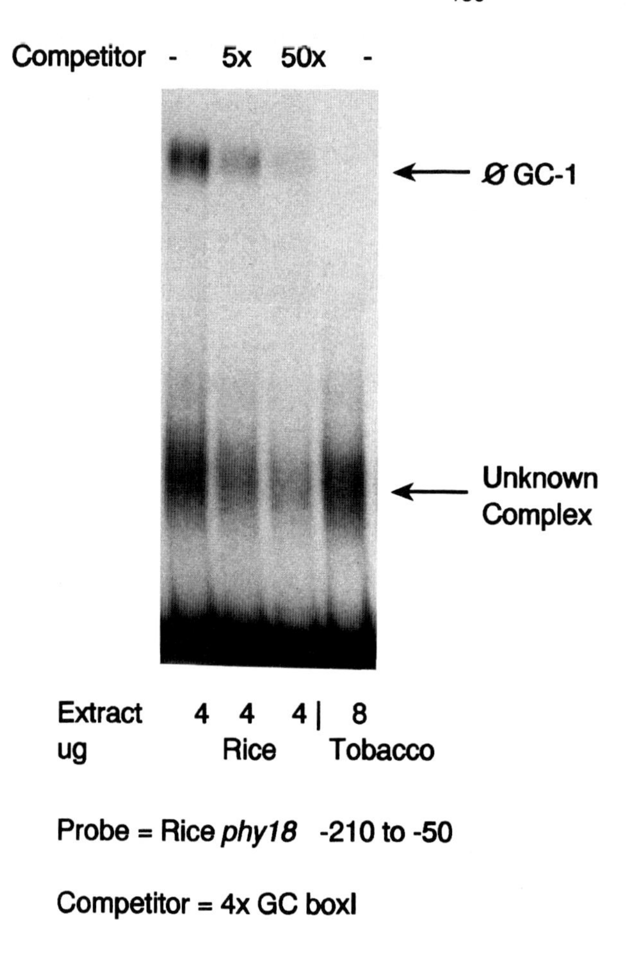

Figure 4. Identification of ϕGC-1 binding to the GC-rich conserved Box I sequence of the rice *phyA* promoter. Rice nuclear extracts were prepared from etiolated tissue and sequentially fractionated on heparin-sepharose and biorex 70 columns. Peak protein fractions were pooled and concentrated. Fractionated extract was used in gel retardation experiments with a probe extending from -210 to -50. In some reactions, cold competitor was added that consisted of a tetramer of the Box I sequence, CCCATCCGCG-CCGG, in which each 14 bp unit is separated by a hexameric restiction site. An equivalent nuclear extract fraction from dark-adapted tobacco was also used.

CONCLUSIONS AND FUTURE PROSPECTS

Studies from Cab and rbcS have revealed the complexity of LREs. Light-inducible promoters can exhibit redundancy in more than one element, and also contain information for correct spatial and temporal expression patterns (Gilmartin *et al.*, 1990; Kuhlemeier *et al.*, 1988). It is therefore likely that the autoregulation of *phyA* genes will also involve complex interactions at the DNA level. However, the identification of factors that bind to conserved motifs, coupled with both transient and transgenic *in vivo* expression assays (Bruce *et al.*, 1989; Shimamoto *et al.*, 1989), provide powerful tools for dissecting the autoregulatory mechanism. Recent success in the cloning of DNA binding proteins for LREs (see Gilmartin *et al.*, this volume) provides access to the terminal steps of the signal transduction pathway. Cloning of genes encoding monocot factors that exhibit a similar binding specificity to cloned dicot factors, such as GT-1, should prove interesting. Knowledge of both conserved and divergent residues outside of the DNA binding domain may be helpful in functional dissection of these proteins.

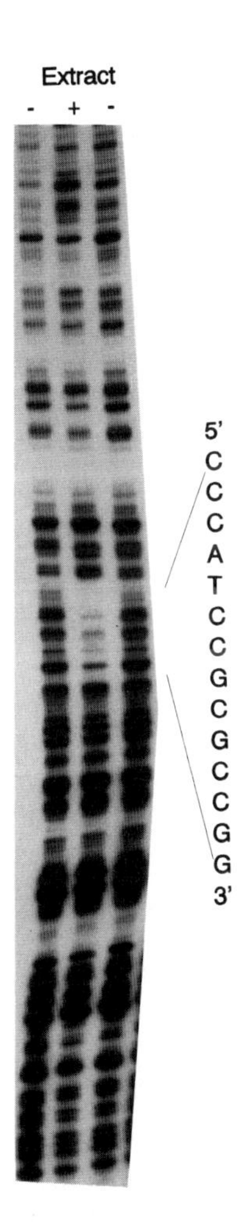

Figure 5. In vitro footprinting of ϕGC-1 on the rice *phyA* promoter. Gel retardation experiments were performed as described above, and the bound and free fraction of DNA was isolated. DNAse I was then added to the isolated complex. The footprinted region was mapped against G-reactions as described in Figure 3.

The autoregulatory loop exhibited by *phyA* genes is an intriguing system to study and several questions still need to be addressed. The recent discovery of different forms of phytochrome in pea (Abe *et al.,* 1989), *Arabidopsis thaliana* (Sharrock and Quail, 1989) and tobacco (S. Kay, M. Deak, R. Kern and Nam-Hai Chua, unpublished data) prompts the question: which form of phytochrome is responsible for down-regulation of the *phyA* genes? One way to address this question may be through the use of photomorphogenic mutants that are defective in only one form of phytochrome. These mutants can be isolated by either classic screening strategies (Adamse *et al.,* 1988) or by overexpression or antisense of specific forms of phytochrome in transgenic plants (Kay *et al.,* 1989b; Boylan and Quail, 1989; Keller *et al.,* 1989). The mechanism of down-regulation of dicot *phyA* genes also requires investigation. In general, dicot *phyA* mRNAs show a less dramatic decrease in response to phytochrome activation than the monocot genes (Tomizawa *et al.,* 1990). The pea *phyA* promoter does not share significant homology with the monocot promoters, and exhibits three distinct start sites, that respond differently to illumination (Sato, 1988; Tomizawa *et al.,* 1989). It is therefore unclear if monocot and dicot *phyA* genes use homologous components for autoregulation. Ultimately, it will require a combination of both molecular and genetic techniques to dissect the chain of events that result in the photophobic expression pattern of higher plant *phyA* genes.

ACKNOWLEDGEMENTS

This work was supported by a grant from the Rockefeller Foundation to Dr. Nam-Hai Chua. We are grateful to Irene Roberson for expert technical assistance, Dr. Gerry Thompson, Harvard University, for expert advice and to Dr. Philip Gilmartin for helpful comments.

REFERENCES

Abe H, Takio K, Titani K, Furuya M (1989) Plant Cell Physiol 30:1089-1097

Adamse P, Kendrick RE, Koornneef M (1988) Photomorphogenetic mutants of higher plants. Photochem Photobiol 48:833-841

Boylan M, Quail PH (1989) Oat phytochrome is biologically active in transgenic tomatoes. Plant Cell 765-773

Bruce WB, Christensen AH, Klein T, Fromm M, Quail PH (1989) Photoregulation of a phytochrome gene promoter from oat transferred into rice by particle bombardment. Proc Natl Acad Sci 86:9692-9696

Christensen AH, Quail PH (1989) Structure and expression of a maize phytochrome encoding gene. Gene 85:381-390

Gilmartin P, Sarokin L, Memelink J, Chua NH (1990) Molecular light switches for plant genes. Plant Cell 2:369-378.

Gilmartin P, Chua NH (1990a) Spacing between GT-1 binding sites within a light-responsive element is critical for transcriptional activity. Plant Cell 2:447-456

Gilmartin P, Chua NH (1990b) Localization of a phytochrome responsive element within the upstream region of pea *rbcS-3A*. Mol Cell Biol in press

Green PJ, Kay SA, Chua NH (1987) Sequence-specific interactions of a pea nuclear factor with light-responsive elements upstream of the *rbcS-3A* gene. EMBO J 6:2543-2549

Green PJ, Yong MH, Cuozzo M, Kano-Murakami Y, Silverstein P, Chua NH (1988a) Binding site requirements for pea nuclear protein factor GT-1 correlate with sequences required for light-dependent transcriptional activation of the *rbcS-3A* gene. EMBO J 7:4035-4044

Green PJ, Kay SA, Lam E, Chua NH (1988) *In vitro* DNA footprinting. In: Gelvin S, Schilperoort R (eds) Plant Molecular Biology Manual. Kluwer Academic Publishers, Dordrecht, B11:1-22

Hershey HP, Barker RF, Idler KB, Murray MG, Quail PH (1987) Nucleotide sequence and characterization of a gene encoding the phytochrome polypeptide from *Avena*. Gene 61:339-348

Kay SA, Keith B, Shinozaki K, Chye ML, Chua NH (1989a) The rice phytochrome gene: structure, autoregulated expression, and binding of GT-1 to a conserved site in the 5′ upstream region. Plant Cell 1:351-360

Kay SA, Keith B, Shinozaki K, Chua NH (1989b) The sequence of the rice phytochrome gene. Nucl Acids Res 17:2865-2866

Kay SA, Nagatani A, Keith B, Deak M, Furuya M, Chua NH (1989c) Rice phytochrome is biologically active in transgenic tobacco. Plant Cell 1:775-782

Keller JM, Shanklin J, Vierstra RD, Hershey HP (1989) Expression of a functional monocotyledonous phytochrome in transgenic tobacco. EMBO J 8:1005-1012

Kuhlemeier C, Cuozzo M, Green P, Goyvaerts E, Ward K, Chua NH (1988) Localization and conditional redundancy of regulatory elements in *rbcS-3A*, a pea gene encoding the small subunit of ribulose-bisphosphate carboxylase. Proc Natl Acad Sci USA 85:4662-4666

Lam E, Chua NH (1990) GT-1 binding site confers light responsive expression in transgenic tobacco. Science 248:471-474

Lissemore JL, Quail PH (1988) Rapid transcriptional regulation by phytochrome of the genes for phytochrome and chlorophyll *a/b* binding protein in *Avena sativa*. Mol Cell Biol 8: 4840-4850

Nagy F, Kay SA, Chua NH (1988) Gene regulation by phytochrome. Trends Genet 4:37-42

Sato N (1988) Nucleotide sequence and expression of the phytochrome gene in *Pisum sativum*: differential regulation by light of multiple transcripts. Plant Mol Biol 11: 697-710

Sharrock RA, Quail PH (1989) Novel phytochrome sequences in *Arabidopsis thaliana*: structure, evolution and differential expression of a plant regulatory photoreceptor family. Genes Dev 3:1745-1757

Shimamoto K, Terada R, Izawa T, Fujimoto H (1989) Fertile transgenic rice plants regenerated from transformed protoplasts. Nature 338:274-276

Tomizawa KI, Sato N, Furuya M (1989) Phytochrome control of multiple transcripts of the phytochrome gene in *Pisum sativum*. Plant Mol Biol 12:295-299

Tomizawa KI, Nagatani A, Furuya M (1990) Phytochrome genes: studies using the tools of molecular biology and photomorphogenetic mutants. Photochem Photobiol in press

DISSECTION OF THE LIGHT-RESPONSIVE ELEMENTS OF PEA *RBCS3A*

P.M. Gilmartin, J. Memelink[1] and N-H. Chua
Laboratory of Plant Molecular Biology,
The Rockefeller University,
1230 York Avenue,
New York. NY 10021-6399,
USA.

LIGHT-RESPONSIVE GENES.

Genes encoding the small subunit of ribulose-1/5-bisphosphate carboxylase-oxygenase (rbcs) and the chlorophyll a/b binding proteins (cab) have been the focus of many studies on the organization and regulation of plant nuclear genes. rbcs and cab genes are typically light-responsive. In etiolated and dark-adapted tissue transcript abundance is low; following illumination of the plant rbcs and cab transcript abundance increases dramatically (see Tobin & Silverthorne 1985; Kuhlemeier *et al.,* 1987a; Jenkins, 1988). There are however exceptions to this rule in which the transcripts of specific rbcs or cab genes are also present in the dark (Manzara & Gruissem, 1988; Sullivan *et al.*, 1989). Light-responsive genes that exhibit a reciprocal pattern of expression, namely, high transcript levels in the dark and low levels in the light, have also been characterized (Batschauer & Apel, 1984; Lissemore & Quail, 1988; kay *et al.*, 1989; Darrah *et al.*, 1990; Tsai & Coruzzi, 1990; Kay, this volume).

Light-responsive expression of rbcs and cab is mediated primarily by red light via the photoreceptor phytochrome. However, other wavelengths of light also play an accessory role in this response. (see Ellis, 1986; Kendrick & Kronenberg, 1986; Cuozzo *et al.*, 1987; Nagy *et al.*, 1988). The increase in rbcs and cab transcript abundance following illumination is regulated at the level of transcription as shown by both nuclear run-on (Gallagher & Ellis, 1982; Silverthorne & Tobin, 1984; Berry-Lowe & Meagher, 1985; Mosinger *et al.*, 1985) and promoter fusion experiments in transgenic plants (see Kuhlemeier *et al.*, 1987a; Silverthorne & Tobin 1987; Benfey & Chua, 1989). Therefore

---

[1]Present address; Department of Molecular Biology, Leiden University, Clusius Laboratory, Lassenaarseweg 64, 2333AL Leiden, The Netherlands.

NATO ASI Series, Vol. H 50
Phytochrome Properties and Biological Action
Edited by B. Thomas and C. B. Johnson

following the perception of distinct wavelengths of light by specific photoreceptors, the resulting signal is transmitted through an as yet unknown transduction pathway to the nucleus where it leads to the transcriptional activation of specific genes.

We have focused on pea *rbcS-3A* as a paradigm for the study of light-responsive transcription. The aim of these studies is to elucidate the signal transduction pathway that links photoperception to a transcriptional response. Our approach has been as follows: 1) *Cis*-element analyses of *rbcS-3A* in order to define specific upstream regions that mediate light-responsive transcription. 2) Identification of discrete sequence elements within these regions that bind nuclear proteins *in vitro*. 3) Definition of the role of these elements in the transcriptional light response by assaying synthetic promoter elements *in vivo*. 4) Cloning of genes encoding the nuclear proteins with the purpose of using them to identify intermediary components of the light signal transduction pathway. Here we summarize these studies to date and describe the isolation of a cDNA clone encoding a DNA-binding protein that interacts with a discrete *cis*-acting through which light-responsive transcription is mediated.

CIS-ELEMENTS OF *RBCS-3A* INVOLVED IN MEDIATING LIGHT-RESPONSIVE TRANSCRIPTION

Pea *rbcS-3A* accounts for 40% of the rbcs transcript levels in light grown plants and is therefore the most highly expressed family member (Fluhr *et al.*, 1986). Following introduction of *rbcS-3A* with 410 bp of upstream sequences into transgenic petunia (Fluhr & Chua, 1986) and tobacco (Kuhlemeier *et al.*, 1987a), high expression levels were observed. Introduction of *rbcS-3A* with an additional 1kb of upstream sequence into transgenic tobacco did not significantly enhance these expression levels (Gilmartin & Chua, unpublished observations). Sequences located downstream of -410 are therefore likely sufficient for maximal expression. Figure 1 summarizes the *cis* element analyses of the *rbcS-3A* upstream regions.

Fusion of the -330 to -50 region of *rbcS-3A* to the -46 cauliflower mosaic virus (CaMV) 35S promoter linked to the chloramphenicol acetyl transferase (CAT) reporter gene results in CAT gene expression that is responsive to phytochrome (Fluhr *et al.*, 1986). Sufficient information for the transcriptional phytochrome-mediated response is therefore located between -330 and -50. These studies were refined by the analysis of

5' deletions of the -410 upstream region in transgenic tobacco exposed to white light. A deletion to -166 retains sufficient sequence information for light-responsive and organ specific expression when assayed in the test gene site of pMON200 (Kuhlemeier *et al.*, 1987a). Furthermore, this -166 *rbcS-3A* promoter retains phytochrome responsiveness (Gilmartin & Chua, 1990b). A phytochrome responsive element is therefore located downstream of -166. A deletion to -50 shows no transcriptional activity (Kuhlemeier *et al.*, 1989). The functional region within the -166 to -50 fragment was further localized to a 58 bp element spanning -169 to -112 (Davis *et al.*, 1990).

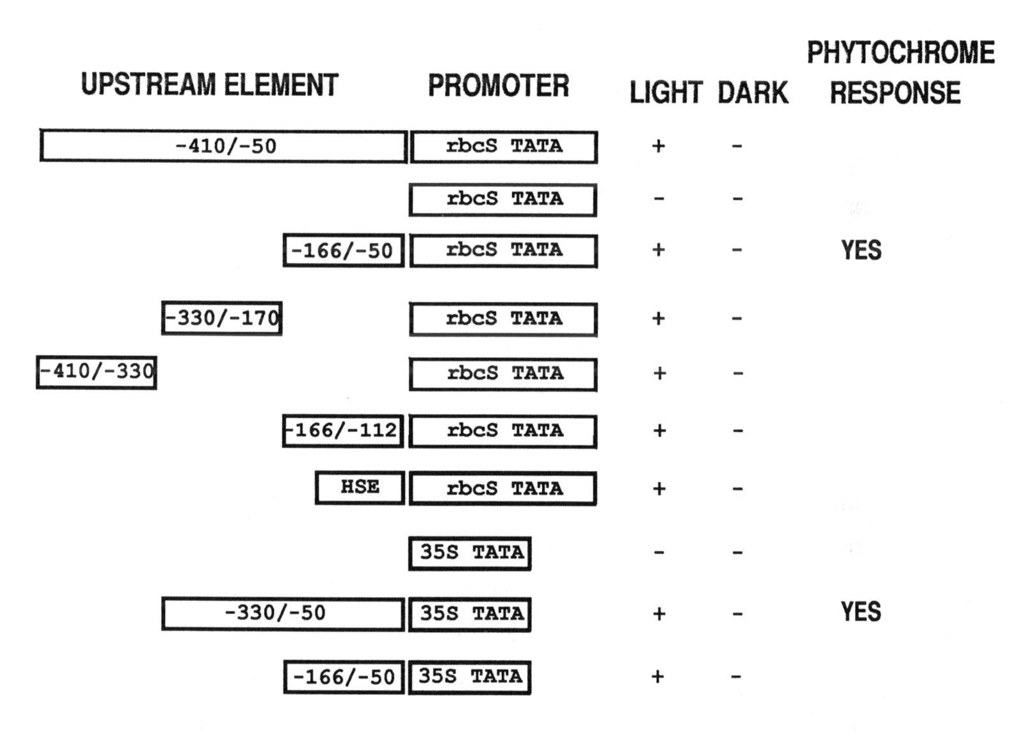

Figure 1. *Cis*-element constructs that define functional upstream regions of *rbcS-3A*. Upstream regions of *rbcS-3A* and their end-points are indicated; HSE, heat shock element. The assay promoters rbcs TATA and 35S TATA refer to the -50 *rbcS-3A* and -46 CaMV 35S promoters respectively. Activity in the light and dark is indicated by the + and - signs. Phytochrome responsive constructs are indicated, those not indicated were not assayed.

An internal deletion of -162 to -117 within -410 *rbcS-3A* does not reduce promoter activity (Kuhlemeier *et al.*, 1987) suggesting a redundancy of function between sequences located upstream -166 and downstream of -162. Using convenient restriction sites within the 410 bp upstream region, DNA fragments extending from -410 to -330 and -330 to -170 were fused to the -50 *rbcS-3A* promoter (Davis *et al.*, 1990). In transgenic tobacco, activity of the -50 *rbcS-3A* promoter is below detectable levels (Kuhlemeier *et al.*, 1989). Both upstream regions can confer light-responsive expression upon the -50 promoter. The -330 to -170 region confers expression primarily in leaf tissue; expression conferred by the -410 to -330 region is too low to assay for organ specificity (Davis *et al.*, 1990). These data demonstrate a reiteration of genetic information within the upstream sequence such that distinct upstream regions can confer similar responses (Kuhlemeier *et al.*, 1988b, Davis *et al.*, 1990).

Studies with pea *rbcS-E9* demonstrated that a 33 bp element overlapping the TATA box confers low levels of light-responsive transcription in transgenic petunia calli (Morelli *et al.*, 1985). Similar observations were also made for the -90 to -2 region of pea *rbcS-3.6* (Timko *et al.*, 1985). In order to uncover a regulatory role for the apparently inactive -50 *rbcS-3A*, a heat shock element (HSE) was fused to this promoter. This element was used as a weak enhancer to assay for -50 *rbcS-3A* activity in both light-grown and dark-adapted tissue and revealed the presence of a light-responsive element (LRE) located between -50 and +15 of *rbcS-3A*. Experiments with the corresponding -46 to +8 region from CaMV 35S show activity in both light-grown and dark-adapted tissue (Kuhlemeier *et al.*, 1989). In order to demonstrate the role of the -166 to -50 region in the light-response observed from the -166 *rbcS-3A* promoter, the light-responsive *rbcS-3A* TATA element was replaced by the light-insensitive CaMV 35S TATA region (Kuhlemeier *et al.*, 1989). These experiments unequivocally demonstrate the presence of two independent LREs downstream of -166, one upstream of -50 and one downstream.

Most studies on *rbcS-3A* have not enabled the separation of light-responsiveness and organ-specificity. However, the activity mediated by the HSE/*rbcS-3A* promoter is light-responsive but not organ-specific (Kuhlemeier *et al.*, 1989). These functions have also been separated within *Nicotiana plumbaginifolia rbcS-8B*. In this case, the far-upstream region confers organ-specificity but is not light-responsive (Poulsen & Chua, 1988). Organ-specificity and light-responsiveness can therefore be dissociated.

These studies have delineated several independent positive elements upstream of *rbcS-3A*. Additionally, they have localized distinct LREs and a phytochrome-responsive element to discrete upstream regions. The definition of nuclear protein factor binding sites within these upstream regions of *rbcS-3A* and several other light-responsive genes have focused our analyses on the role of these elements in light-responsive transcription.

TRANS-ACTING FACTORS THAT INTERACT WITH THE UPSTREAM REGION OF *RBCS-3A*

Sequence comparisons between many light-responsive genes from several species reveal the presence of conserved sequence motifs within the upstream regions of these genes (see Manzara & Gruissem,1988; Dean *et al.*, 1989). Several nuclear proteins that specifically interact with distinct sequence elements within these upstream sequences have been identified (see Gilmartin *et al.*, 1990). The locations of factor binding sites upstream of *rbcS-3A* are shown in Figure 2. Conserved motifs within other light-responsive promoters but not yet shown to interact with nuclear proteins within *rbcS-3A* are also indicated. Comparison of the regions delineated by *cis*-element analyses in transgenic tobacco (Fig.1) with the locations of factor binding sites within these regions (Fig.2), reveals that several of the binding sites are located within functionally defined regions. Within each of the active upstream regions [-410 to -330, -330 to -170 and -166 to -50] (Kuhlemeier *et al.*, 1988; Davis *et al.*, 1990) are paired binding sites for the nuclear factor GT-1 (Green *et al.*, 1987, 1988).

Mutation of either box II ($^{-151}$GTGTGGTTAATATG^{-138}) or box III ($^{-125}$ATCATTTTCACT^{-114}) in -410 *rbcS-3A* does not affect transcriptional activity suggesting a role for these elements in the functional redundancy observed between the three upstream regions (Kuhlemeier *et al.*, 1987b; 1988b). However, mutation of either box II or box III within the -170 *rbcS-3A* promoter (Kuhlemeier *et al.*, 1988b), or within the 58 bp element fused to -50 (Davis *et al.*, 1990) dramatically reduces transcriptional activity. Mutation of box II* ($^{-224}$GTGAGGTAATAT^{-213}) within the -330 to -170 region does not affect activity (Davis *et al.*, 1990) suggesting that *cis*-acting elements other than the GT-1 binding sites can play a role in mediating light-responsive transcription. Within this region we have identified a binding site for the factor GAF-1 (Memelink, Gilmartin & Chua, manuscript in preparation, see Gilmartin *et al.*, 1990). Factors binding to similar sequences within other light-responsive promoters have been identified (GA-1 [see Donald & Cashmore, 1990; Donald *et al.*, 1990] and LRF-1 [Busby *et al.*, 1990). Additional putative binding sites for GBF (Giuliano *et al.*, 1988), AT-1 (Datta & Cashmore, 1989) and GAF-1 (Memelink, Gilmartin & Chua, manuscript in preparation) are also present between -330 and -170 of *rbcS-3A* (see Figure 2 and Gilmartin *et al.*, 1990). The role of these elements within this fragment has not been addressed. However, a positive role has been demonstrated for these elements within upstream regions of other light-responsive genes (Castresana *et al.*, 1988; Gidoni *et al.*, 1989 Datta & Cashmore, 1989; Donald & Cashmore, 1990).

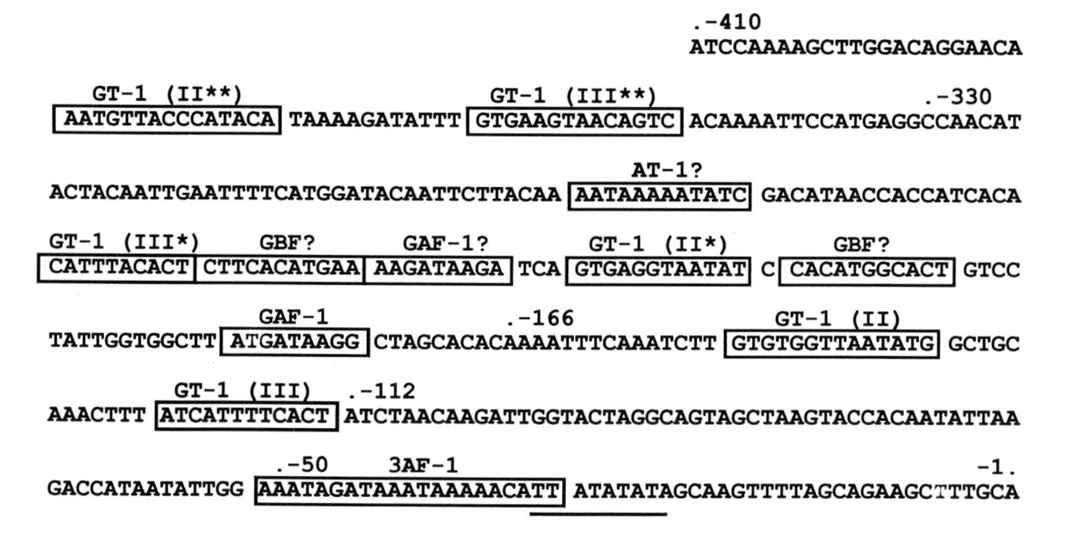

Figure 2. Upstream sequence of pea *rbcS-3A*. The nucleotide sequence between -410 and -1 is presented. The GT-1 binding sites boxes II, III, II*, III*, II** and III**, (Green *et al.*, 1987; 1988) as well as the defined binding sites for 3AF-1 (Lam, Kano-murakami, Gilmartin, Niner and Chua, submitted) and GAF-1 Memelink, Gilmartin and Chua, in preparation) are shown. Putative binding sites for the factors AT-1 (Datta and Cashmore, 1989), GBF (Giuliano *et al.*, 1988) , and GAF-1 (Memelink, Gilmartin and Chua, in preparation) based on sequence homology to their defined binding sites are indicated (?). The TATA element is underlined. Indicated nucleotide positions correspond to those of Figure 1.

The critical role of box II within the -170 *rbcS-3A* promoter was further defined by the introduction of a 2 bp (GG-CC) mutation into this element (Kuhlemeier *et al.*, 1988b). This mutation not only severely reduces GT-1 binding *in vitro* (Green *et al.*, 1987) but abolishes transcriptional activity *in vivo* (Kuhlemeier *et al.*, 1988b). We have demonstrated that critical spacing requirements exist between boxes II and III for transcriptional activity *in vivo* (Gilmartin & Chua, 1990a) and that these two elements differ in their ability to interact with GT-1 *in vitro* (Gilmartin & Chua 1990b). These combined data define a pivotal role for the GT-1 binding sites in mediating light-responsive transcription.

Sequences containing boxes II and III can, in a different context from the *rbcS-3A* promoter, repress expression from the CaMV 35S enhancer in the dark. (Kuhlemeier *et al.*, 1987b). However, it is not known whether the ability of *rbcS-3A* promoter fragments to reduce activity of the 35S enhancer in the dark correlate with GT-1 binding. The presence of negative elements within the upstream sequences of cab genes from pea (Simpson *et al.*, 1986) and *N. plumbaginifolia* has been reported (Castresana *et al.*, 1988). These silencer elements reduce transcriptional activity in the roots and in light-grown leaves respectively. However no correlation was made between these activities and defined factor binding sites.

SYNTHETIC PROMOTER ELEMENTS DEFINE A ROLE FOR THE GT-1 BINDING SITE BOX II

Our understanding of the role of the GT-1 binding sites upstream of *rbcS-3A* has developed from the localization of light-responsive elements to defined upstream regions and mutational analyses of these regions. These studies have demonstrated a requirement for both boxes II and III in the transcriptional response to a light signal and correlated GT-1 binding affinity *in vitro* to the level of this response *in vivo*. Deletion studies with other genes have demonstrated that binding sites for the nuclear factors ASF-2 (Castresana *et al.*, 1988; Gidoni *et al.*, 1989; Fang *et al.*, 1989; Lam & Chua, 1989), GBF (Donald & Cashmore, 1990), and GA-1 (Donald & Cashmore, 1990) play a role in the activation of rbcs and cab. However, gain-of-function experiments are required to define the role of an element within the molecular light switch as a regulatory component or merely as a positive non regulated module of the switch. The ASF-2 binding site, as-2, from the CaMV 35S promoter has been shown by such gain-of-function experiments to confer transcriptional activity primarily in leaf tissue that is not responsive to light (Lam & Chua, 1989). Such studies demonstrate a positive non regulatory role for this element within light-responsive promoters.

To demonstrate the role of the GT-1 binding site box II within the light switch, several chimeric promoter elements have been constructed (Fig.3). Fusion of a tetramer of box II to either the -50 *rbcS-3A* or the -46 CaMV 35S promoter does not lead to transcriptional activation *in vivo* (Davis *et al.*, 1990). Both promoters have been used previously to identify *cis*-acting elements within the *rbcS-3A* upstream region (see Figure 1). However, fusion of the box II tetramer to the -90 CaMV 35S promoter results in a synthetic promoter element capable of conferring light-responsive transcription primarily in chloroplast containing cells in response to light (Lam & Chua, 1990). A tetramer of box II containing the GG to CC mutation within each box II element fused to the same -90 35S promoter is inactive (Lam & Chua, 1990) demonstrating a requirement for GT-1 binding.

Located between -90 and -46 of the 35S promoter is a binding site, as-1, for the nuclear factor ASF-1 (Lam *et al.*, 1989). It is presumably an interaction of GT-1 with ASF-1 within this context that results in transcriptional activity from the -90 but not the -46 CaMV 35S promoter. In striking contrast, a tetramer of the ASF-1 binding site, as-1, fused to the -90 CaMV 35S promoter confers expression in the leaves of transgenic plants but is not light-responsive (Lam & Chua, 1990). A tetramer of the mutant as-1 site fused to -90 35S is not active in leaves. Additionally, Fusion of a tetramer of the ASF-2 binding site, as-2, to the same truncated promoter confers expression primarily in the leaves but is also not responsive to light. A tetramer of the mutant as-2 element fused to -90 35S is inactive in leaves (Fig.3). These observations demonstrate three principles: 1) Sequence motifs required for high level expression in the light are not necessarily light regulatory elements. 2) Box II is a regulatory component of the molecular light switch. 3) GT-1 binding to box II is necessary but not sufficient for the formation of a light-responsive promoter.

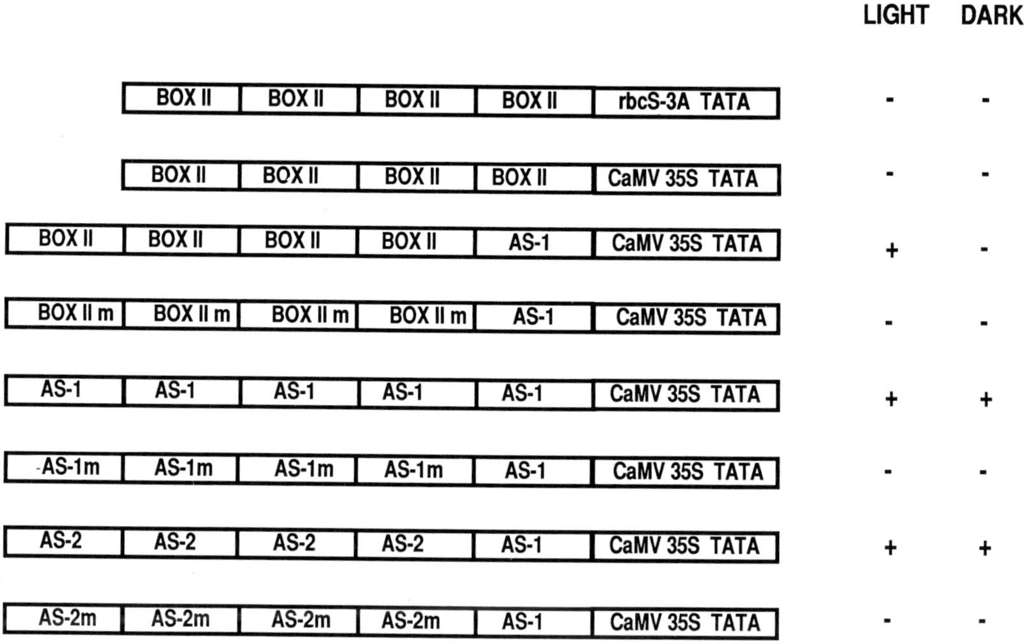

Construct						LIGHT	DARK
	BOX II	BOX II	BOX II	BOX II	rbcS-3A TATA	-	-
	BOX II	BOX II	BOX II	BOX II	CaMV 35S TATA	-	-
BOX II	BOX II	BOX II	BOX II	AS-1	CaMV 35S TATA	+	-
BOX II m	BOX II m	BOX II m	BOX II m	AS-1	CaMV 35S TATA	-	-
AS-1	AS-1	AS-1	AS-1	AS-1	CaMV 35S TATA	+	+
AS-1m	AS-1m	AS-1m	AS-1m	AS-1	CaMV 35S TATA	-	-
AS-2	AS-2	AS-2	AS-2	AS-1	CaMV 35S TATA	+	+
AS-2m	AS-2m	AS-2m	AS-2m	AS-1	CaMV 35S TATA	-	-

Figure 3. Structure of synthetic promoter elements. Tetramers of the GT-1 binding site, BOX II (GTGTGGTTAATATG), the ASF-1 binding site, AS-1 (ATGACGTAAGGGATGACGCAC), the ASF-2 binding site (GTGGATTGATGTGATATCTCC) and their mutant derivatives BOX II^{m} (GTGTCCTTAATATG), AS-1^{m} (CTGCTGTAAGGGATCTCGCAC) and AS-2^{m} (GTGGATTCATGTAATATCTCC) (Green *et al.*, 1987; Lam *et al.*, 1989; Lam and Chua, 1989) were fused to the test promoters indicated. *RbcS-3A* TATA, -50 *rbcS-3A* promoter; CaMV 35S TATA, -46 CaMV 35S promoter; AS-1/CaMV 35S TATA, -90 CaMV 35S promoter. Transcriptional activity in light-grown and dark-adapted leaves is indicated.

A cDNA ENCODING A PROTEIN WITH SPECIFICITY FOR THE GT-1 BINDING SITE BOX II

Our analyses of the *cis*-elements and *trans*-acting factors involved in mediating a transcriptional response to a light stimulus have focused on GT-1 and its binding sites boxes II and III. A combination of mutational analyses and gain-of-function experiments have led to the definition of box II and GT-1 as regulatory components of the transcriptional light-response. With the aim of using these terminal components of the phytochrome signal transduction chain as a basis for the definition of intermediary components of the pathway, we undertook to isolate cDNA clones encoding DNA binding proteins with specificity for box II.

Following a slightly modified method of Singh *et al.* (1989) we have isolated a cDNA encoding a DNA binding protein with specificity for the GT-1 binding site box II (Gilmartin, Memelink & Chua, manuscript in preparation). A lambda zap etiolated tobacco seedling cDNA library was screened. Approximately 600,000 pfu of phage (20,000 pfu per 150 cm petri dish) were plated on LB agar plus 10 mM $MgSO_4$ on *E.coli* XL-1 blue in LB top agarose plus 10 mM $MgSO_4$. As a control, phage B9, encoding a non-specific DNA binding protein (provided by Dr. S. McKnight), was plated on *E.coli* Y1090. Plates were incubated at 37°C for 5 h. Nitrocellulose filters were soaked in 10 mM IPTG, blotted dry and placed onto the plates which were incubated for a further 5 h at 37°C. The nitrocellulose filters were removed from the plates and incubated plaque side up in 20 ml B-buffer (20 mM HEPES-KOH pH7.5, 40 mM KCl, 1 mM EDTA, 10% glycerol, 1 mM DTT, 0.8 mM PMFS) supplemented with 5% non-fat dry milk in 150 mm petri dishes with gentle rocking for 1 h at room temperature. Filters were washed twice in 25 ml B-buffer and incubated for 1 h in B-buffer containing 2 ng/ml of labelled probe and 5 μg/ml sonicated and denatured salmon testis DNA with gentle rocking at room temperature.

As a probe, the 84 bp tetramer of box II was purified by gel electrophoresis and end-labelled with all four radiolabelled nucleotides and Klenow polymerase to a specific activity of approximately 10^8 cpm/μg. Filters were washed four times with 25 ml B-buffer for a total of 1 h at room temperature, blotted dry and exposed to X-ray film for 12 h. A shorter exposure, 2 h was used to distinguish between pin-prick signals and halo-like signals (Singh *et al.*, 1989). Several putative positives were identified from the first screen. These were re-screened at a density of 4000-5000 pfu on 150 mm petri dishes. After one further round of screening a single positive clone was identified. A single plaque was isolated and its binding positive confirmed by a 'pizza test' filter binding assay.

Phage carrying cDNAs encoding putative GT-1 and the non-specific DNA-binding protein (B9) were plated and grown as described above on 100 mm petri dishes. Following removal of the nitrocellulose filters, they were cut into six sections. One of each section with bound B9 and putative GT-1 were incubated in pairs with one of six probes. Probes were, 4x Box II, 4x Box IIm, 4x as-2, 4x as-2m, 4x ga-1, 4x ga-1m (Fig.4).

These probes are the wild-type and mutant binding sites for GT-1 (Green *et al.*, 1987), ASF-2 (Lam & Chua, 1989) and GAF-1 (Memelink, Gilmartin & Chua, in preparation) respectively. As can be clearly seen in Figure 4, the isolated cDNA encodes for a protein with specificity for box II but does not bind to any of the other five probes. This DNA binding protein is therefore a good candidate for GT-1.

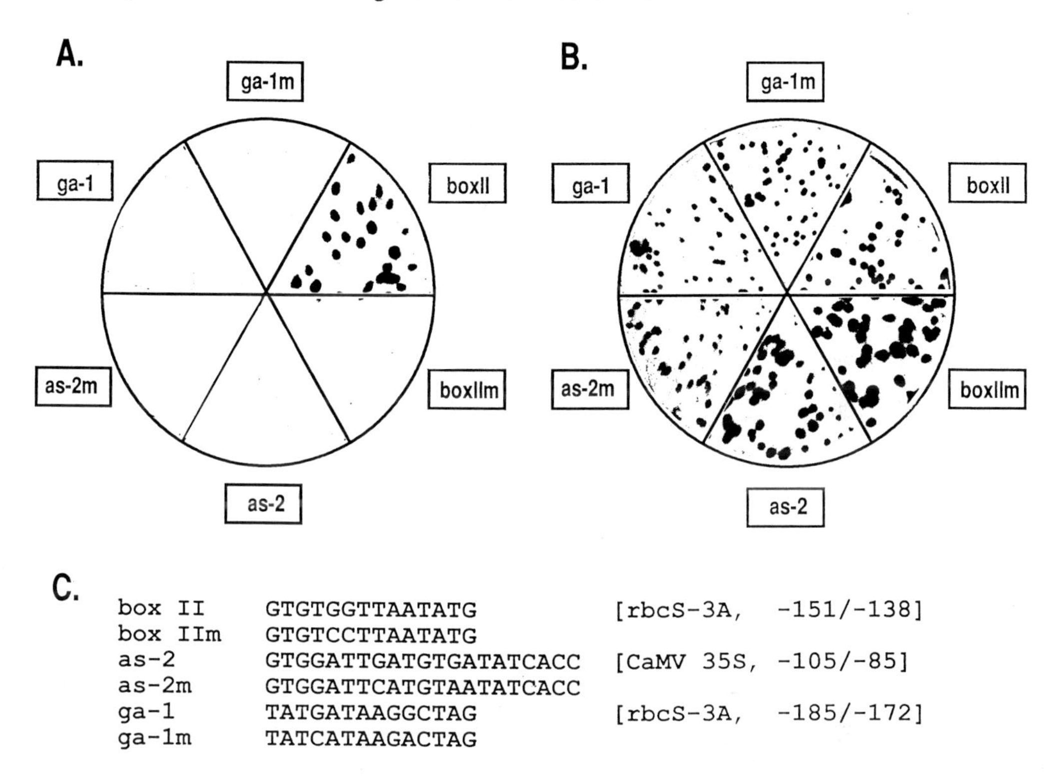

Figure 4. Pizza-test for recombinant factor binding site specificity. A. Specificity of putative GT-1. Filter binding assay with phage carrying recombinant GT-1 using either box II, box II^m, as-2, as-2^m, ga-1 and ga-1^m; the wild-type and mutant binding sites for the factors GT-1, ASF-2 and GAF-1 respectively (Green *net al.*, 1987; Lam and Chua, 1989; Memelink, Gilmartin and Chua, manuscript in preparation). B. Binding of non-specific factor. As A but phage B9, encoding a non-specific DNA binding protein was used. C. Sequence and location within the cognate promoters of the wild-type and mutant binding sites used.

SUMMARY AND FUTURE PROSPECTS

Through a combination of approaches, the upstream region of *rbcS-3A* has been delineated into an array of separate regulatory sequences with overlapping functions.

These upstream sequences have been further dissected into an assembly of discrete *cis*-acting elements. Several of these elements have been defined within *rbcS-3A* in terms of specific interactions with nuclear proteins. Elements conserved within promoters of other light-responsive genes are also present. We have focused on the GT-1 binding sites upstream of *rbcS-3A*. These sequences have been defined as critical for light-responsive gene transcription by mutational analyses within the cognate promoter, and by the creation of synthetic promoter elements. A phytochrome responsive element has been localized downstream of -166 within the *rbcS-3A* promoter and the paired GT-1 binding sites boxes II and III have been shown to be critical for this response. It is not yet known whether wavelengths of light other than red can mediate a response through these binding sites. The presence of additional *cis*-acting elements upstream of *rbcS-3A* and the observation that GT-1 binding to a promoter element is not necessarily sufficient for light-responsive transcription suggest roles for factors bound to these other elements. These roles may overlap those of the GT-1 binding sites or may be distinct (see Gilmartin *et al.*, 1990).

The isolation of a cDNA encoding a protein with specificity for box II will enable further dissection of the phytochrome mediated signal transduction chain from the terminus of the pathway. These studies may also enable the definition of light-responsive transcription in terms of whether it is mediated via activation in the light, repression in the dark or a combination of both. Additionally, molecular analyses of GT-1 may clarify the role of the GT-1 binding sites present in both light-activated genes and those expressed primarily in the dark (see Kay, this volume). Future studies may lead to the identification of intermediary components of the transduction chain linking photoperception by phytochrome to a transcriptional response.

ACKNOWLEDGEMENTS

We thank Dr. M. Deak for providing the tobacco cDNA library and Dr. S. McKnight for the non-specific DNA binding protein clone B9. Thanks to Dr. Steve Kay and other members of our laboratory for helpful comments on the manuscript and numerous stimulating conversations. P.M.G is supported by a post-doctoral fellowship from the Winston Foundation and J.M. was supported by a NATO post-doctoral fellowship.

REFERENCES

Batschauer A, Apel K (1984) An inverse control by phytochrome of the expression of two nuclear genes in barley. Eur J Biochem 143:593-597

Benfey PN, Chua N-H (1989) Regulated genes in transgenic plants. Science 244:174-181

Berry-Lowe SL, Meagher RB (1985) Transcriptional regulation of a gene encoding the small subunit of ribulose-1,5-bisphosphate carboxylase in soybean tissue is linked to the phytochrome response. Mol Cell Biol 5:1910-1917

Buzby JS, Yamada T, Tobin EM (1990) A light-regulated DNA binding activity interacts with a conserved region of a *Lemna gibba rbcS* promoter. Plant Cell:In Press

Castresana C, Garcia-Luque I, Alonso E, Malik VS, Cashmore AR (1988) Both positive and negative regulatory elements mediate expression of a photoregulated CAB gene from *Nicotiana plumbaginifolia.* EMBO J 7:1929-1936

Cuozzo M, Kay SA, Chua N-H (1987) Regulatory circuits of light-responsive genes. in Temporal and Spatial Regulation of Plant Genes eds Verma, DPS and Goldberg:RB Springer-Verlag pp 131-153

Darrah P, Kay S, Teakle G, Griffiths T (1990) Cloning and sequencing of protochlorophyllide reductase. Biochem J 265:789-798

Datta N, Cashmore AR (1989) Binding of a pea nuclear protein to promoters of certain photoregulated genes is modulated by phosphorylation. Plant Cell 1:1069-1077

Davis MC, Yong M-H, Gilmartin PM, Goyvaerts E, Kuhlemeier C, Sarokin L, Chua N-H (1990) Minimal sequence requirements for the regulated expression of *rbcS-3A* from *Pisum sativum* in transgenic tobacco plants. Photochem Photobiol in press

Dean C, Pichersky E, Dunsmuir P (1989) Structure, evolution and regulation of rbcs genes in higher plants. Ann Rev Plant Physiol 40:415-439

Donald RGK, Cashmore AR (1990) Mutation of either G box or I box sequences profoundly affects expression from the *Arabidopsis rbcS-1A* promoter. EMBO J 9:1717-1726

Donald RGK, Schindler U, Batschauer A, Cashmore AR (1990) The plant G box promoter sequence activates transcription in *Saccharomyces cerevisiae* and is bound *in vitro* by a yeast activity similar to GBF, the plant B box binding factor. EMBO J 9:1727-1735

Ellis RJ (1986) Photoregulation of plant gene expression. Bioscience Reports 6:127-135

Fang RX, Nagy F, Sivasubramaniam S, Chua N-H (1989) Multiple *cis*-regulatory elements for maximal expression of the cauliflower mosaic virus 35S promoter in transgenic plants. Plant Cell 1:141-150

Fluhr R, Chua N-H (1986) Developmental regulation of two genes encoding ribulose-bisphosphate carboxylase small subunit in pea and transgenic petunia plants: Phytochrome response and blue-light induction. Proc Natl Acad Sci 83:2358-2362

Fluhr R, Moses P, Morelli G, Coruzzi G, Chua N-H (1986) Expression dynamics of the pea rbcs multigene family and organ distribution of the transcripts. EMBO J 5:2063-2071

Gallagher TF, Ellis RJ (1982) Light-stimulated transcription of genes for two chloroplast polypeptides in isolated pea leaf nuclei. EMBO J 1:1493-1498

Gidoni D, Brosio P, Bond-Nutter D, Bedbrook J, Dunsmuir P (1989) Novel *cis*-acting elements in petunia cab gene promoters. Mol Gen Genet 215:337-344

Gilmartin PM, Chua N-H (1990a) Spacing between GT-1 binding sites within a light-responsive element is critical for transcriptional activity. Plant Cell 2:369-378

Gilmartin PM, Chua N-H (1990b) Localization of a phytochrome responsive element within the upstream region of pea *rbcS-3A*. Mol Cell Biol In Press

Gilmartin PM, Sarokin L, Memelink J, Chua N-H (1990) Molecular light switches for plant genes. Plant Cell 2:369-378

Giuliano G, Pichersky E, Malik VS, Timko MP, Scolnik PA, Cashmore AR (1988) An evolutionarily conserved protein binding sequence upstream of a plant light-regulated gene. Proc Natl Acad Sci USA 85:7089-7093

Green PJ, Kay SA, Chua N-H (1987) Sequence-specific interactions of a pea nuclear factor with light-responsive elements upstream of the *rbcS-3A* gene. EMBO J 6:2543-2549

Green PJ, Yong M-H, Cuozzo M, Kano-Murakami Y, Silverstein P, Chua N-H (1988) Binding site requirements for pea nuclear protein factor GT-1 correlate with sequences required for light-dependent transcriptional activation of the *rbcS-3A* gene. EMBO J 7:4035-4044

Jenkins G (1988) Photoregulation of gene expression in plants. Photochem Photobiol 48:821-832

Kay SA, Keith B, Shinozaki K, Chye M-L, Chua N-H (1989) The rice phytochrome gene: structure, autoregulated expression, and binding of GT-1 to a conserved site in the 5' upstream region. Plant Cell 1:351-360

Kendrick RE, Kronenberg GMH eds (1986) Photomorphogenesis in plants, Martinus Nijhoff

Kuhlemeier C, Green PJ, Chua N-H (1987a) Regulation of gene expression in higher plants. Ann Rev Plant Physiol 38:221-257

Kuhlemeier C, Fluhr R, Green P, Chua N-H (1987b) Sequences in the pea *rbcS-3A* gene have homology to constitutive mammalian enhancers but function as negative regulatory elements. Genes Dev 1:247-255

Kuhlemeier C, Fluhr R, Chua N-H (1988a) Upstream sequences determine the difference in transcript abundance of pea rbcs genes. Mol Gen Genet 212:405-411

Kuhlemeier C, Cuozzo M, Green P, Goyvaerts E, Ward K, Chua N-H (1988b) Localization and conditional redundancy of regulatory elements in *rbcS-3A*, a pea gene encoding the small subunit of ribulose-bisphosphate carboxylase. Proc Natl Acad Sci 85:4662-4666

Kuhlemeier C, Strittmatter G, Ward K, Chua N-H (1989) The pea *rbcS-3A* promoter mediates light responsiveness but not organ specificity. Plant Cell 1:471-478

Lam E, Chua N-H (1989) ASF-2: A factor that binds to the cauliflower mosaic virus 35S promoter and a conserved GATA motif in cab promoters. Plant Cell 1:1147-1156

Lam E, Chua N-H (1990) GT-1 binding site confers light-responsive expression in transgenic tobacco. Science 248:471-474

Lam E, Benfey PN, Gilmartin PM, Fang R-X, Chua N-H (1989) Site-specific mutations alter *in vitro* factor binding and change promoter expression pattern in transgenic plants. Proc Natl Acad Sci USA 86:7890-7894

Lissemore JL, Quail PH (1988) Rapid transcriptional regulation by phytochrome of the genes for phytochrome and chlorophyll a/b/-binding protein in *Avena sativa*. Mol Cell Biol 8:4840-4850

Manzara T, Gruissem W (1988) Organization and expression of the genes encoding ribulose-1,5-bisphosphate carboxylase in higher plants. Photosynth Res 16:117-139

Morelli G, Nagy F, Fraley RT, Rogers SG, Chua N-H (1985) A short conserved sequence is involved in the light-inducibility of a gene encoding ribulose 1,5-bisphosphate carboxylase small subunit of pea. Nature 315:200-204

Mosinger E, Batschauer A, Schafer E, Apel K (1985) Phytochrome control of *in vitro* transcription of specific genes in isolated nuclei of barley (*Hordeum vulgare*). Eur. J Biochem 147:137-142

Nagy F, Kay SA, Chua N-H (1988) Gene regulation by phytochrome. Trends Genet 4:37-42

Poulsen C, Chua N-H (1988) Dissection of 5' upstream sequences for selective expression of the *Nicotiana plumbaginifolia rbcS-8B* gene. Mol Gen Genet 214:16-23

Silverthorne J, Tobin E (1984) Demonstration of transcriptional regulation of specific genes by phytochrome action. Proc Natl Acad Sci USA 81:1112-1116

Silverthorne J, Tobin E (1987) Phytochrome regulation of nuclear gene expression. Bioessays 7:18-23

Simpson J, Schell J, Van Montagu M, Herrera Estrella L (1986) The light-inducible and tissue specific expression of a pea LHCP gene involves an upstream element combining enhancer and silencer properties. Nature 323:551-553

Singh H, Clere RG, Lebowitz JH (1989) Molecular cloning of sequences-specific DNA binding proteins using recognition site probes. Biotechniques 7:252-261

Sullivan TD, Christensen AH, Quail PH (1989) Isolation and characterization of a maize chlorophyll a/b binding protein gene that produces high levels of mRNA in the dark. Mol Gen Genet 215:431-440

Timko MP, Kausch AP, Castresana C, Fassler J, Herrera Estrella L, Van den Broeck G, Van Montagu M, Schell J, Cashmore AR (1985) Light regulation of plant gene expression by an upstream enhancer-like element. Nature 318:579-582

Tobin EM, Silverthorne J (1985) Light regulation of gene expression in higher plants. Ann Rev Plant Physiol 36:569-593

Tsai F-Y, Coruzzi GM (1990) Dark-induced and organ-specific expression of two asparagine synthetase genes in *Pisum sativum*. EMBO J 9:323-332

AN *ARABIDOPSIS THALIANA* LEUCINE ZIPPER PROTEIN THAT BINDS TO G-BOX PROMOTER SEQUENCES

U. Schindler, J.R. Ecker and A.R. Cashmore
Plant Science Institute,
Department of Biology,
University of Pennsylvania,
Philadelphia, PA 19104, USA.

INTRODUCTION

We have shown previously that promoters for RbcS genes are often characterized by conserved sequences that we have referred to as L-, I- and G-boxes (Giuliano *et al.*, 1988). These sequences commonly reside from 200 to 300 bp upstream from the start site of transcription. In keeping with this strong conservation of sequence we have demonstrated that there is a clear requirement for both the G- and I-boxes for expression from the *Arabidopsis thaliana rbcS-1A* promoter (Donald and Cashmore, 1990). These studies were performed by examining GUS expression driven by a "full-length" 1.7 kb *rbcS-1A* promoter in transgenic tobacco plants.

The G-box sequence is not restricted to RbcS promoters. A G-box-like sequence is found in the *cab-E* gene from *Nicotiana plumbaginifolia* (Castresana *et al.*, 1988) and mutation of this sequence also results in a substantial loss in expression (Bringmann and Cashmore, unpublished results). Similarly, chalcone synthase promoters are often characterized by G-box sequences (Schulze-Lefert *et al.*, 1989; Staiger *et al.*, 1989) and, at least in the case of the parsley promoter, these sequences are required for UV light-induced expression. The *Arabidopsis thaliana adh* promoter, which mediates anaerobically-enhanced root expression, also contains a G-box (Ferl and Laughner, 1989). ABA-induced genes have also been shown to contain functional G-box sequences within their promoters (Guiltinan *et al.*, 1990). We have characterized a factor, GBF, from nuclear extracts of tomato and *Arabidopsis* that binds specifically to G-box promoter sequences (Giuliano *et al.*, 1988; Schindler and Cashmore, 1990). Also, GBF-like activity has been characterized in extracts from *Antirrhinum*, *Petunia*, and *Nicotiana tabacum* (Staiger *et al.*, 1989). *In vivo* footprinting studies demonstrate that a GBF-like factor binds to the chalcone synthase G-box sequence and this binding is altered by

NATO ASI Series, Vol. H 50
Phytochrome Properties and Biological Action
Edited by B. Thomas and C. B. Johnson

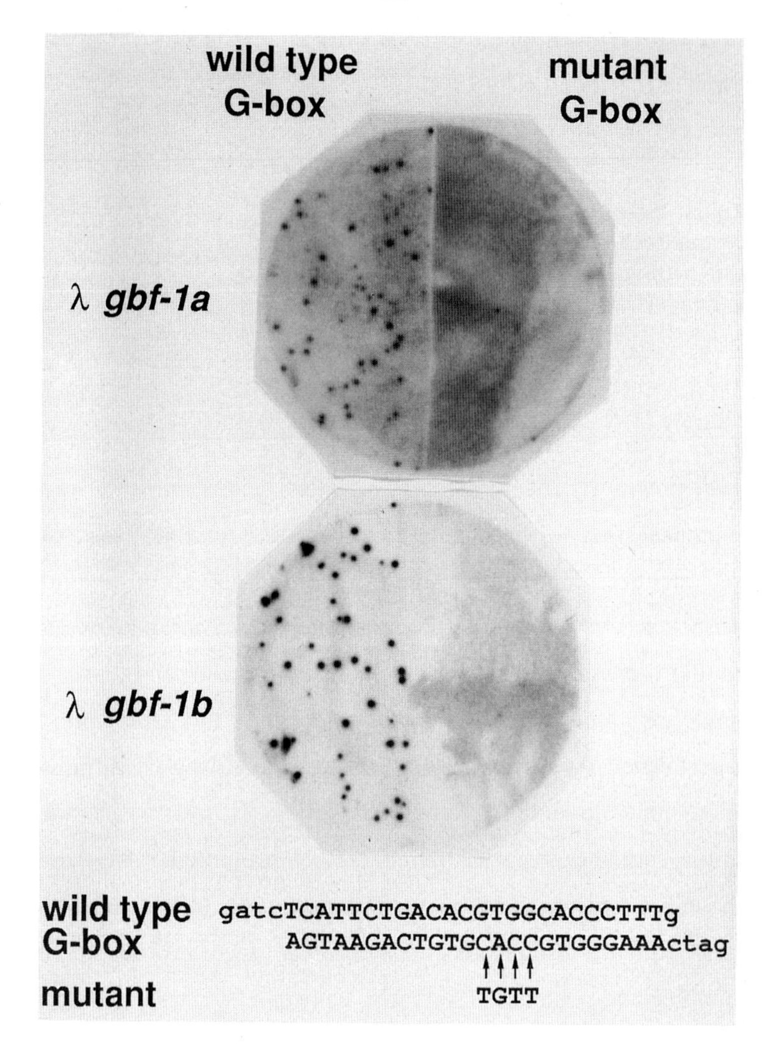

Fig. 1. Oligonucleotide binding properties of proteins encoded by λgbf recombinant phage. An *Arabidopsis thaliana* cDNA expression library in λZAP was screened with a ^{32}P-labeled catenated DNA fragment containing eight copies of a synthetic oligonucleotide representing the tomato *rbcS-3A* G-box motif. Two positive phage (λgbf-1a and λgbf-1b) were isolated from a primary screen of $4x10^5$ phage plaques. The proteins encoded by the two purified recombinant phage were shown to bind specifically to the G-box sequence (left) but not the mutant derivative (right).

UV-induction (Schulze-Lefert *et al.*, 1989). Similarly, *in vivo* footprinting was used to characterize a factor bound to the *Arabidopsis adh* promoter (Ferl and Laughner, 1989). We have demonstrated that G-box sequences are active as upstream activating sequences in yeast and we have characterized a yeast factor with *in vitro* binding properties similar to GBF (Donald *et al.*, 1990).

In view of the essential role that GBF plays in the photoregulated expression of certain plant genes and the widespread occurrence of GBF-like factors in plant tissues and other species, it is obviously of interest to further characterize this factor. Questions of interest include: i), what is the sequence of this factor; ii), are there multiple GBF-like factors present in different tissues and mediating quite distinct expression characteristics and iii), what is the nature of associated factors that we presume are involved, along with GBF, in mediating specific expression.

A RECOMBINANT PHAGE EXPRESSES A PROTEIN THAT BINDS SPECIFICALLY TO G-BOX PROMOTER SEQUENCES

We screened an *Arabidopsis thaliana* λZAP cDNA expression library for recombinant phage expressing proteins that bound oligomerized G-box sequences. Two positive phage were plaque-purified and the specificity of their DNA binding properties was examined. Protein produced by these phage was shown to bind strongly to the synthetic oligonucleotide containing the tomato *rbcS-3A* G-box sequence ACACGTGG but not to the mutant G-box oligonucleotide containing the sequence ACACtgtt (Fig. 1).

GBF-1 IS A LEUCINE ZIPPER PROTEIN

In vivo excision yielded two recombinant Bluescript plasmids containing the cDNAs *gbf-1a* and *gbf-1b*. Southwestern studies confirmed the G-box-binding specificities of the proteins encoded by these cDNAs (Fig. 2). Restriction analysis of the plasmids showed that the two cDNA sequences were very similar but differed in size; *gbf-1a* was 1.42 kb and *gbf-1b* was 0.82 kb. Sequence studies showed that the two clones contained identical sequences differing only in length. The sequences were indicative of a leucine zipper protein - which we called GBF-1 - containing a basic domain adjacent to a heptameric repeat of leucines (Fig. 3).

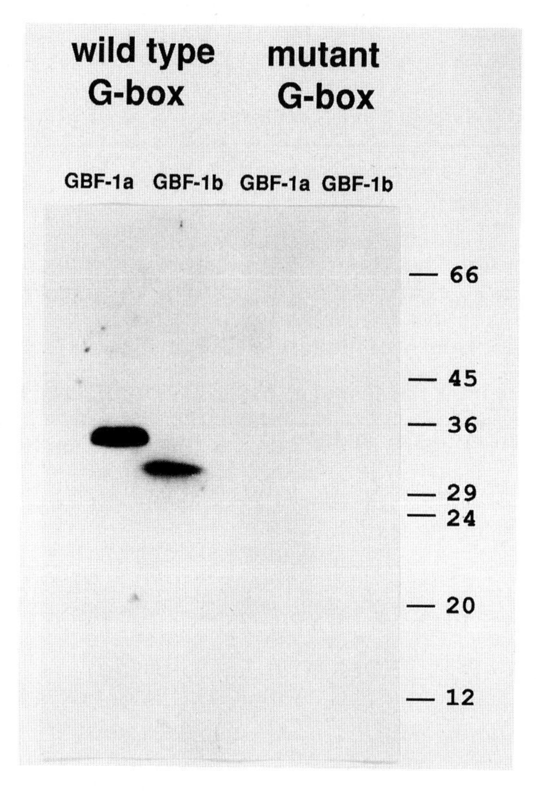

Fig. 2. South western analyses. E. coli cells (DH5α) transformed with recombinant plasmids containing the *gbf-1a* and *gbf-1b* cDNA fragments were grown to an OD_{600} of 0.5. Protein synthesis was induced by adding 5 mM IPTG and growth was continued for an additional hour. The cells (1 ml aliquots) were pelleted, washed and after another centrifugation resuspended in SDS loading buffer. After sonication the proteins were electrophoretically separated on a 15% SDS-PAGE and electroblotted on to nitrocellulose filters. The filters were treated as described by Vinson *et al.* (1988). DNA binding was performed with a DNA fragment containing eight copies of the synthetic oligonucleotide representing the wildtype or mutant *rbcS-3A* G-box motif.

THE BASIC DNA BINDING DOMAIN OF GBF-1 SHOWS STRIKING SEQUENCE HOMOLOGY TO A PROTEIN PREVIOUSLY IDENTIFIED AS HBF-1

We compared the basic putative DNA-binding domain of GBF-1 with other plant leucine zipper proteins (Fig. 3). In a region of 30 amino acids GBF-1 showed some similarity

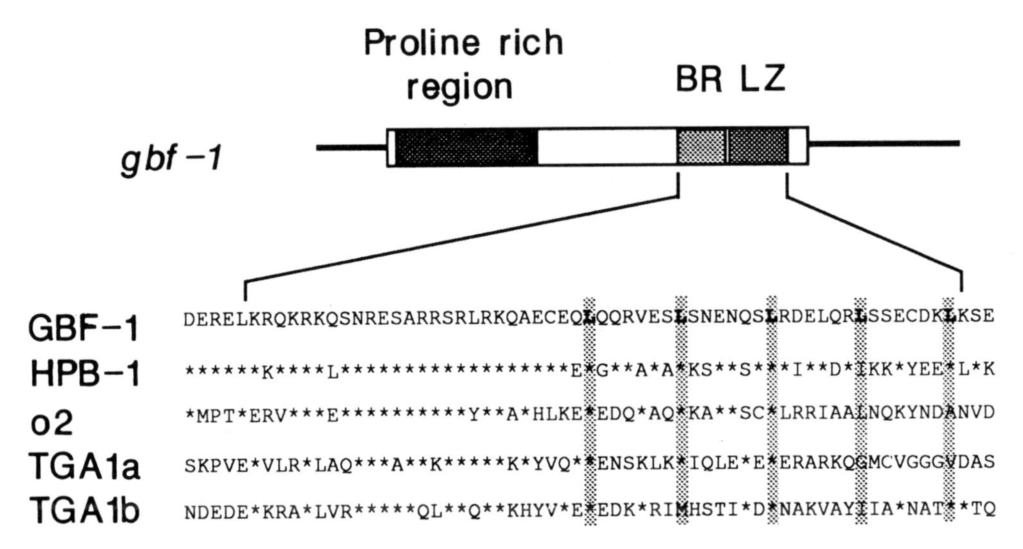

Fig. 3. Comparison of putative DNA binding domains of GBF-1 and other plant leucine zipper proteins. A schematic representation of the cDNA encoding GBF-1 is shown at the top of the figure. The proline rich region, the basic putative DNA binding domain (basic region, BR), and leucine zipper (LZ) are indicated. The amino acid sequence within the basic region is aligned with the corresponding regions of other plant DNA binding proteins (Katagiri *et al.*, 1989; Tabata *et al.*, 1989; Schmidt *et al.*, 1990).

(43% and 40% respectively) to TGA1a and TGA1b (Katagiri *et al.*, 1989) and was even more similar (63%) to o2 (Schmidt *et al.*, 1990). Both TGA1a and TGA1b bind to cauliflower mosaic virus 35S promoter sequences containing the element TGACG; this sequence has some similarity to the core ACGT sequence of the G-box. The most striking similarity was observed on comparing GBF-1 with a protein encoded by a wheat cDNA sequence. This protein has been identified as HBP-1, a nuclear factor that binds to the hexameric ACGTCA sequence found in many histone gene promoters (Tabata *et al.*, 1989). The complement of this hexamer sequence (TGACGT) is also similar to the core G-box sequence. GBF-1 and the sequence identified as HBP-1 exhibited 93% homology in the 30 amino acid basic putative DNA-binding domain. This striking sequence similarity raises the possibility that these two proteins may be evolutionary homologues and this prospect in turn raises questions concerning the true function of these related proteins. It is quite clear that the *in vivo* role of any transcription factor is difficult to assign in the absence of genetic studies. However, in evaluating the relationship between these closely-related DNA-binding proteins, we note that the oligonucleotide used to screen for the wheat clone identified as HBP-1 somewhat fortuitously contained an overlapping partial G-box sequence (in the form TGACGTGG). In the mutant oligonucleotide used for these studies (Tabata *et al.*, 1989) the G-box

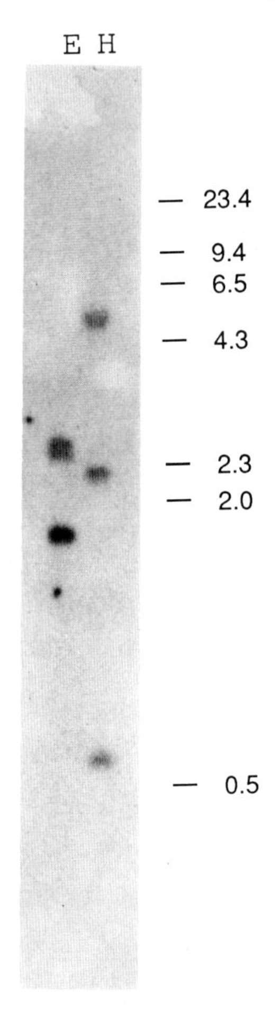

Fig. 4. Southern blot analyses with *gbf-1. Arabidopsis thaliana* genomic DNA (6 ug) was digested with HindIII (H) and EcoRI (E). The DNA fragments were electro-phoretically separated and blotted on to nitrocellulose filters. Hybridization was performed with the ^{32}P-labelled EcoRI cDNA fragment of λgbf-1a under the following conditions: 5xSSPE, 0.2 % PVP, 0.2 % BSA, 0.2 % Ficoll, 0.5 % SDS, 10 μg ml^{-1} sonicated salmon sperm DNA for 12 h at 42°C. The filters were washed in 0.5 x SSC at 65°C and autoradiographed. Lambda DNA digested with HindIII was used as a molecular weight standard.

sequence, as well as the hexamer sequence, was altered (TtACtTGG) and thus these experiments did not distinguish binding of the two sequences. We have repeated these studies with GBF-1 and obtained similar results. Furthermore, we have observed that binding does not occur to a mutant oligonucleotide containing the sequence TGACGTtt in which the G-box sequence, but not the hexamer sequence, has been destroyed (data not shown). In view of these observations we conclude that GBF-1 binds a G-box sequence and not the hexamer sequence. Furthermore, as the putative DNA-binding domain of the protein encoded by the wheat cDNA is so similar to GBF-1, we consider it likely that it is also a G-box-binding protein and that it is not HBP-1. Here it should be noted that we are using the term HBP-1 as it was originally used to describe the protein present in wheat nuclear extracts - this DNA-binding protein has been clearly demonstrated to bind to a hexamer sequence distinct from the G-box (Mikami *et al.*, 1989). We predict that the sequence of the true HBP-1 is likely to be less similar to GBF-1 than it is to the hexamer binding proteins TGA1a and TGA1b.

GENOMIC SOUTHERN BLOTS SHOW THAT *GBF-1* IS PRESENT AS A SINGLE OR LOW-COPY SEQUENCE

In order to examine the complexity of the DNA sequences encoding GBF-1, we performed southern blot hybridization studies with genomic *Arabidopsis* DNA (Fig. 4). A simple pattern of hybridization was observed indicating that *gbf-1* was present as a single or low-copy sequence. The isolation of genomic clones will be required to determine the exact copy number of this sequence.

CONCLUDING REMARKS

We have identified a cloned DNA sequence encoding a "leucine zipper" DNA-binding protein which we have designated GBF-1. The binding properties of GBF-1 are similar to those of GBF, the activity that we have previously characterized in nuclear extracts of tomato and *Arabidopsis thaliana* (Giuliano *et al.*, 1988; Schindler and Cashmore, 1990). A determination of the precise relationship between these two activities will require further studies. The sequence of the putative DNA-binding domain of GBF-1 is strikingly similar to the protein encoded by a wheat cDNA sequence and identified as the histone promoter binding factor HBP-1 (Tabata *et al.*, 1989). Both the sequence and the published binding properties of the factor encoded by this wheat cDNA prompt us to suggest that it may have been misidentified as HBP-1 and that it may be a homologue of GBF-1.

ACKNOWLEDGEMENTS

This work was supported by grants to ARC from NIH (GM-38409) and DOE (DE-FG02-87ER13680) and grants to J.R.E. from NIH (GM-38894 and HG-00322).

REFERENCES

Castresana C, Garcia-Luque I, Alonso E, Malik VS, Cashmore AR (1988) Both positive and negative regulatory elements mediate expression of a photoregulated CAB gene from *Nicotiana plumbaginifolia*. EMBO J. 7:1929-1936

Donald RGK, Cashmore, AR (1990) Mutation of either G-box or I-box sequences profoundly affects expression from the *Arabidopsis rbcS-1A* promoter. EMBO J. 9:1717-1726

Donald RGK, Schindler U, Batschauer A, Cashmore AR (1990) The plant G-box promoter sequence activates transcription in *Saccharomyces cerevisiae* and is bound *in vitro* by a yeast activity similar to GBF, the plant G-box binding factor. EMBO J. 9:1727-1735

Ferl RJ, Laughner BH (1989) *In vivo* detection of regulatory factor binding sites of *Arabidopsis thaliana Adh*. Plant Mol. Biol. 2:357-366

Giuliano G, Pichersky E, Malik VS, Timko MP, Scolnik PA, Cashmore AR (1988) An evolutionarily conserved protein binding sequence upstream of a plant light-regulated gene. Proc. Natl. Acad. Sci. USA. 85:7089-7093

Guiltinan MJ, Marcotte WR, Quatrano RS (1990) A plant leucine zipper protein that recognizes an abscisic acid response element. Science. (in press)

Katagiri F, Lam E, Chua N-H (1989) Two tobacco DNA-binding proteins with homology to the nuclear factor CREB. Nature. 340:727-730

Mikami K, Nakayama T, Kawata T, Tabata T, Iwabuchi M (1989) Specific interaction of nuclear protein HBP-1 with the conserved hexameric sequence ACGTCA in the regulatory region of wheat histone genes. Plant Cell Physiol. 30:107-119

Schindler U, Cashmore AR (1990) Photoregulated gene expression may involve ubiquitous DNA binding proteins. EMBO J. 11 (in press)

Schmidt RJ, Burr FA, Aukerman MJ, Burr B (1990) Maize regulatory gene opaque-2 encodes a protein with a "leucine-zipper" motif that binds to zein DNA. Proc. Natl. Acad. Sci. USA. 87:46-50

Schulze-Lefert P, Dangl JL, Becker-Andre M, Hahlbrock K, Schulz W (1989) Inducible *in vivo* DNA footprints define sequences necessary for UV light activation of the parsley chalcone synthase gene. EMBO J. 8:651-656

Staiger D, Kaulen H, Schell J (1989) A CACGTG motif of the *Antirrhinum majus* chalcone synthase promoter is recognized by an evolutionarily conserved nuclear protein. P.N.A.S., USA. 86:6930-6934

Tabata T, Takase H, Takayama S, Mikami K, Nakatsuka A, Kawata T, Nakayama T, Iwabuchi M (1989) A protein that binds to a cis-acting element of wheat histone genes has a leucine zipper motif. Science. 245:965-967

Vinson CR, LaMarco KL, Johnson PF, Landschulz WH, McKnight SL (1988) *In situ* detection of sequence-specific DNA binding activity specified by a recombinant bacteriophage. Genes and Dev. 2:801-806

PHYTOCHROME REGULATION OF TRANSCRIPTION: BIOCHEMICAL AND GENETIC APPROACHES

E.M. Tobin, J.A. Brusslan, J.S. Buzby, G.A. Karlin-Neumann, D.M. Kehoe, P.A. Okubara, S.A. Rolfe, L. Sun, and T. Yamada
Biology Department,
University of California,
Los Angeles, CA 90024, USA.

INTRODUCTION

Phytochrome action has been demonstrated to affect the transcription of a number of different genes in many different species (see reviews by Tobin and Silverthorne, 1985; Kuhlemeier *et al.*, 1987). The effect may be either a positive or negative one. Other light receptors, circadian rhythms, and tissue type have also been shown to have effects on the transcription of phytochrome regulated genes. Particular short "light responsive elements" that can interact with protein factors have been identified upstream of a number of rbcS and cab genes encoding, respectively, the small subunit (SSU) of ribulose 1,5-bisphosphate carboxylase/oxygenase (Rubisco) and the major apoproteins of the photosystem II light-harvesting chlorophyll a/b-protein complex (LHCII). The evidence suggests these sequences play an important role in the overall response to light/dark conditions (reviewed in Silverthorne and Tobin, 1987; Benfey and Chua, 1989). There is also evidence that phytochrome action can alter RNA levels by effects on additional, post-transcriptional processes (Colbert, 1988; Thompson, 1988; Elliott *et al.*, 1989), as well as influence many other processes, such as membrane permeability, that may not involve altered gene expression (Kendrick and Kronenberg, 1986). Although the phytochrome chromoprotein has itself been the subject of biochemical studies for many years, to date there is no clear understanding of the chain of events by which the phototransformation of phytochrome leads to specific transcriptional changes.

In *Lemna gibba* we have studied the transcriptional regulation of both rbcS and cab genes in response to phytochrome action (Silverthorne and Tobin, 1984). We have also recently identified genes that have a negative transcriptional response to phytochrome action in this species (Okubara and Tobin, 1990). We are particularly interested in

NATO ASI Series, Vol. H 50
Phytochrome Properties and Biological Action
Edited by B. Thomas and C. B. Johnson

defining the DNA sequences and interacting protein factors important in mediating both positive and negative responses. Because *Lemna* is a monocot that has not been transformed with *Agrobacterium tumefaciens*-based vectors, we are exploring the possibility of using an homologous transient expression system to investigate the regulatory DNA elements. We have also been able to demonstrate appropriate expression of reporter genes driven by a *Lemna* promotors in a heterologous system (tobacco).

Analysis of mutant plants has offered some interesting observations and insights into the action of phytochrome (e.g. Adamse *et al.*, 1988; Chory *et al.*, 1989a, 1989b; Koornneef *et al.*, 1980; Parks *et al.*, 1989; Sharrock *et al.*, 1988). *Arabidopsis thaliana* offers a particularly good opportunity to utilize a genetic approach to understanding the phytochrome signal transduction pathway. The three characterized cab genes of this species (Leutwiler *et al.*, 1986) show phytochrome regulation of their RNA levels, and one of these, cab140, is the most responsive to brief red illumination of etiolated seedlings (Karlin-Neumann *et al.*, 1988). Existing long hypocotyl (hy) mutants, including ones potentially affected in the phytochrome signal transduction pathway (Koornneef *et al.*, 1980), have been examined and found to have normal phytochrome regulation (Chory *et al.*, 1989; Sun and Tobin, 1990) of their cab RNA levels. Therefore, we are trying to make other mutants that will alter the cab transcription increase seen in response to phytochrome action and represent mutations in components of the signal transduction pathway.

THE rbcS GENE FAMILY OF *LEMNA GIBBA*

The rbcS gene family of *Lemna gibba* comprises 12-14 genes (Tobin *et al.*, 1985); we have isolated six of these as genomic clones and a seventh as a cDNA clone (Stiekema *et al.*, 1983; Tobin *et al.*, 1985). Portions of the 3′ untranslated (3′UT) regions of all seven genes have been sequenced and subcloned for use as gene-specific probes to distinguish relative expression levels and phytochrome responsiveness of individual genes (Tobin *et al.*, 1985; Silverthorne *et al.*, 1990). These genes have been designated SSU1, SSU5A, SSU5B, SSU13, SSU26, SSU40A, and SSU40B. Expression of SSU13 could not be detected in total RNA from light-grown plants, but levels of transcripts corresponding to each of the other clones could be regulated by phytochrome action. Of the transcripts represented by isolated genomic clones, SSU5B showed the highest level of phytochrome responsiveness and was chosen for further study. Transcriptional activity assayed in isolated nuclei by *in vitro* "run-on" experiments using a gene-specific probe confirmed that the transcription of the SSU5B gene increases within two hours in response to a single minute of red light (P.A. Okubara, unpublished work).

A LIGHT-REGULATED NUCLEAR FACTOR THAT BINDS UPSTREAM OF *LEMNA GIBBA* rbcS PROMOTER

Sequence comparison of the upstream regions of SSU5A, SSU5B, and SSU13 revealed a number of regions of sequence conservation, including three in which the sequence identity is extremely high. These three, which occur within 150nt of the transcription starts, have been designated boxes X, Y, and Z (Buzby *et al.*, 1990). Box X, which has 15 identical nucleotides out of 16 in the three genes, proved to be particularly interesting. It stretches from position -149 to -134 (from transcription start) in the SSU5B gene, and it includes a GATAAG motif that is present in a number of other rbcS genes (e.g. see Manzara and Gruissem, 1988). A DdeI-HindIII restriction fragment of the SSU5B gene (-208 to -123) was used in gel retardation assays with extracts from *Lemna gibba* nuclei, and it was found to interact specifically with a protein factor (Buzby *et al.*, 1990). Footprint analysis was used to determine the region of the promoter that is involved in the interaction, and the protected bases on each strand are indicated by lines in Figure 1. Figure 1 also shows the extent of an oligonucleotide (B16) that was found to compete for binding with the 86 bp DdeI-HindIII fragment.

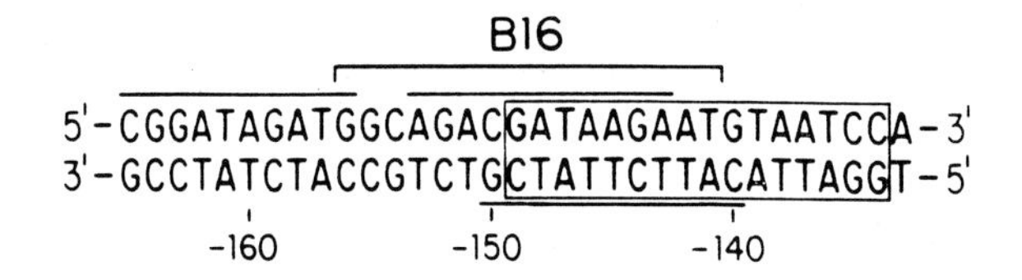

Figure 1. SSU5B upstream sequence that interacts with a nuclear protein factor. Nucleotides protected from DNase I action in the DNA-protein complex are indicated by lines. Numbers at the bottom show the distance from the start of transcription. The highly conserved region designated Box X is indicated by a box. The sequence of a synthesized oligonucleotide, B16, is also shown.

The level of activity was considerably lower in nuclei from dark- treated plants (1-7 day) than in nuclei from plants grown in the light or returned to light after a 24 h dark treatment. The activity was measured on the basis of nuclear protein amount; neither the amount of nuclear protein nor DNA were changed significantly by the light treatments. Furthermore, if plants were given a 1 day dark treatment, then a single 2 min red illumination, the level of activity showed a transient increase of about two-fold at two hours after the red treatment. For this reason, we have designated the activity as light-regulated factor 1 (LRF-1) (Buzby *et al.*, 1990). *Lemna* extracts also contained a GT-1-like activity that could bind to the pea Box II sequence (Green *et al.*, 1988); as was found in pea, this binding activity was not higher in extracts from light-grown plants. Taken together, the evidence suggests that phytochrome may be involved in the regulation of LRF-1 binding.

Although there is sequence identity between a portion of the sequence footprinted by LRF-1 and conserved regions upstream of other rbcS genes, e.g. Box I of Giuliano *et al* (1988), our evidence suggests that additional sequences are involved in the binding of LRF-1 (Buzby *et al.*, 1990). In order to be able to see whether the DNA region that interacts with LRF-1 plays an important role *in vivo*, it will be necessary to test its effect in the plant itself. However, since *Lemna gibba*, a monocot, cannot yet be stably transformed, we have taken other approaches to test *Lemna* promoters *in vivo*.

LEMNA PROMOTER-REPORTER GENE CONSTRUCTS IN TRANSFORMED TOBACCO

In an initial set of experiments, fusion constructs were made that contained *Lemna* rbcS promoter regions (either 640 bp upstream of SSU5A or 960 bp upstream of SSU5B) and a 1.6 kbp promoter region from the CaMV 35S gene upstream and in inverse orientation to the rbcS promoter fragment. This 35S sequence was used because it had been reported to enhance expression of a monocot promoter in a dicot (Ellis *et al.*, 1987). These constructs were inserted upstream of an NPTII gene with an ocs 3′ region in the plasmid pLGVneo2103B (Hain *et al.*, 1985) and mobilized to an *Agrobacterium tumefaciens* strain containing pGV3850 (Hain *et al.*, 1985) using a helper plasmid R64 *drd*11 as described by Van Haute *et al.* (1983). Tobacco leaf disks were transformed by the procedures described by Rogers *et al.* (1986), and regenerated tobacco plants were tested for NPT II activity (Reiss *et al.*, 1984). Dark treatment of transgenic plants lowered the level of NPT II activity, and returning them to light for three days resulted in an increase in this activity. Thus, the constructs could confer on the NPTII gene a difference in expression in white light and darkness. Seedlings obtained from these plants showed a similar response, but no clear evidence of phytochrome regulation of the NPT II activity could be obtained. Furthermore, the role of the CaMV 35S sequences was uncertain, and experiments with these constructs was not pursued further.

In a second set of experiments, a 1.6 kb fragment upstream of the cab gene AB19 (Karlin-Neumann *et al.*, 1985) was fused (in both positive and negative orientations) to the bacterial gus gene (Jefferson *et al.*, 1987) containing a nos 3′ region and a 30 bp fragment of the Kunitz trypsin inhibitor 2 gene (containing 21 bp of 5′ untranslated region to provide an appropriate plant translation start site; Jofuku and Goldberg, 1989) at its 5′ end. These *AB19:gus* constructs were inserted into the polylinker site of pGV1501A (derived from pGV1501 from Plant Genetics Systems, Ghent, Belgium, C. Reeves and R.B. Goldberg, personal communication) in the opposite direction to the NPT II gene, and transformed into tobacco plants (Rogers et al, 1986). At least ten independently transformed lines of each construct were used to generate homozygous transformed progeny. Of these, at least six different lines of each construct showed segregation ratios consistent with a single insertion site, and seeds from these were used for further analysis.

We tested whether etiolated seedlings from the transformed lines showed phytochrome regulation of GUS activity by germinating the seeds in darkness and giving various light-treatments for 10 days. Figure 2 shows the results of these experiments for the construct with the AB19 promoter in the positive orientation. Growth in intermittent red light can increase the GUS activity by two fold compared to the dark level, and the effect of red can be reversed by far-red light. However, the dark level of activity was substantial and was present in dry seeds. The lines with the AB19 promoter in the negative orientation showed about ten-fold lower levels of GUS activity and gave no evidence of phytochrome regulation.

Lines containing the positively oriented promoter were further examined for changes in GUS activity in several organ types in response to growth in the intermittent red illumination. For cotyledons and hypocotyls, a four-fold increase in GUS activity compared to dark-grown seedlings was seen. The relative increase in the roots was only 1.7-fold.

Thus, we have shown that a transgenic dicot containing a phytochrome regulated monocot promoter can confer phytochrome regulation on a reporter gene. We have also found that some degree of organ specificity of expression exists for the construction in tobacco. These results indicate that the processes that are involved in phytochrome signal transduction leading to transcriptional responses are conserved between these two major plant groups. However, because it is difficult to quantitate the phytochrome responsiveness over the substantial dark level that is present already in the seeds and increases slightly during germination in complete darkness, and in order to be able to test promoters in an homologous system, we are developing a transient assay system in *Lemna* fronds.

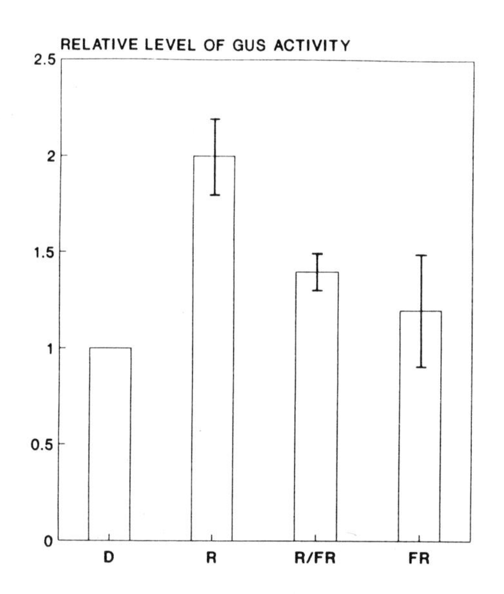

Figure 2. Phytochrome regulation of *AB19:gus* in transgenic tobacco seedlings. Values are the average normalized level of GUS activity relative to the dark level on a protein basis. Bars show the standard errors (n=6). Seedlings were germinated in the dark and grown for 10 days with the illumination indicated: D, no illumination; R, 2min red/8 h; R/FR, 2min red followed by 10min FR/8 h; FR, 10min FR/8 h.

TRANSIENT EXPRESSION IN *LEMNA GIBBA* FRONDS

A transient expression system in which reporter gene fusion constructs can be assayed for phytochrome responsiveness has been reported by Bruce et al. (1989) using an oat phytochrome promoter construct expressed in rice seedlings. We are using the Biolistics/DuPont Particle Acceleration System (Klein *et al.*, 1988) to deliver DNA to a monolayer of intact *Lemna* fronds which can then be assayed for the reporter gene activity. As controls, we have used maize ubiquitin promoter-reporter genes that have been shown to be constitutively active in the rice transient expression system and to result in a higher level of activity than CaMV 35S promoter constructs (Bruce et al., 1989). We have made a translational fusion that includes 965nt of the SSU5B sequence upstream of the transcription start, the 82nt 5′ untranslated region, and 135nt of coding sequence fused to a bacterial gene for chloramphenicol acetyl transferase (CAT) with a nos 3′UT to give a construct designated *SSU5B:cat*.

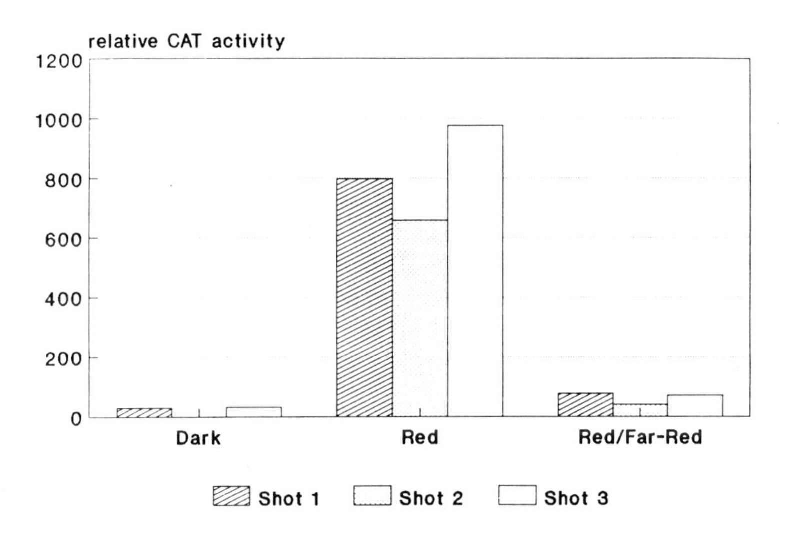

Figure 3. Phytochrome regulation of *SSSU5B:cat* in a transient assay in *Lemna* fronds. Etiolated *Lemna* fronds were maintained on sucrose in darkness for two days prior to bombardment with 1.2μ tungsten particles coated with *SSU5B:cat* DNA. Experiments were performed under a dim green safelight and all samples were exposed to 2min FR light immediately after bombardment. Samples were then exposed to 2min R light or 2min R followed by 2min FR light, placed in darkness for 16h, and then assayed for chloramphenicol acetyl transferase (CAT) activity. Data for three different samples (Shots 1-3) are shown for each treatment.

The *SSU5B:cat* construct directed expression of CAT activity that continued to increase for about 16 hours after the DNA was introduced into the *Lemna* fronds. This increase was evident if the fronds were incubated after the shot in continuous white light, but very little activity could be detected if the fronds were maintained (on sucrose) in the dark. The control construct using the maize ubiquitin promoter fused to CAT was equally expressed in fronds maintained in the light and in darkness (Tobin *et al.*, 1990).

Figure 3 demonstrates that phytochrome can regulate expression of the *SSU5B:cat* construct in this transient assay system. A single 2min red (R) illumination given immediately after the shot introducing the DNA was sufficient to greatly increase the CAT activity measured 16h later. This effect of red was reversed by 2min far-red (FR) light given immediately after the red treatment. The average level of CAT activity seen in these red samples reached about half the level seen if the fronds were maintained in white light.

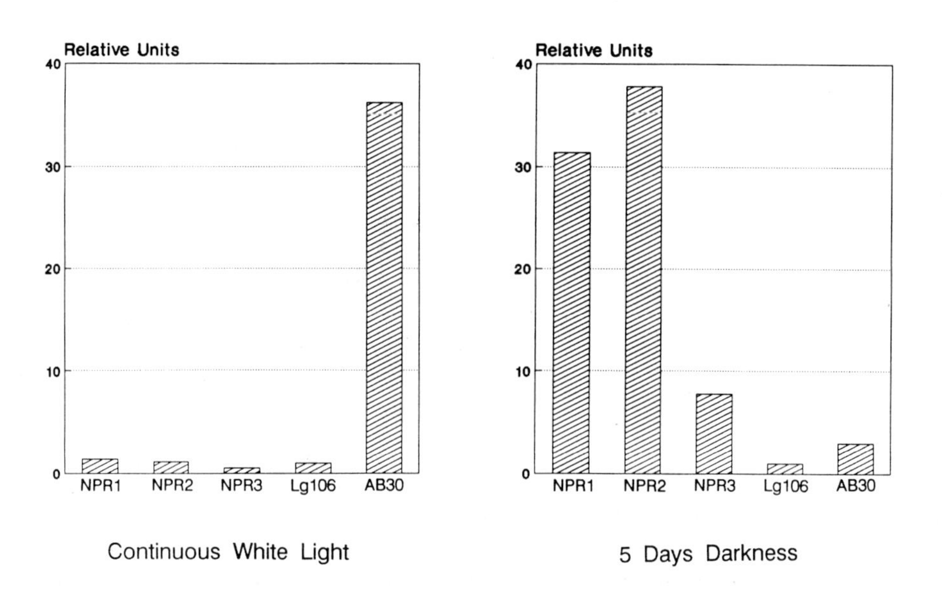

Figure 4. Relative transcription levels in white light and darkness. Nuclei were isolated from plants grown in continuous white light or after a five day dark treatment and used for "run-on" transcription experiments. The labeled transcripts were hybridized to ZetaProbe (BioRad) blots of DNA probes excised from genomic clones indicated. The amounts of hybridized transcripts were quantititated by liquid scintillation counting of the NPR-specific radioactivity visualized previously by autoradiography. The relative units shown are normalized to the value for Lg106, the constitutively transcribed gene.

IDENTIFICATION OF GENES NEGATIVELY REGULATED BY PHYTOCHROME ACTION IN *LEMNA*

We have isolated three different cDNAs that represent genes whose transcripts accumulate during dark treatment. These genes were originally designated NR11, NR18, and NR300 (Tobin *et al.*, 1990), but have been renamed NPR1, NPR2, and NPR3, respectively. The increased RNA levels can be demonstrated to involve increases in the relative levels of transcription of these genes using the *in vitro* "run-on" isolated nuclei assay (Silverthorne and Tobin, 1984). Figure 4 shows the relative levels of transcription of these three NPR genes, the cab gene family (AB30, Kohorn *et al* 1986) and Lg106 (Okubara *et al.*, 1988), a constitutively transcribed *Lemna* gene, in light-grown and dark-treated plants.

The increase in NPR transcription in the dark could be seen by the first 6-12 hours, and has been seen as early as 2h for NPR1 and NPR2. Figure 5 shows that if these plants

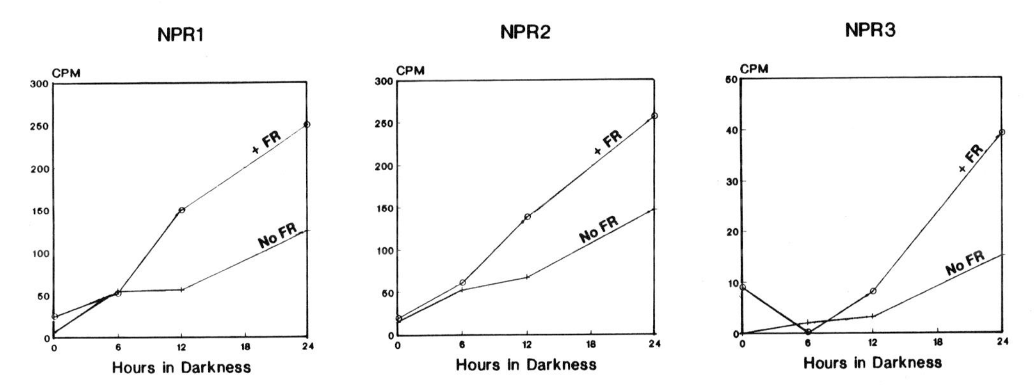

Figure 5. Effect of far-red light treatment on the increased transcription of NPR genes in darkness. Light-grown *Lemna* plants were put into darkness for the times indicated with (+FR) or without (No FR) a preceding 10min FR illumination. The transcription was quantitated as described for Figure 4, and the counts shown are those that hybridized from aliquots of transcripts containing 1.2 x 10^7 cpm (No FR) and 1.6 x 10^7 cpm (+FR).

were given a FR illumination at the end of their growth in white light, the increase was accelerated, suggesting that Pfr is involved in decreasing the transcription of these genes.

We have shown that phytochrome is involved in regulating the transcription of the three negatively phytochrome regulated (NPR) genes. When nuclei were isolated from plants put into darkness for 4 days (D), and given a 2min red (R) light treatment 4h before harvest, a phytochrome-induced decrease in transcription of all three NPR genes could be demonstrated. This response was reversible by immediate far-red illumination (Okubara and Tobin, 1990). Sequencing of the cDNAs revealed that none of the three are homologous to either phytochrome or protochlorophyllide reductase genes, other genes that have been demonstrated to be negatively regulated by phytochrome action in some species. The functions of the encoded proteins are unknown. Genomic clones corresponding to the cDNAs have been isolated from a library made from *Lemna* DNA partially digested with BamHI and Sau3AI and cloned into Charon 35; these clones are currently being characterized.

ARABIDOPSIS cab GENE EXPRESSION AND MUTANT SCREENS

Existing long hypocotyl mutants of *Arabidopsis* (Koornneef *et al.*, 1980; Chory *et al.*, 1989) fail to exhibit the phytochrome mediated suppression of hypocotyl growth, but none of these has substantially altered cab gene responsiveness to phytochrome (Chory *et al.*, 1989; Sun and Tobin, 1990). We are attempting to isolate *Arabidopsis* mutants in which this transcriptional response is affected by two different strategies. In the first of these we have mutagenized wild type seeds and are screening M2 seeds for those that do not germinate in response to red illumination, but can be rescued by being induced to germinate in the presence of gibberellin. We have a number of candidates that have shown reduced germination in response to red light over three generations and these lines will be tested to see if phytochrome-regulated transcription of cab genes is also aberrant.

In a more directed approach we are using a "suicide" gene selection scheme that will allow survival of mutants with disruptions in the phytochrome transduction pathway activating nuclear gene transcription, while plants with the normal pathway will be severely affected. For this purpose, we have created homozygous lines of *Arabidopsis* transformed with promoter fusion constructs containing the *tms2* gene (Klee *et al.*, 1984) of *Agrobacterium tumefaciens*. The product of this gene is an auxin amide hydrolase that catalyzes the ultimate step in the formation of auxin in the bacterial pathway. The plant auxin biosynthetic pathway is different, and the wild type plants lack this activity. Thus, plants expressing the *tms2* gene would be expected to be sensitive to otherwise non-toxic levels of auxin amides because they would convert them to toxic levels of auxins. We have found that the transgenic *tms2+* seedlings do show strikingly higher sensitivity to auxin amides compared with wild type seedlings.

In order to determine the usefulness of this marker for selection of phytochrome response chain mutants, the *tms2* gene was placed under the control of either the *Arabidopsis* cab140 promoter (1 kb upstream of the start of transcription) or the CaMV 35S promoter and transformed into *Arabidopsis* (Lloyd *et al*., 1986). It was important to demonstrate that the expression of the introduced *tms2* gene fused to the cab promoter is under phytochrome regulation. Indeed, we have been able to show phytochrome regulation of the *tms*2 transcript levels, the level of auxin amide hydrolase activity, and the increased growth inhibition in intermittent red light in comparison to complete darkness in such transformed lines; lines transformed with the 35S fusion construct do not show such regulation (Karlin-Neumann et al., 1990). Seeds of the homozygous lines with the cab 140:tms2 constructs have been mutagenized, and M2 seeds are being selected for survivors growing under inductive light conditions on medium containing naphthalene acetamide. The M3 progeny that show the auxin amide insensitive phenotype will be tested for alterations in phytochrome regulation of the endogenous cab genes.

ACKNOWLEDGMENTS

This research was supported by N.I.H. grant GM-23167 (E.M.T.) and by N.I.H. BRSG grant RR07009-23 to U.C.L.A. We thank Ruth Doxsee and Lu Huang for technical assistance.

REFERENCES

Adamse P, Kendrick RE, Koornneef M (1988) Photomorphogenetic mutants of higher plants. Photochem Photobiol 48:833-841

Benfey PN, Chua N-H (1989) Regulated genes in transgenic plants. Science 244:174-181

Bruce WB, Christensen AH, Klein T, Fromm M, Quail PH (1989) Photoregulation of a phytochrome gene promoter from oat transferred into rice by particle bombardment. Proc Natl Acad Sci USA 86:9692-9696

Buzby J, Yamada T, Tobin EM (1990) A light-regulated DNA-binding activity interacts with a conserved region of a *Lemna gibba* rbcS promoter. Plant Cell 2:805-814

Chory J, Peto CA, Ashbaugh M, Saganich R, Pratt L, Ausubel F (1989a) Different roles for phytochrome in etiolated and green plants deduced from characterization of *Arabidopsis thaliana* mutants. Plant Cell 1:867-880

Chory J, Peto CA, Feinbaum R, Pratt L, Ausubel F (1989b) *Arabidopsis thaliana* mutant that develops as a light-grown plant in the absence of light. Cell 58:991-999

Colbert JT (1988) Molecular biology of phytochrome. Plant Cell Environ 11:305-318

Elliott RC, Dickey LF, White MJ, Thompson WF (1989) Cis-acting elements for light regulation of pea ferredoxin gene expression are located within transcribed sequences. Plant Cell 1:691-698

Ellis JG, Llewellyn DL, Dennis ES, Peacock, WJ (1987) Maize adh-1 promoter sequences control anaerobic regulation: addition of upstream promoter elements from constitutive genes is necessary for expression in tobacco. EMBO J 6:11-16

Giuliano G, Pichersky E, Malik VS, Timko MP, Scolnik PA, Cashmore AR (1988) An evolutionarily conserved protein binding sequence upstream of a plant light regulated gene. Proc Natl Acad Sci USA 85:7089-7093

Green PJ, Yong M-H, Cuozzo M, Kano-Murakami Y, Silverstein P, Chua N-H (1988) Binding site rerequirements for pea nuclear protein factor GT-1 correlate with sequences required for light-dependent transcriptional activation of the *rbcS-3A* gene. EMBO J 6:2543-2549

Hain R, Stabel P, Czernilofsky AP, Steinbiss HH, Herrera-Estrella L, Schell J (1983) Uptake, integration, expression and genetic transmission of a selectable chimaeric gene by plant protoplasts. Mol Gen Genet 199:161-168

Jefferson RA, Kavanagh TA, Bevan MW (1987) Gus fusions: ß-glucoronidase as a sensitive and versatile gene fusion marker in higher plants. EMBO J 6:3901-3907

Jofuku KD, Goldberg RB (1989) Kunitz trypsin inhibitor genes are differentially expressed during the soyabean life cycle and in transformed tobacco plants. Plant Cell 1:1079-1093

Karlin-Neumann GA, Brusslan J, Tobin EM (1990) Manuscript in preparation

Karlin-Neumann GA, Kohorn BD, Thornber JP, Tobin EM (1985) A chlorophyll a/b-protein encoded by a gene containing an intron with characteristics of a transposable element. J Mol Appl Genet 3:45-61

Karlin-Neumann GA, Sun L, Tobin EM (1988) Expression of light harvesting chlorophyll a/b-protein genes is phytochrome regulated in etiolated *Arabidopsis thaliana* seedlings. Plant Physiol 88:1323-1331

Kendrick RE, Kronenberg GHM (eds) (1986) Photomorphogenesis in Plants, Martinus Nijhoff, Boston

Klee H, Montoya A, Horodyski F, Lichtenstein C, Garfinkel D, Fuller S, Flores C, Peschon J, Nester E, Gordon M (1984) Nucleotide sequence of the EMS genes of the pTiA6NC octopine Ti plasmid: two gene products in plant tumorigenesis. Proc Natl Acad Sci USA 81:1728-1732

Klein TM, Fromm M, Weissinger A, Tomes D, Schaaf S, Sletten M, Sanford JC (1988) Transfer of foreign genes into intact maize cells with high-velocity microprojectiles. Proc Natl Acad Sci 85:4305-4309

Kohorn BD, Harel E, Chitnis PR, Thornber JP, Tobin EM (1986) Functional and mutational analysis of the chlorophyll a/b protein of thylakoid membranes. J Cell Biol 102:972-981

Koornneef M, Rolff E, Spruit CJP (1980) Genetic control of light-inhibited hypocotyl elongation in *Arabidopsis thaliana* (L.) Heynh. Z Pflanzenphysiol 100:147-160

Kuhlemeier C, Green PJ, Chua N-H (1987) Regulation of gene expression in higher plants. Annu Rev Plant Physiol 38:221-257

Leutwiler LS, Meyerowitz EM, Tobin EM (1986) Structure and expression of three light harvesting chlorophyll a/b binding protein genes in *Arabidopsis thaliana.* Nucl Acids Res 14:4051-4064

Lloyd AM, Barnason AR, Rogers SG, Byrne MC, Fraley RT, Horsch RB (1986)

Tansformation of *Arabidopsis thaliana* with *Agrobacterium tumefaciens.* Science 234:464-466

Manzara T, Gruissem W (1988) Organization and expression of the genes encoding ribulose-1,5-bisphosphate carboxylase in higher plants. Photosynth Res 16:117-139

Nagy F, Kay SA, Chua N-H (1988) Gene regulation by phytochrome. Trends Genet 4:37-42

Okubara PA, Flores S, Tobin EM (1988) Characterization of a negatively light-regulated mRNA from *Lemna gibba.* Plant Mol Biol 11:673-681

Okubara PA, Tobin EM (1990) Manuscript in preparation

Parks BM, Shanklin J, Koornneef M, Kendrick RE, Quail PH (1989) Immunochemically detectable phytochrome is present at normal levels but is photochemically nonfunctional in the *hy 1* and *hy 2* long hypocotyl mutants of *Arabidopsis.* Plant Mol Biol 12:425-437

Reiss B, Sprengel R, Will H, Schaller H (1984) A new sensitive method for qualitative and quantitative assay of neomycin phosphotransferase in crude cell extracts. Gene 30:217-223

Rogers SG, Horsch RB, Fraley RT (1986) Gene transfer in plants: Production of transformed plants using Ti plasmid vectors. Meth Enz 118:627-640

Sharrock RA, Parks BM, Koornneef M, Quail PH (1988) Molecular analysis of the phytochrome deficiency in an aurea mutant of tomato. Mol Gen Genet 213:9-14

Silverthorne J, Tobin EM (1984) Demonstration of transcriptional regulation of specific genes by phytochrome action. Proc Natl Acad Sci USA 81:1112-1116

Silverthorne J, Tobin EM (1987) Phytochrome regulation of nuclear gene expression. BioEssays 7:18-23

Silverthorne J, Wimpee CF, Yamada T, Rolfe S, Tobin EM (1990) Differential expression of individual genes encoding the small subunit of ribulose-1,5-bisphosphate carboxylase in *Lemna gibba.* Plant Mol Biol 15:49-58

Stiekema WJ, Wimpee CF, Tobin EM (1983) Nucleotide sequence encoding the precursor of the small subunit of ribulose-1,5-bisphosphate carboxylase from *Lemna gibba.* Nucl Acids Res 11:8051-8061

Sun L, Tobin E M (1990) Phytochrome-regulated expression of genes encoding light-harvesting chlorophyll a/b-protein in two long hypocotyl mutants and wt plants of *Arabidopsis thaliana.* Photochem Photobiol 52:51-56

Thompson WF (1988) Photoregulation: diverse gene responses in greening seedlings. Plant Cell Environ 11:319-328

Tobin EM, Silverthorne J (1985) Light regulation of gene expression in higher plants. Annu Rev Plant Physiol 36:569-593

Tobin EM, Wimpee CF, Karlin-Neumann GA, Silverthorne J, Kohorn BD (1985) In: Arntzen CJ, Bogorad L, Bonitz S, Steinback KE (eds) The Molecular Biology of the Photosynthetic Apparatus Cold Spring Harbor Laboratory, Cold Spring Harbor, NY, pp373-380

Tobin EM, Brusslan JA, Buzby JS, Karlin-Neumann, GA, Kehoe DM, Okubara PA, Rolfe SA, Sun L (1990) In: Herrmann R (ed) Plant Molecular Biology, NATO/ASI Series Plenum, New York, In press

Van Haute E, Joos H, Maes M, Warren G, Van Montagu M, Schell J (1983) Intergeneric transfer and exchange recombination of restriction fragments cloned in PBR322: A novel strategy for the reversed genetics of the Ti plasmids of *Agrobacterium tumefaciens.* EMBO J 2:411-417

LIGHT AND CELL SPECIFIC GENE EXPRESSION

A. Batschauer, B. Ehmann, M. Furuya[1], R. Grimm, E. Hofmann, K. Harter, T. Kretsch, A. Nagatani[1], B. Ocker, V. Speth and E. Schäfer
Institut für Biologie II,
Universität Freiburg,
Schänzlestrasse 1,
D-7800 Freiburg, Germany.

INTRODUCTION

There is general agreement that photomorphogenesis is the synopsis of signal perception and transduction on one side and the temporal and spatial patterns of competence on the other side. To analyse photoregulation of cell specific gene expression the following topics will be discussed: 1. The cellular and intracellular localisation of the photoreceptors; 2. The photoregulation of gene expression and finally; 3. The tissue and cell specific gene expression of light-regulated genes. In the context of this article we will concentrate on the photoreceptor phytochrome and chalcone synthase as a candiate for a light-regulated gene.

CELLULAR AND INTRACELLULAR LOCALISATION OF PHYTOCHROME

Unfortunately, our knowledge about the localisation of phytochromes in plant tissue is very poor although around 1970, Pratt and co-workers had already started to analyse cellular distribution of type I (etiolated or "phy A") phytochrome by immunological methods. Because the localisation of the other types of phytochrome has not been analysed so far, we will concentrate on the analysis of type I phytochrome.

Up to now the studies of cellular localisation of phytochrome is restricted mainly to grass seedlings (Coleman and Pratt, 1974) with the exceptions of pea (Saunders *et al.*, 1983) and most recently soybean (Cope and Pratt, 1989). The general view is that phytochrome is highly concentrated in meristematic tissues and parenchymatic cells of the coleoptile. In dicotyledonous seedlings the low levels of phytochrome in the epidermal

[1]Laboratory of Plant Biological Regulation, Frontier Research Program, RIKEN Institute, Wako City, Saitama, Japan 251-01.

NATO ASI Series, Vol. H 50
Phytochrome Properties and Biological Action
Edited by B. Thomas and C. B. Johnson

cells seems to be remarkable. Unfortunately, the number of species which have been analysed is so small that no general statements about the localisation are possible. Intracellular localisation of phytochrome has been studied extensively only in parenchyma cells of oat coleoptiles (Speth *et al.*, 1986, 1987; McCurdy and Pratt, 1986 a,b). The common view of these studies is that the *detectable* phytochrome is homogenously distributed in the cytosol in etiolated seedlings. After a brief red light pulse a rapid formation of sequestered areas of phytochrome (SAP's) is detectable (Fig. 1). The SAP's disappear slowly (2-3 h) after conversion of Pfr back to Pr. The reformation of SAP's can be induced by a second irradiation with a red light pulse (Hofmann *et al.*, 1990a). If phytochrome remains in its Pfr form the disappearance of the SAP's is slower than in the Pr form (Speth *et al.*, 1986). SAP's seem to contain ubiquitin (Speth *et al.*, 1987), are not associated with any known structure and their formation seems to require active steps which show partial adaptation (Hofmann *et al.*, 1990a). The

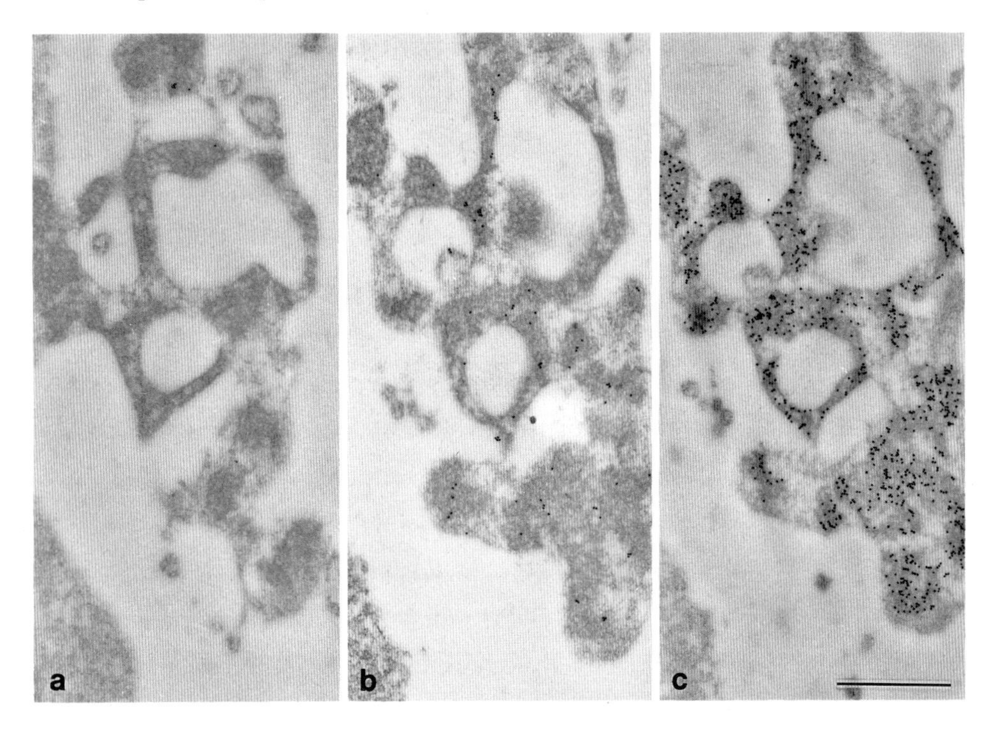

Figure 1a-c. Serial sections of an *Avena* coleoptile cell from material fixed 2 h after a 5 min red light pulse. Pfr is sequestered in large electron-dense cytoplasmic areas (c), which are also positive for ubiquitin (b), and are lot labelled by unspecific serum (a) x 20000, bar 21μm.

ubiquitination of phytochrome seems to be an important step in the phytochrome destruction pathway (Jabben *et al.*, 1989).

There now seems to be general agreement that Pfr-induced pelletability, Pfr destruction and sequestering reflect the same process (Schäfer, 1981; McCurdy and Pratt, 1986a). On the other side sequestering is a general process and therefore we will study it in more detail.

Is pelletability of type II phytochrome observable? Using polyclonal antibodies against pea phytochrome and the monoclonal antibody mAP 5, which recognizes both pea type I and II phytochrome, it was possible to discriminate between type I and II not only in pea but also in other dicotyledonous seedlings. The general result is that type II phytochrome is so far never observed as pelletable phytochrome and therefore probably does not undergo sequestering.

Partial purification of SAP's. To understand more about the mechanism of sequestering, purification of SAP's was important, especially with respect to the question whether other proteins are present in the SAP's. A partial purification of SAP's from oat leading to a 200-fold enrichment of phytochrome has recently been described (Hofmann *et al.*, 1991). This method can possibly also be used for dicotyledonous systems. Based on immunological methods it could be demonstrated that SAP's contain no other proteins in major amounts beside phytochrome (Hofmann *et al.*, 1991). In native PAGE phytochrome from SAP's shows a different mobility than conventionally purified phytochrome (Grimm and Rüdiger, 1986). This indicates some possible modifications (Fig. 2).

Beside phosphorylation and methylation, glycosylation was analysed in great detail as a possible post-translational modification. Based on antibody and lectin experiments it is concluded that phytochrome is a glycoprotein of the complex type (Fig. 3). This was confirmed by gas chromatography and mass spectroscopy. By competition assay and microsequencing of glycosylated proteolytic fragments of phytochrome a glycosylation site could be narrowed to a stretch of 40 amino acids nearby the chromophore binding domain. Up to now no differences in the glycosylation pattern between soluble Pr and phytochrome purified as Pfr or Pr from SAP's has been detected.

Therefore the mechanism of SAP formation remains obscure. Nevertheless SAP formation might be a relevant step in signal transduction, because SAP's seem to be the places of phytochrome destruction (Speth *et al.*, 1987; Jabben *et al.*, 1989). Physiological

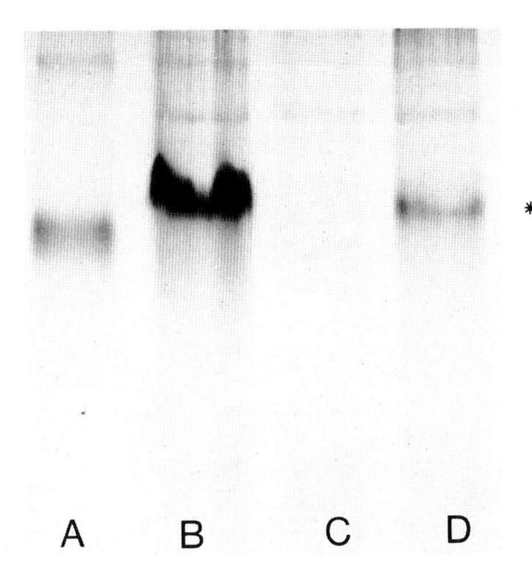

Figure 2. Retardation of pelleted phytochrome. SAP's enriched from red-irradiated seedlings were solubilized in 5% glycerol, 14 mM 2-ME, 2 mM PMSF and supernatants subjected to 10% native gels using the Laemmli-system. Gel was stained with Coomassie. A = conventionally purified phytochrome, C = extract from unirradiated seedlings, B and D = two different concentrations of phytochrome from red-light irradiated seedlings, * = phytochrome band.

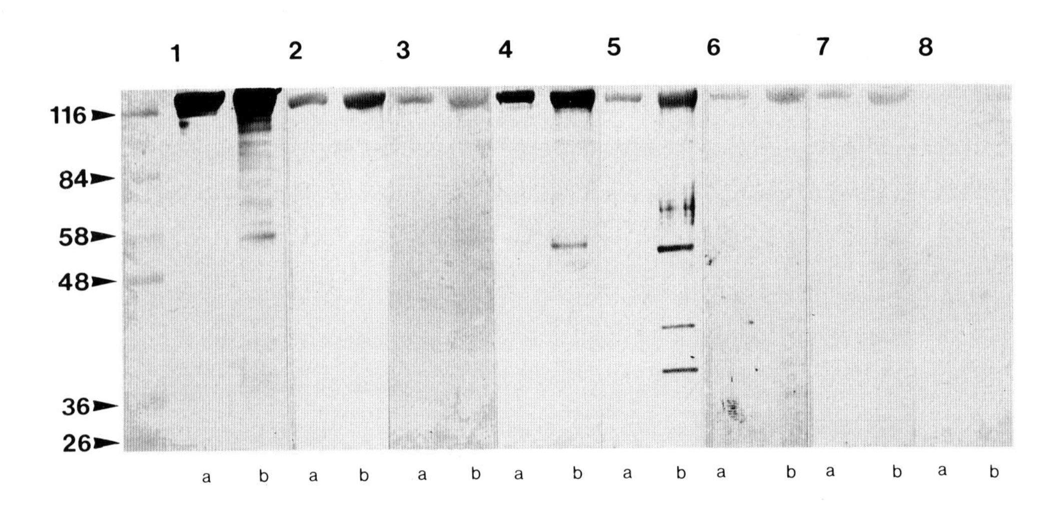

Figure 3. Detection of carbohydrate binding to Avena phytochrome using sugar-specific antisera and biotinylated lectins.

(a) Conventionally purified phytochrome
(b) Phytochrome partially purified from sap's
1: Anti-phytochrome antibody (MAC 198)
2: Anti-patatin antiserum (N-glycan)
3: Anti-xylose antiserum
4: Tetragonolobus purpureas lectin (fuc)
5: Pisum sativum lectin (man)
6: Abrus precatorius lectin (gal)
7: Helix pomatia lectin (galNAc)
8: Wheat germ lectin (gluNAc)

experiments of loss of reversibility after a light pulse have generally been misinterpreted to reflect the kinetics of coupling reaction (see Fukshansky *et al.*, 1983 for discussion). Simple calculations show that this assay is a measurement of the inactivation of Pfr and thus mainly reflects the time constant of phytochrome destruction. Therefore, we must conclude that the active phytochrome is inactivated with a half life similar if not identical with the half life of destruction of the spectroscopically and immunologically measured phytochrome. This implies that either a: the signal transduction starts from Pfr of SAP's b: another destruction mechanism with the same half life occurs at a different place in the cell or c: the active phytochrome interacts with SAP's.

PHOTOREGULATION OF CHS GENE EXPRESSION

In parsley cell cultures we could demonstrate that CHS expression is under control of a UV-B receptor, blue light receptor and phytochrome (Bruns *et al.*, 1986; Ohl *et al.*, 1989). In the cotyledons of mustard seedlings both CHS genes tested so far are under the control of phytochrome and the HIR. Both genes also showed a low expression in the dark starting from 36 h after sowing. In contrast, light-induced increases in CHS are already observed 27 h after sowing. Surprisingly, the accumulation of CHS mRNA levels can be decreased by a far-red light pulse given 6 h after sowing. This light effect is red/far-red reversible indicating that the dark expression of the genes is a phytochrome type II response; i.e. the increase in darkness is due to remaining Pfr from the seed. This also implies that the competence points for phytochrome type I and type II regulation are different. Whereas in cotyledons CHS expression is induced by Pfr, we see in primary leaves a strong blue/UV-light response (Figs. 4 and 5). These data do not exclude the possibility that in cotyledons also blue and UV-light might play a role in CHS expression and that in primary leaves phytochrome might regulate CHS expression. Despite their complication, clearly we can conclude from these results that there is a change in using different photoreceptors for light regulated gene expression during development.

TISSUE AND CELL SPECIFIC EXPRESSION OF LIGHT REGULATED GENES

An antiserum raised against a CHS fusion protein (Kretsch *et al.*, in preparation) allowed us to analyse tissue and cell specific expression of CHS in mustard. CHS can be detected

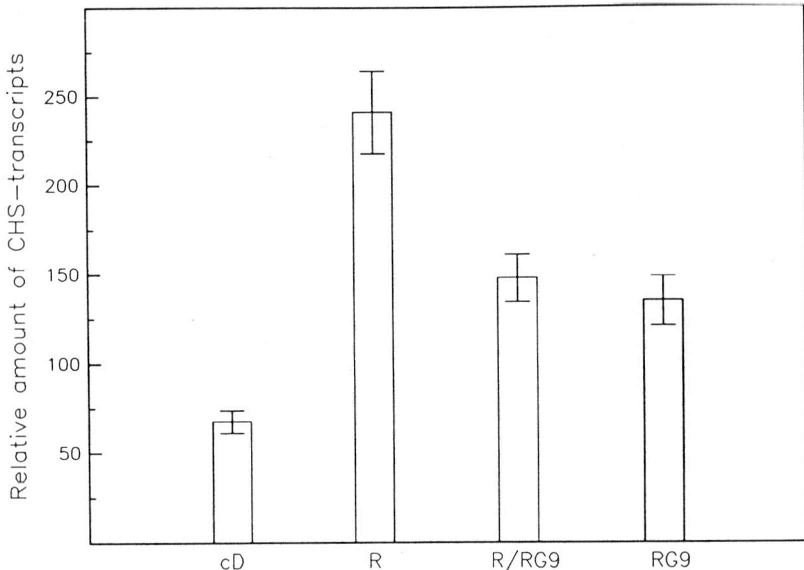

Figure 4. Analysis of the phytochrome control of transcript accumulation of CHS mRNA in cotyledons of mustard seedlings. 42 h dark grown mustard seedlings obtained 5 min red, 5 min RG9 or 5 min red followed by 5 min RG9 light pulse and were harvested 48 h after sowing. cD 48 dark; R 5 min red light 6.7 Wm^{-2}; RG9 5 min R69 light 10 Wm^{-2}.

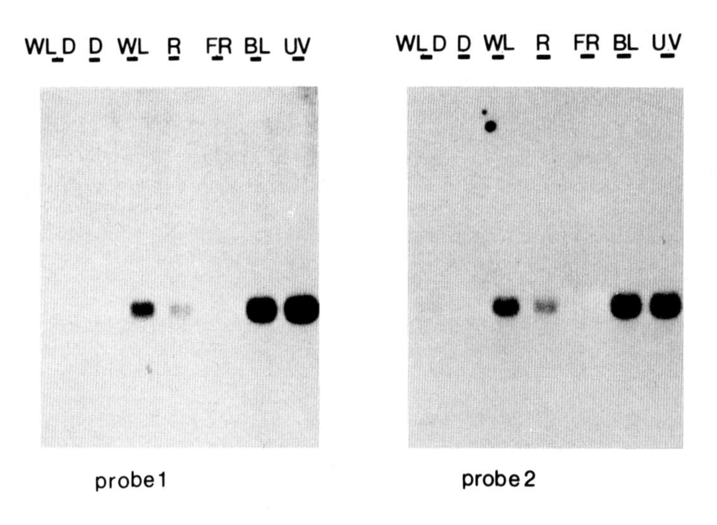

Figure 5. Plants were grown for 23 d in a light/dark cycle (16 hrs white-light/8 hrs darkness: Wl/D) and then transferred to darkness for 3 d (D). Afterwards, plants were treated with different lights for 16 hrs. 15 μg per lane of total RNA was separated on 1.2% agarose-formadehyde gels, transferred to nitrocellulose filters and hybridized with gene-specific probes: probe 1 was from SCHS1 and probe 2 from SCHS2. W1/D: White-light/dark cycle for 23 d (samples taken at the end of light phase); D: Plants kept in darkness for 3 d after W1/D cycle; W1: White-light; R: Red-light; FR: Far-red light; Bl: Blue-light; UV: UV-light.

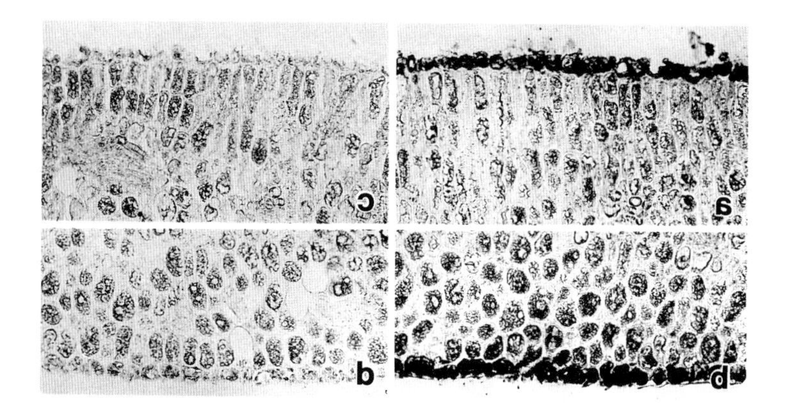

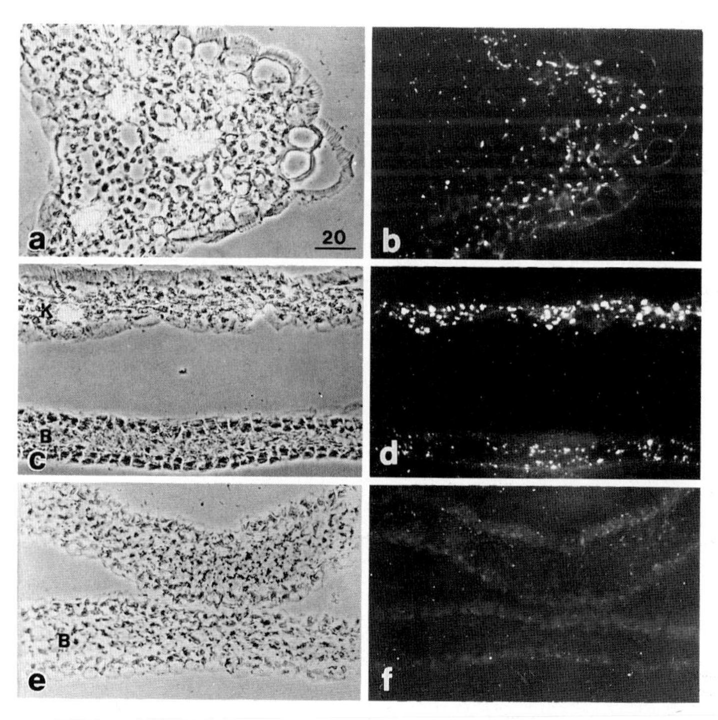

Figure 6. Immunolocalisation of CHS in mustard seedlings. Upper panel a-d, cotyledons of mustard seedlings grown for 54 h in continous far red light. a, b: Primary antibody: Polyclonal antibodies, secondary antibody: coloidal gold coupled goat antirabbit antibody using the silver-enhancement technique a,c,d controls using pre-immune serum. Lower panel a-f, sepals and petals of flowers of 21 days light-grown mustard plants. As secondary antibody a FITC compled goat antirabbit antibody was used a,b sepals; c, d, sepals, e,f controls using pre-immune serum.

by Western blotting in extracts from hypocotyls, cotyledons, epicotyls, primary leaves and flowers. With immunocytological techniques we localized the CHS expression mainly in epidermal and subepidermal layers of hypocotyls and epicotyls, the upper and lower epidermal layer of cotyledons, epidermal and mesophyll cells of primary leaves, sepals and petals (Fig. 6). Using gene specific probes for two of the four CHS genes from mustard we demonstrated that both genes are expressed in both epidermal layers of the cotyledons (Fig. 7).

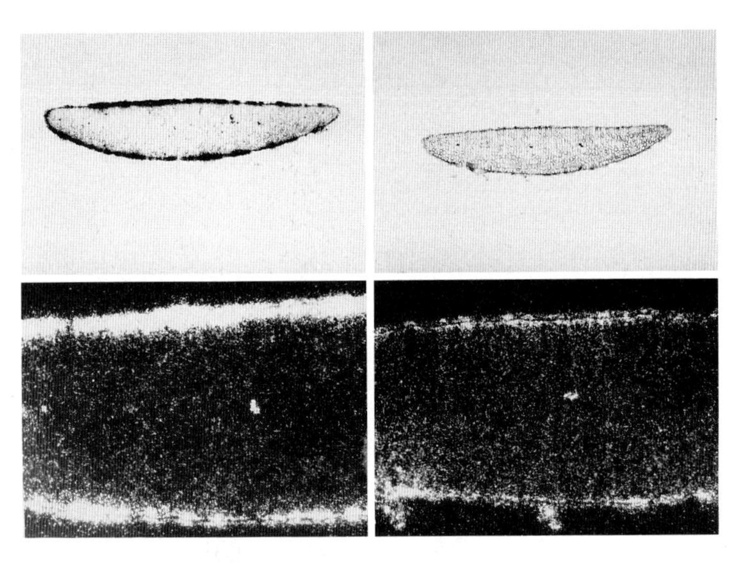

Figure 7. *In situ* hybridization with section of a cotyledon grown in 42 h continous far-red light. Hybridization with SCHS1 (SCHS2) are shown on the left (right) half of the picture (light microscopy); details out of the sections are shown in the lower half of the picture (epifluorescence technique). In case of SCHS1 the sections were made proximal and in case of SCHS2 distal from the central vascular vein.

Expression of CHS in epidermal and subepidermal cells of hypocotyls and in the upper and lower epidermis of cotyledons is under phytochrome control. The expression pattern of both CHS mRNAs in the dark and after continuous light is different in the cotyledons indicating possibly that the response to type I and type II phytochrome shows cell specific differences.

To test whether a mustard CHS gene can also be active in parsley protoplasts, a CHS promotor GUS reporter fusion was constructed and tested in a transient gene expression assay. To our surprise this construct showed only strong responsiveness to

UV-B containing light (and weak response to blue light). Its photoregulation was in this cellular background therefore, similar if not identical to that of the endogenous parsley CHS or a parsley CHS promotor GUS reporter fusion construct (Fig. 8).

These observations demonstrate that the photoregulation of gene expression depends strongly, not only on the photoreceptor present and the promotor of the genes, but also on the developmental state of the cell in which the expression of the gene will be studied.

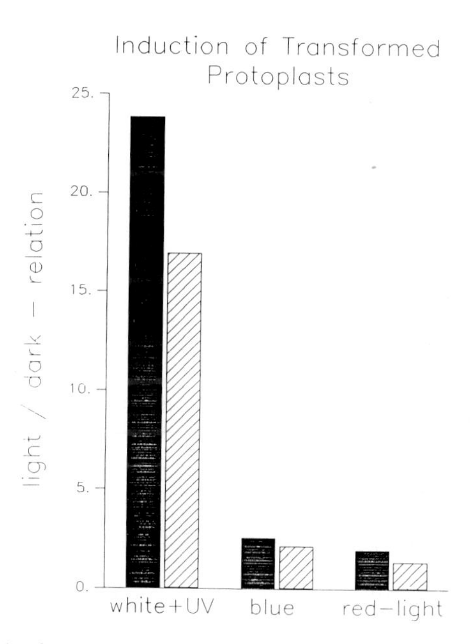

Figure 8. Photocontrol of parsley and mustard CHS promoter GUS fusion constructs tested in a transient gene expression assay in transformed parsley protoplasts. The transformed protoplasts were irradiated for 9 h and GUS was extracted after the end of the irradiation period. The dark levels were 43 and 37 (pmol $MU/mgP_{tot}/min$) for the parsley and mustard CHS promotor, respectively. The fluence rates were 4.2, 3.7 and 5 Wm^{-2} for UV-containing white light, blue and red light, respectively.

Within the mustard seedling we observed that the endogenous CHS in cells of the primary leaf is also mainly under control of UV-B and blue light. We therefore must conclude that photoregulation of a specific gene is quantitatively and qualitatively different in different cellular backgrounds, irrespective whether this is a homologous or heterologous system. These differences are probably not due to differences in cellular phytochrome content, because we know that parsley cells contain relatively large amounts of phytochrome. It should be mentioned that this phytochrome is probably type II phytochrome (Harter and Schäfer, unpublished).

To evaluate photoregulation of cell specific gene expression clearly much more must be known about the cellular contents and distribution of the pigments and more detailed photobiological studies of cell specific gene expression must be carried out.

REFERENCES

Bruns B, Hahlbrock K, Schäfer E (1986) Fluence dependence of the ultraviolet-light-induced accumulation of chalcone synthase mRNA and effects of blue and far-red light in cultured parsley cells. Planta 169:393-398

Coleman RA, Pratt LH, (1974) Subcellular localization of the red-absorbing form of phytochrome by immunocytochemistry. Planta 121:119-131

Cope M, Pratt LH (1989) Subcellular localozation of phytochrome in soybean (*Glycine max* L.). Europ Symp Photomorphogenesis in Plants Freiburg S 5

Grimm R, Rüdiger W (1986) A simple and rapid method for isolation of 124 kDA oat phytochrome. Z Naturforsch 41c:988-992

Hofmann E, Speth V, Schäfer E (1990) Intracellular localisation of phytochrome in oat coleoptiles by electron microscopy. Planta 180:372-377

Hofmann E, Grimm R, Harter K, Speth V, Schäfer E (1991) Partial purification of sequestered particles of phytochrome (SAP's) from oat (*Avena sativa* L.) seedlings. Planta 183: 265-273

Jabben M, Shanklin J, Vierstra RD (1989) Ubiquitin-phytochrome conjugates: Pool dynamics during *in vivo* phytochrome degradation. J Biol Chem 264:4998-5005

Kretsch T, Ehmann B, Ocker B, Speth V, Schäfer E (1990) in press

McCurdy DW, Pratt LH (1986a) Kinetics of intracellular redistribution of phytochrome in *Avena* coleoptiles after its photoconversion to the active, far-red-absorbing form. Planta 167:330-336

McCurdy DW, Pratt LH (1986b) Immunogold electron microscopy of phytochrome in *Avena*: Identification of intracellular sites responsible for phytochrome sequestering and enhanced pelletability. J Cell Biol 103:2541-2550

Ohl S, Hahlbrock K, Schäfer E (1989) A stable blue-light-derived signal modulated ultraviolet-light-induced activation of the chalcone-synthase gene in cultured parsley cells. Planta 177:228-236

Saunders MJ, Cordonnier M-M, Palevitz BA, Pratt LH (1983) Immunofluorescence visualization of phytochrome in *Pisum sativum* L. epicotyls using monoclonal antibodies. Planta 159:545-553

Schäfer E (1981) Phytochrome and day-light. In: Smith, H (ed) Plants and the Daylight Spectrum. Academic Press, London, pp 461-580

Speth V, Otto V, Schäfer E (1986) Intracellular localisation of phytochrome in oat coleoptiles by electron microscopy. Planta 168:299-304

Speth V, Otto V, Schäfer E (1987) Intracellular localisation of phytochrome and ubiquitin in red-light-irradiated oat coleoptiles by electron microscopy Planta 171:332-338

CIRCADIAN CLOCK AND LIGHT REGULATED TRANSCRIPTION OF THE WHEAT CAB-1 GENE IN WHEAT AND IN TRANSGENIC TOBACCO PLANTS

E. Adam, M. Szell, A. Pay, E. Fejes and F. Nagy[1]
Institute of Plant Physiology,
Biological Research Center of the Hungarian Academy of Sciences,
Szeged, P.O.Box 521, H-6701,
Hungary.

INTRODUCTION

Light plays an indispensable role throughout the life span of higher plants. It provides energy for photosynthesis and has profound effects on gene expression. The alterations in the activity of specific genes triggered by fluctuations in ambient light quality and quantity culminate in a variety of developmental responses (Kuhlemeier *et al.*, 1987).

It has also been established that numerous light-induced developmental responses of higher plants are further modulated by endogeneous circadian rhythm(s) (Vince-Prue, 1983). Processes such as flowering or the activity of various enzymes have been found to be under circadian and/or diurnal control and in many cases light has been shown to reset the phase.

The mechanism(s) by which light and endogeneous rhythm(s) regulate gene expression is the subject of considerable interest. To perceive changes in light conditions, higher plants developed a number of photoreceptors including a blue/UV-A absorbing cryptochrome, a UV-B absorbing pigment and the red/far-red absorbing phytochrome. Among these, phytochrome is the most extensively characterized. This photoreceptor exists in two forms: Pr (= phytochrome absorbing red light, max = 660 nm) and Pfr (= phytochrome absorbing far-red light, max = 730 nm) which are interconvertible by the appropriate irradiation. This photoreversibility endows phytochrome with its regulatory function (Tobin and Silverthorne, 1985).

Recent studies demonstrated that photosynthetic genes such as genes encoding the chlorophyll a/b binding protein (Cab) are regulated by light and that their induction by light is mediated by phytochrome (Nagy *et al.*, 1988). In addition, it has also been

[1]Also Friedrich-Miescher Institute, P.O. Box 2543, H-4002 Basel, Switzerland

NATO ASI Series, Vol. H 50
Phytochrome Properties and Biological Action
Edited by B. Thomas and C. B. Johnson

established that the expression of these genes is further modulated by an endogeneous rhythm (Kloppstech, 1985). The dual regulation of Cab gene expression by phytochrome and by the circadian clock is particularly interesting and provides a relatively simple experimental model to study these regulatory mechanisms on the molecular level (Tavladoraki *et al.*, 1989).

To this end we have recently initiated a series of experiments to characterize the expression pattern of wheat Cab genes under various conditions and define regulatory elements for phytochrome and/or circadian clock responsiveness. We have previously reported that steady state mRNA levels of the wheat Cab genes and particularly that of the *Cab-1* gene oscillated with a periodicity of approximately 24 hours in plants grown under a 16L/8D light-dark regime. The transfer of plants to constant light or dark altered the level of the Cab-1 specific transcript but did not eliminate the rhythmic oscillation, indicating that the fluctuation (timing) is controlled by a circadian clock. In addition we found that the light induction of *Cab-1* gene expression was mediated by phytochrome in etiolated and green wheat seedlings. Moreover we demonstrated that the *Cab-1* gene mainatined its fluctuating and phytochrome regulated expression in transgenic tobacco plants (Nagy *et al.*, 1988).

The dual regulation of Cab gene expression raises the following questions:

1. How does phytochrome and the clock interact?
2. What is the relative contribution of phytochrome and the circadian rhythm to gene expression?
3. Does phytochrome regulate the expression pattern of the Cab genes exclusively through or by the circadian clock, or are parallel signal transducing chains involved in fine-tuning Cab gene expression?

To answer at least some of these questions, we first carried out nuclear run-off experiments and demonstrated that the circadian clock as well as phytochrome regulate the expression of the Cab genes on the level of transcription. To define individual *cis*-regulatory sequences (if any) for phytochrome and/or circadian regulation, we constructed a series of 5' deletion mutants and chimaeric genes and analyzed their expression in transgenic tobacco plants. The analysis of transgenic plants allowed us to conclude that the *Cab-1* promoter contains several *cis*-regulatory elements for maximal and regulated expression and that there are individual, well-defined *cis*-regulatory sequences for phytochrome mediated light response and circadian rhythm.

THE CIRCADIAN CLOCK REGULATES THE EXPRESSION OF THE WHEAT CAB GENES AT THE LEVEL OF TRANSCRIPTION

The oscillation and the increase of the steady state mRNA level after illumination could be due to changes of transcription rate or mRNA stability or both. It is generally accepted that *in vitro* nuclear systems from plant cells only elongate transcripts initiated prior to the isolation of nuclei. As a consequence, it is also assumed that the results of these assays reflect the transcription initiation rate at that time point. It has been previously reported that phytochrome regulates Cab gene expression at the level of transcription (Tobin and Silverthorne, 1985). To determine how the circadian clock affects the transcriptional status of the Cab genes during a 24-hour period, we performed *in vitro* transcription assays. Wheat seedlings were grown for 6 days under a 16 hours light and 8 hours dark photoperiodic cycle (16L/8D). On the 7th day half of the seedlings were transferred to constant light and grown for 2 additional days. On the 9th day seedlings were harvested at 4-hour intervals, nuclei were isolated and preparative *in vitro* transcription assays were carried out using identical numbers of nuclei (1.5×10^7).

Based on three independent experiments we found that the transcription initiation rate of the Cab genes is at least 10-15 times higher in the early morning hours (with a peak at around 8 a.m.) than in the late evening hours. In contrast to the Cab genes, SSU specific mRNA levels did not show significant changes during the 24 hour period studied. Transcription initiation rates of the SSU genes were generally 5-fold lower at 8 a.m. than those of the Cab genes and showed at most 2-fold variation among the different time points; however,this variation lacked a characteristic pattern. The transcription initiation rate of the rRNA genes was nearly identical at different time points (similarly to the SSU genes) and showed no significant variation under different light regimes. These data clearly show that the transcription rate of the wheat Cab genes fluctuates in a very specific manner during a 24-hour period. This specific pattern is maintained even under constant illumination. These results strongly suggest that the endogeneous clock, similarly to phytochrome, regulates wheat Cab gene expression, at least partially, at the level of transcription.

IDENTIFICATION OF *CIS*-REGULATORY ELEMENTS REQUIRED FOR MAXIMUM AND REGULATED EXPRESSION OF THE WHEAT CAB-1 GENE IN TRANSGENIC TOBACCO PLANTS

In order to define *cis*-acting elements that mediate the circadian clock-controlled transcription of the wheat *Cab-1* gene, we performed two complementary lines of experiments. First we analyzed in transgenic plants the expression of a series of 5' deletion mutants retaining various lengths of upstream sequences from -1816 to -127. We found that the -244 mutants still maintain maximum expression level. Further removal of upstream sequences from -244 dramatically decreased the expression level of the transgenes. Deletion to -230 resulted in a tenfold decrease; additional deletion from -211 to -190 lowered the expression below detection level in the majority of samples. One plant out of 15, however, still expressed the -190 deletion mutant at a low but detectable level. Moreover, we found that all of the deletion mutants from -1816 down to -211 (plus the one -190 mutant with detectable expression) exhibited circadian clock responsive gene expression regardless of the level of transcription.

Several conclusions can be drawn from these results:

1. Sequences upstream of -244 are not required for maximum transcription level in transgenic tobacco plants.
2. A 54 bp sequence located between -244 and -190 contains at least two regulatory elements for maximal gene expression. The major *cis*-acting element for maximal expression resides between -244 and -230. An auxiliary *cis*-acting element is located between -211 and -190. Circumstantial evidence indicates that a third regulatory element is positioned between -190 and -160. The existence of this putative third element, however, has yet to be directly proven. We have found only one transgenic plant supporting this statement.
3. The 5' boundary of the promoter region that is sufficient to maintain circadian clock responsive *Cab-1* gene expression is at -211.

We also attempted to determine the 3' boundary of the clock-responsive *cis*-acting element. To this end we analyzed the expression of chimaeric genes containing various regions of the *Cab-1* promoter. We found that a chimaeric gene containing a 1.8 kb region of the *Cab-1* gene (-1816 to +31) confers clock responsive expression to the bacterial GUS gene. Removal of upstream sequences from -1816 to -357 did not change the level or pattern of expression. An internal deletion of the *Cab-1* promoter sequences between -357 and -127, however, completely eliminated the expression of this chimaeric gene. We report here that sequences located between -357 and -90 have a specific function in the clock-regulated expression of the *Cab-1* gene. We found that this 268 bp promoter fragment endows a transcription unit containing a truncated 35S promoter and the bacterial GUS gene with fluctuating transcription. These data, together with those obtained from the analysis of the 5' deletion mutants, indicate that:

1. The *cis*-acting element that mediated circadian clock regulated expression is located within an enhancer-like region.
2. The 3' boundary of this element can be placed at -90 (Fejes *et al.*, 1990).

DOES THE CAB-1 PROMOTER CONTAIN INDEPENDENT *CIS*-REGULATORY ELEMENTS FOR LIGHT-INDUCED AND CIRCADIAN CLOCK CONTROLLED EXPRESSION?

As described above, the transcription of the wheat *Cab-1* gene is regulated by phytochrome and by an endogeneous circadian rhythm. We prepared several internal deletions in the promoter region downstream of -357 and analyzed their expression in transgenic tobacco plants. Up to now we have completed the analysis of the following three constructs:

1. (-357 - -179) (-127 Cab/Cab/Cab)
2. (-306 - -179) (-127 Cab/Cab/Cab)
3. (-267 - -179) (-127 Cab/Cab/Cab)

We found that the expression of these internal deletion mutants was clearly regulated by phytochrome in etiolated transgenic seedlings. Transcript levels of these mutants, as well as that of the -127 5' deletion mutant, were invariably below detection in dark-grown transgenic seedlings. We showed that the mRNA level of these three transgenes (but not that of the -127 5' deletion mutant) increased at least 5-fold upon a short red illumination. Moreover, the red light dependent induction was clearly reversed by far-red light.

We observed that, consistant with the low-level expression in etiolated seedlings, mRNA levels were also below detection in dark-adapted (3 days in constant darkness) mature transgenic plants. Interestingly, however, the mRNA level of the internal deletion mutants did not show the characteristic fluctuating pattern in transgenic plants grown under a 16L/8D cycle or in constant light. The level of *Cab-1* specific mRNA was invariably high during a 24-hour period, while the endogeneous tobacco Cab genes exhibited the usual oscillatory expression pattern. In contrast to the internal deletion mutants, the expression of the -127 5' deletion mutant was below detection level in all transgenic plants. It should also be noted that the expression level of the internal deletion mutants was 5-fold lower than that of the -267 5' deletion mutant.

We drew three conclusions from these findings:

1. Promoter sequences of the *Cab-1* gene located between -267 and -179 are responsible for the high-level, light inducible expression of these transgenes.
2. *Cis*-regulatory elements positioned between -179 and -127 are needed for maximum level expression of the *Cab-1* gene.
3. This latter *cis*-acting element is absolutely necessary for circadian rhythm responsive expression in transgenic plants.

It is tempting to conclude from these data that the wheat *Cab-1* promoter contains separate *cis*-regulatory elements for light-induced and circadian clock responsive gene expression. There is, however, an alternative interpretation for these data. One could assume the presence of multiple regulatory elements for the optimal (light and rhythm responsive) expression of the *Cab-1* gene. Deletion of one or more regulatory element(s) would still result in a detectable difference between expression in light and in dark (in spite of the relatively low level expression of these mutants). The more subtle differences between time points during a 24-hour period are more difficult to observe and may have escaped detection at this expression level. In an effort to obtain unambiguous data we are presently undertaking a very detailed analysis of the expression of these mutants.

Based on our preliminary data, however, we suggest that:

1. Two independent signal transduction chains are involved in controlling the expression of the *Cab-1* gene.
2. Alternatively light initiated signal transduction pathway branches before reaching the "circadian clock".

We are presently mutagenizing the potentially critical region of the *Cab-1* promoter and hope that our data will enable us to distinguish between these alternative models.

THE CIRCADIAN OSCILLATOR IN WHEAT IS CONTROLLED BY PHYTOCHROME AND EXHIBITS VERY LOW FLUENCE RESPONSE

Oscillation of the wheat Cab mRNA level is a well documented phenomenon in plants grown under various light/dark regimes. In addition, Piechulla (1989) has shown that the oscillator can be reset by shortening the light period. These data indicated that the light/dark transition plays a critical role in setting the clock but did not point out the photoreceptor involved. On the other hand, we previously showed that phytochrome controls the transcript level of Cab genes in green wheat seedlings, however we were not able to reset the circadian clock by short red or far-red light pulses (Nagy *et al.*, 1988). The oscillation of Cab mRNA level in etiolated wheat seedlings has not yet been studied. To understand the circadian clock regulated expression of the Cab gene at this developmental stage, we carried out the following experiments. Wheat seedlings were germinated and grown in constant dark for 7 days and on the 7th day various light treatments were applied. After light treatment the seedlings were again kept in constant dark and the expression of the Cab genes was monitored throughout a 72-hour period.

Our results clearly showed that:

1. After a short (5 min.) white or red light pulse, the mRNA level of the Cab genes and particularly that of the *Cab-1* gene fluctuates throughout the 72-hour period studied;
2. A short (5 min.) far-red light pulse is sufficient to initiate the fluctuating expression of the *Cab-1* gene.
3. A far-red pulse applied immediately after the red pulse decreases the expression level but does not prevent fluctuation.
4. Red or far-red pulses applied after various lengths of time following the initial light treatment can reset the circadian rhythm.

To determine whether the circadian oscillator shows a similar type of regulation in dicots, we are currently performing experiments on transgenic tobacco seedlings. Irrespective of the results of the ongoing experiments, our present data clearly indicate that

1. The circadian oscillator is controlled by phytochrome.
2. It exhibits a "very low fluence response" type regulation at least in monocot plants.

CONCLUSIONS

We have shown that a short region of the *Cab-1* promoter (between -357 and -90) contains multiple *cis*-regulatory elements that arc required for maximum level, regulated expression in transgenic tobacco. In addition, our data also indicate the presence of separate *cis*-regulatory elements for light induced and circadian rhythm responsive gene expression. Based on the analysis of several other Cab promoters from various plant species, the complex structure of the wheat *Cab-1* promoter is not unexpected. Multiple regulatory elements have been found in the promoter region of Cab genes in tobacco (Castresana *et al.*, 1988), *Petunia* (Gidoni *et al.*, 1988) and *Arabidopsis* (Ha and An, 1989). The contribution of these various *cis*-acting elements to regulated gene expression is not fully understood. *In vitro* footprinting experiments combined with the analysis of gene expression in transgenic plants have already yielded valuable information about the fine structure of various plant promoters (Lam *et al.*, 1989; Green *et al.*, 1987; Giuliano *et al.*, 1988). We hope that a similar approach will allow us:

1. To determine the contribution of the identified *cis*-acting elements to the regulated transcription of the wheat *Cab-1* gene.
2. To identify the exact position of the *cis*-acting element that mediates circadian clock responsive gene expression of this wheat gene.

The characterization of the clock responsive element could be the first step towards elucidating the molecular mechanism by which an endogeneous oscillator controls gene expression in higher plants.

REFERENCES

Castresana C, Garcia-Luque I, Alonso E, Malik VS, Cashmore AR (1988) Both positive and negative regulatory elements mediate expression of a photoregulated Cab gene from *Nicotiana plumbaginifolia*. EMBO J 7:1929

Fejes E, Pay A, Kanevski I, Szell M, Kay SA, Nagy F (1990) submitted

Gidoni D, Brosio P, Bond-Nutter D, Bedbrook J, Dunsmuir P (1988) Novel *cis*-acting element in *Petunia* Cab gene promoters. Mol Gen Genet 215:337

Giuliano G, Piechersky E, Malik E, Timko VS, Scolnik PA, Cashmore AR (1988) An evolutionarily conserved protein binding sequence upstream of a plant light regulated gene. Proc Natl Acad Sci USA 85:7089

Green PJ, Kay SA, Chua N-H (1987) Sequence specific interactions of a pea nuclear factor with light-responsive elements upstream of a plant light regulated gene. EMBO J 6:2543

Ha SB, An G (1989) Identification of upstream regulatory elements involved in the developmental expression of the *Arabidopsis* Cab gene. Proc Natl Acad Sci USA 85:8017

Kloppstech K (1985) Diurnal and circadian rhythmicity in the expression of light-induced plant nuclear messenger RNAs. Planta 165:502

Kuhlemeier C, Green PJ, Chua N-H (1987) Regulation of gene expression in higher plants. Ann Rev Plant Physiol 38:221

Lam E, Benfey P, Gilmartin P, Fang RX, Chua N-H (1989) Site-specific mutations alter *in vitro* binding and change promoter expression pattern in transgenic plants. Proc Natl Acad Sci USA 86:7890

Nagy F, Kay SA, Chua N-H (1988) A circadian clock regulates transcription of the wheat *Cab-1* gene. Genes Dev 2:376

Nagy F, Kay SA, Chua N-H (1988) Gene regulation by phytochrome. Trends in Genetics 4:37

Piechulla B (1989) Changes of the diurnal and circadian (endogeneous) mRNA oscillations of the chlorophyll a/b binding protein in tomato leaves during altered day/night (light/dark) regimes. Plant Mol Biol 12:317

Tavladoraki P, Kloppstech K, Argyroudi-Akoyunoglou J (1989) Circadian rhythm in the expression of the mRNA coding for the apoprotein of the light-harvesting complex of photosystem II. Plant Physiol 90:665

Tobin EM, Silverthorne J (1985) Light regulation of gene expression in higher plants. Ann Rev Plant Physiol 36:569

Vince-Prue D (1983) Photomorphogenesis and flowering. In: Shropshire W, Mohr H (eds) Encyclopedia of plant physiology, Vol 16B, p 458

UNUSUAL FEATURES OF THE LIGHT RESPONSE SYSTEM REGULATING FERREDOXIN GENE EXPRESSION

W.F. Thompson, R.C. Elliott[1], L.F. Dickey, M. Gallo, T.J. Pedersen and D.A. Sowinski
Departments of Botany and Genetics,
North Carolina State University,
Raleigh, NC 27695,
USA.

INTRODUCTION

The ferredoxin gene system. Recently we identified a cDNA clone encoding ferredoxin I and showed that the level of the corresponding transcript in pea seedlings is increased by white light (Sagar *et al.*, 1988a, b) and by red light acting through the phytochrome system (Dobres *et al.*, 1987; Kaufman *et al.*, 1985, 1986). The light responses of this gene - which we have designated *Fed-1* - differ strikingly from those of RbcS and Cab which have so far served as the main models for light responsive genes, providing us with an opportunity to significantly broaden our understanding of light regulatory mechanisms.

Ferredoxin is also of considerable inherent interest. Ferredoxins play a pivotal role in photosynthesis, both in noncyclic electron transport and in cyclic electron flow via cytochrome b, as well as in redox systems regulating enzyme activity such as the thioredoxin system (Anderson, 1986; Buchanan, 1980; Crawford *et al.*, 1989). In addition, ferredoxins are important redox carriers in nitrogen metabolism, sulfur metabolism, and lipid biosynthesis. For a review of the many biochemical roles of ferredoxin see Cammack (Cammack *et al.*, 1985). Understanding the structure and regulation of a gene with so many important functions will certainly be of value to those attempting to improve crop yields by genetic engineering.

We have now isolated and sequenced a gene encoding ferredoxin I from *Pisum sativum*. In contrast to other well-studied light-responsive genes, *Fed-1* appears to be a single copy gene in pea (Dobres, *et al.*, 1987; Elliott *et al.*, 1989b) as well as in *Silene* and

[1]Present address: John Innes Institute, Colney Lane, Norwich NR4 7UH, U.K.

NATO ASI Series, Vol. H 50
Phytochrome Properties and Biological Action
Edited by B. Thomas and C. B. Johnson

Arabidopsis (Smeekens, *et al.*, 1985; Somers *et al.*, 1990; Vorst *et al.*, 1990). Although other forms of ferredoxin are known at the protein level (Dutton *et al.*, 1980; Kimata and Hase, 1989; Sakihama and Shin, 1987; Takahashi *et al.*, 1983; Wada *et al.*, 1989) the DNA sequences encoding them are evidently sufficiently diverged that they do not cross hybridize at normal stringency. The absence of cross hybridizing gene copies will greatly facilitate many experiments, as will the fact that *Fed-1* is a small gene (about 800 bp) with no introns (Elliott *et al.*, 1989b).

Light responses of Fed-1 mRNA. The present project derives from our earlier work, reviewed in (Thompson, 1988; Thompson *et al.*, 1985), in which we studied thirteen cDNA clones representing different light-responsive mRNAs. The responses of *Fed-1* were particularly striking (Kaufman *et al.*, 1986). In time course experiments, *Fed-1* transcripts increased rapidly, with virtually all of the response occurring within the first hour after induction by a red light pulse. Most other mRNAs accumulated much more slowly.

A particularly striking feature of the *Fed-1* response was seen in photoreversal ('escape') experiments, in which the increase in *Fed-1* mRNA induced by a red light pulse was shown to be reversible by far red light for at least 7 hours - that is, for at least 6 hours after the original accumulation is complete. Again, *Fed-1* is unusual in this respect, with most other genes showing substantial escape from phytochrome control by 7 hours. These results indicate that *Fed-1* mRNA is degraded following the far red light treatment. The data do not establish whether RNA synthesis or degradation is being affected by light, although the difference between the behavior of *Fed-1* and other mRNAs known to be regulated at transcription may be suggestive in this regard. An attractive hypothesis is that light affects the abundance of *Fed-1* transcripts mostly or entirely at a post-transcriptional level, while transcriptional regulation is more important for genes such as RbcS and Cab.

EXPRESSION OF CHIMAERIC GENES

To further analyze the expression of *Fed-1*, we turned to transgenic plants in an effort to localize the *cis*-acting elements involved in its light responses and the organ specificity of its expression. We were particularly interested in the extent to which these elements might or might not resemble those which have been characterized from other light regulated gene systems such as the RbcS and Cab gene families.

For transfer, the *Fed-1* gene and various chimaeric constructs were incorporated into the binary vector pBIN19 (Bevan, 1984) which contains a plant-selectable marker conferring kanamycin resistance (neomycin phosphotransferase; nptII) between the left and right T DNA border sequences. These recombinant plasmids were conjugated from *E. coli* into the *Agrobacterium* strain LBA 4404 (Hoekema *et al.*, 1983), which contains a plasmid encoding the trans acting functions needed for transforming plants, but no T-DNA sequences. The resulting strain was used to inoculate tobacco leaf discs, and kanamycin-resistant shoots were obtained as described (Horsch *et al.*, 1985).

Our initial results with this system showed that the pea *Fed-1* gene is expressed and responds normally to light signals in transgenic tobacco plants. Nuclease mapping and primer extension studies showed that the transcript begins and ends at the same points as it does in pea, while Western blot analysis showed that transgenic *Fed-1* mRNA is

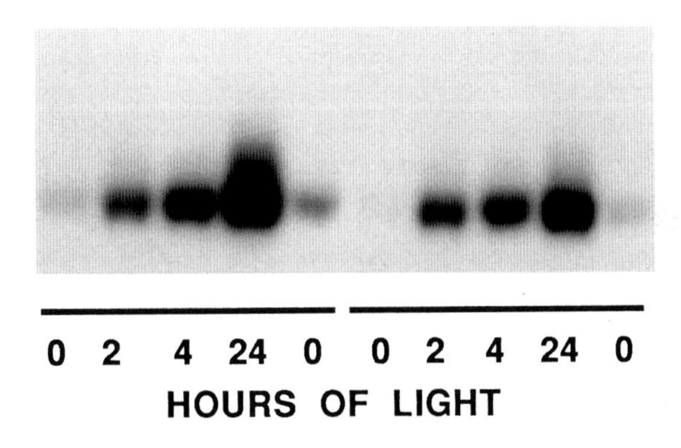

Figure 1. Northern blot showing the time course of accumulation of *Fed-1* mRNA in etiolated transgenic tobacco seedlings. Seedlings containing an intact pea *Fed-1* gene were grown for 6 days in the dark and either harvested immediately or exposed to white light in a growth chamber for the indicated number of hours. Total RNA was extracted from whole seedlings and analyzed after denaturation with glyoxal and electrophoresis on an agarose gel. Results are shown for two different primary transformants. The leftmost lane in each case is RNA from zero time controls, while the right hand lane contains RNA from plants kept for an additional 24 hours in the dark.

translated to produce a protein which is correctly processed and taken up into the tobacco chloroplasts (Elliott *et al.*, 1989a). We also showed that *Fed-1* mRNA levels in mature leaves declined when wild type or transgenic plants were transferred to darkness, and increased again within a few hours when the same plants were transferred back to white light (Elliott 1989a, and unpublished data).

Figure 1 shows that accumulation of *Fed-1* mRNA is also light-dependent in etiolated transgenic tobacco seedlings. As expected from the rapid light response of this gene in pea seedlings (Kaufman *et al.*, 1986), substantial accumulation occurs within the first two hours. Overall, the time course for accumulation in tobacco seedlings compares quite well with that in peas, or with that for the endogenous tobacco transcript (data not shown).

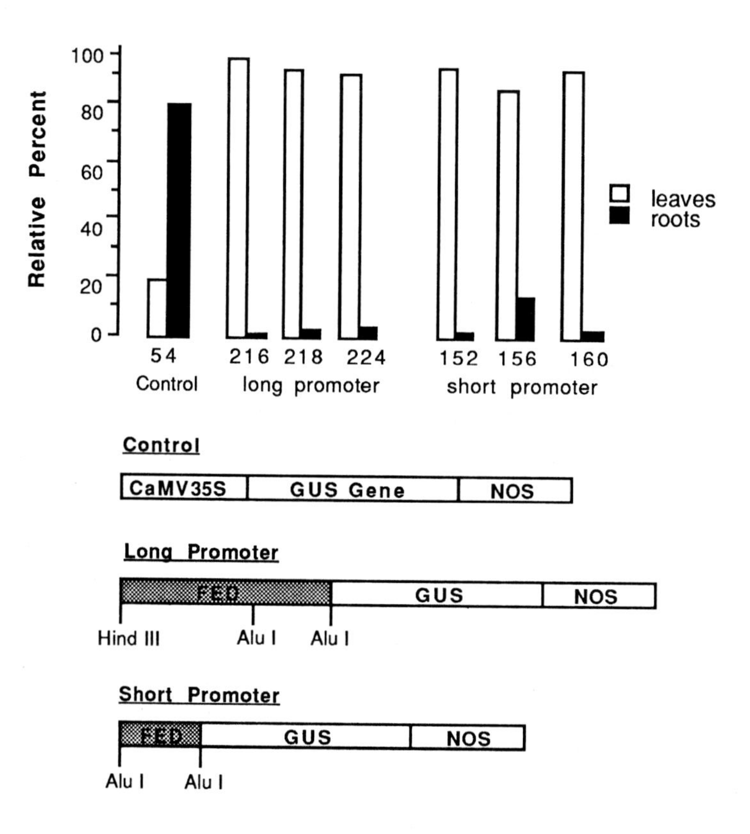

Figure 2. Organ specific expression of GUS activity in plants containing chimaeric Fed-GUS genes. Fluorometric assays were used to measure enzyme activity in extracts from leaves and roots of transgenic plants containing GUS constructs driven by *Fed-1* or CaMV 35S promoters. The constructs used are illustrated below.

ORGAN SPECIFICITY

In addition to exhibiting normal light responses, the pea *Fed-1* gene shows normal organ specificity in transgenic tobacco, with transcripts abundant in leaves but undetectable in roots. This pattern of expression is typical of nuclear genes encoding chloroplast proteins (Benfey and Chua, 1989; Fluhr *et al.*, 1986; Kuhlemeier *et al.*, 1987; Simpson and Herrera-Estrella, 1990), and is generally determined by transcriptional control mechanisms mediated by *cis*-acting elements in the promoter. Figure 2 shows that *Fed-1* conforms to this pattern, since its sequences confer leaf-specific expression on a GUS reporter gene. In a separate series of experiments, we have also shown that high levels of *Fed-1* mRNA accumulate in roots when the gene is placed under control of the constitutive CaMV 35S promoter, so the absence of *Fed-1* mRNA from the roots of plants containing the intact gene must reflect a lack of synthesis rather than instability of the message. We conclude that leaf-specific expression of *Fed-1* is determined primarily by 5′ *cis*-acting elements regulating transcriptional initiation.

DNA BINDING PROTEINS

Comparison of the sequences 5′ to *Fed-1* with those of putative regulatory elements from RbcS and Cab genes reveals a number of similarities, some of which may reflect the presence of common regulatory elements involved in determining leaf-specific expression (Elliott *et al.*, 1989b). To begin an investigation of regulatory elements in the *Fed-1* promoter, we carried out *in vitro* gel-shift assays using probes from *Fed-1* and nuclear protein extracts prepared by techniques similar to those of Green *et al.* (1987). These experiments revealed two types of complexes, which we have called BC (for 'bound complex') 1 and 2. Both complexes are sensitive to detergent and protease treatments, but BC-1 is unusual in that it is heat stable. Normal BC-1, but not BC-2, activity is retained after heating nuclear extracts for 10min at 90°C prior to the binding assays.
Deletion experiments (Elliott *et al.*, 1989b) indicated that an AT-rich region approximately 250 to 350 bp upstream from the start of transcription was required for formation of BC-1. Similar AT-rich regions have been noted in the 5′ flanking regions of a number of other plant genes (Bustos *et al.*, 1989; Datta and Cashmore, 1989; Guiltinan *et al.*, 1989; Holdsworth and Laties, 1989; Jacobsen *et al.*, 1990; Jensen *et al.*, 1988; Jofuku *et*

al., 1987; Marcotte *et al*., 1989); in some of these cases regions containing such sequences have been shown to be important for gene expression. In yeast, somewhat longer AT tracts are thought to increase the level of expression from constitutive promoters (Russell *et al*., 1983; Struhl, 1985), while in animal systems tracts of 6 or more AT residues have been shown to bind HMG-I, a member of a family of low molecular weight nonhistone chromosomal proteins known as high mobility group proteins (Solomon *et al*., 1986).

HMG proteins have been extensively studied in mammalian systems, where they were first discovered. In addition to HMG-I, there are four main classes, HMGs 1, 2, 14, and 17. All are relatively small (10-25kDa) proteins which can be extracted in 350 mM NaCl and are soluble in 2% TCA because of their high content of both positive and negatively charged residues. These charged residues are arranged in blocks to produce acidic and basic domains similar to those of many transcription factors. Although thought to be associated with nucleosomes, HMGs are normally present at much lower levels than would be expected for structural proteins uniformly distributed in chromatin. In mammalian systems, it is thought that only about 10% of the nucleosomes are associated with HMGs at any given time, and that these nucleosomes are derived largely from regions transcriptionally active, or potentially active, chromatin (reviewed by Goodwin and Bustin, 1988; Spiker, 1988a; Spiker, 1988b)

Plant nuclei do not contain proteins exactly like animal HMGs, but they do contain a set of low molecular weight proteins with many of the same properties. HMG-like proteins from wheat have been extensively studied by Dr. Steven Spiker and his colleagues at North Carolina State University. Four major species have been designated HMG a, b, c, and d (Spiker, 1984). Wheat HMGs are released by light DNase I treatments which digest transcriptionally active chromatin (Spiker *et al*., 1983), and thus may be preferentially associated with such regions in plant nuclei.

Using purified wheat HMG proteins obtained from Dr. Spiker, we were able to show that both HMGa and HMGb (but not HMGs c and d) could bind to *Fed-1* promoter fragments containing the AT-rich region, producing complexes with the a mobility similar to that of the BC-1 complex we observed with nuclear extracts. Further experiments with a partially purified preparation of HMGs from peas produced complexes with mobilities identical to BC-1. Since the HMGs also showed the same resistance to heat denaturation that we observed for BC-1 we believe that BC-1 probably involves HMG proteins.

```
                          GAGCTCTTTGTGACATATGTCAAAGAGGGAGTTCAACGTAGCTGTTCTATA
-620  ACTCTTGATGATGTAATCTGATTAAGATAAAGATACCTTAAGCCCCCATAAGTCAGATAATTCAAAATATCTAGTTAAA
-541  GTGTCAATCAATCATTGAATCAAGCAGATCATCTTGAAGTTTGAAAGCTCACATAGTCATGTGCTAAATATTTTAGAGT
-462  GACCATCTTAAATTGTAAAAGGTAAAGAGTATGTCTTGAAGTACTAAGACAACCACACGCACTTAAAATATTTTCAAAC
-383  CTTCTCTTCCTTTATTAAAACTTATTAAAAACCTATCCAGACATTTATTTATGTAATTGATTAAAAAGTATTTCTAATC
-304  AATATTTTAATTAAGTTAATTTAATAAAAATGTAATATCTTATATCAATATTATTGTATTTTTTTATCTATAAAAAAAC
-225  CTTAAACACAGAATGAGTATTGTATATGAAGACTTGAGTAATTAGTTACCACAGTGCACAATACTTCATTTTTAAATAC
-146  TATAAGGTGAAGATGGTGCCAACAAGGATTCACCAACAATATGTGGTGATAAAAGAGATAGATAACCTGAGGCATAAAA
-67   CAATGCCACCTGGCAGATAGGGTTGCATGCAGTTCATATAGCAGCTTGTTCCAACCACACTATTTCC           +1
```

Figure 3. Sequence of the *Fed-1* 5′ flanking region. Potential HMG binding sites (uninterrupted blocks of 6 or more AT base pairs; see Solomon *et al*., 1986) are shaded. Nucleotide positions are numbered relative to the transcription start site (Elliott *et al*., 1989b).

Repeated attempts to footprint the HMG binding site failed to demonstrate sharply defined binding sites. Instead, we believe that HMGs bind to multiple sites within the AT-rich region (Fig. 3) in such a way that no one site is fully occupied on all the molecules. This suggestion is based in part on the observation that plant HMGs produce BC-1 like complexes with a poly dA•poly dT probe and the fact that poly(dA)•poly(dT) and poly(dAdT)•poly(dAdT) are effective competitors for HMG binding to the *Fed-1* probe (Pedersen and Thompson, unpublished). This conclusion is at variance with a recent report by Grasser et al. [Grasser, 1990 p406] that both poly(dA)•(dT) and poly(dG)•(dC) compete for HMG binding. Our binding assays are carried out in the presence of high levels (500 μg/ml) of poly(dI-dC) to suppress non-specific interactions between DNA and positively charged regions of HMG proteins. However, we can find no reference in the Grasser et al. paper to the use of a nonspecific competitor DNA, such as poly(dI-dC). Thus, we argue that the competition they observe with poly(dG)•(dC) is simply a reduction in nonspecific binding. Indeed, their data show that poly(dG)•(dC) affects only the most slowly migrating (presumably nonspecific) complexes; complexes resembling our BC-1 are still present at the highest levels of poly(dG)•(dC), but are eliminated by similar concentrations of poly(dA)•(dT). We have confirmed this interpretation in our system by comparing poly(dI-dC) and poly(dG)•(2C) competitors, showing that they behave similarly and that neither is able to compete for specific binding to AT-rich probes. Grasser et al. also showed footprinting data to suggest that maize HMGs bind to the CAAT and TATA boxes of a zein gene. We have not seen such footprints under our conditions, and we suggest that other factors in the partially purified HMG preparations used by Grasser et al. may have been responsible for the footprints they observed.

It remains to be determined whether or not HMGs also participate in the formation of BC-2. Deletion experiments show that BC-2 requires DNA sequences other than (or in addition to) those required for BC-1 formation (Elliott *et al.*, 1989b), although we have not yet defined this requirement very precisely. In addition, BC-2 formation requires a heat labile factor, since we do not see BC-2 in reactions with heat-denatured extract. We are currently exploring the possibility that BC-2 is formed by interactions between HMGs and other DNA binding factors.

It also remains to be determined whether BC-1 or BC-2 play a role in *Fed-1* transcription *in vivo*. Experiments to test the effect of mutations in these regions of the promoter are in progress. It is possible that such complexes could be important for transcriptional controls such as that leading to organ-specific expression of *Fed-1*, but data from mammalian systems are more consistent with the hypothesis that HMG complexes help to maintain a high rate of constitutive transcription and/or define a chromatin structure open to interaction with more specific transcription factors.

LOCATION OF LIGHT RESPONSE ELEMENTS

To determine which portion of the *Fed-1* gene confers light responsiveness, we have prepared five constructs which we have introduced into tobacco plants by *Agrobacterium*--mediated transformation. Different portions of the *Fed-1* gene were used to replace the corresponding regions of a control gene whose expression should not be affected by light. We then compared the light responses of different chimaeric gene constructs using a simple light/dark assay. The results (Elliott *et al.*, 1989a) are summarized in Figure 4. Constructs containing 5′ flanking sequences from *Fed-1* attached to a GUS reporter and NOS terminator showed no light responses, in dramatic contrast to previous results with other light-regulated genes. Both the 2kb and 0.45kb promoter fragments from the *Fed-1*

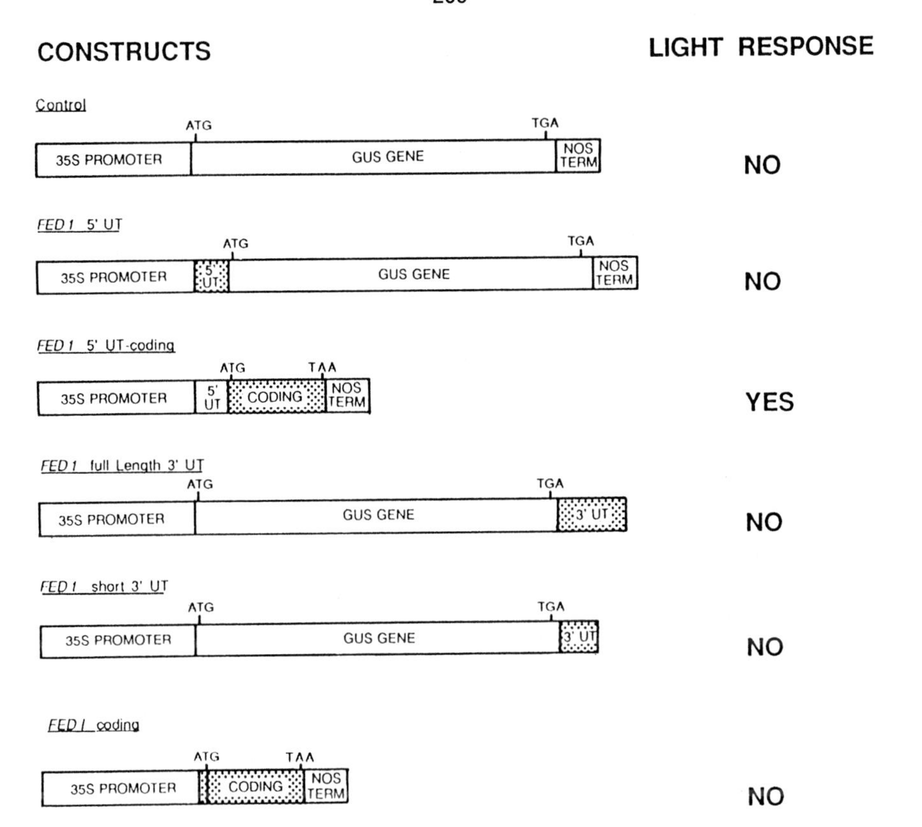

Figure 4. 'First generation' constructs introduced into tobacco plants to examine light responsiveness of different segments of the *Fed-1* gene. Each of the 6 constructs illustrated were incorporated into tobacco plants and *Fed-1* mRNA levels measured in clonally duplicated transgenic plants grown in the light or dark-adapted for 3 days (Elliott *et al.*, 1989a). Each determination is based on results from 5 to 16 independent transformants.

gene were able to support considerable transcription, as evidenced from the accumulation of GUS mRNA in these plants, but no consistent light responses were observed with either of these constructs. An additional series of experiments showed that 3′ flanking sequences from *Fed-1* also fail to confer light responsiveness.

However, a chimaeric gene driven by the CaMV 35S promoter, and containing only transcribed sequences from *Fed-1*, exhibits light effects on mRNA levels identical to those observed with the intact *Fed-1* gene in tobacco, or with the endogenous *Fed-1* gene in

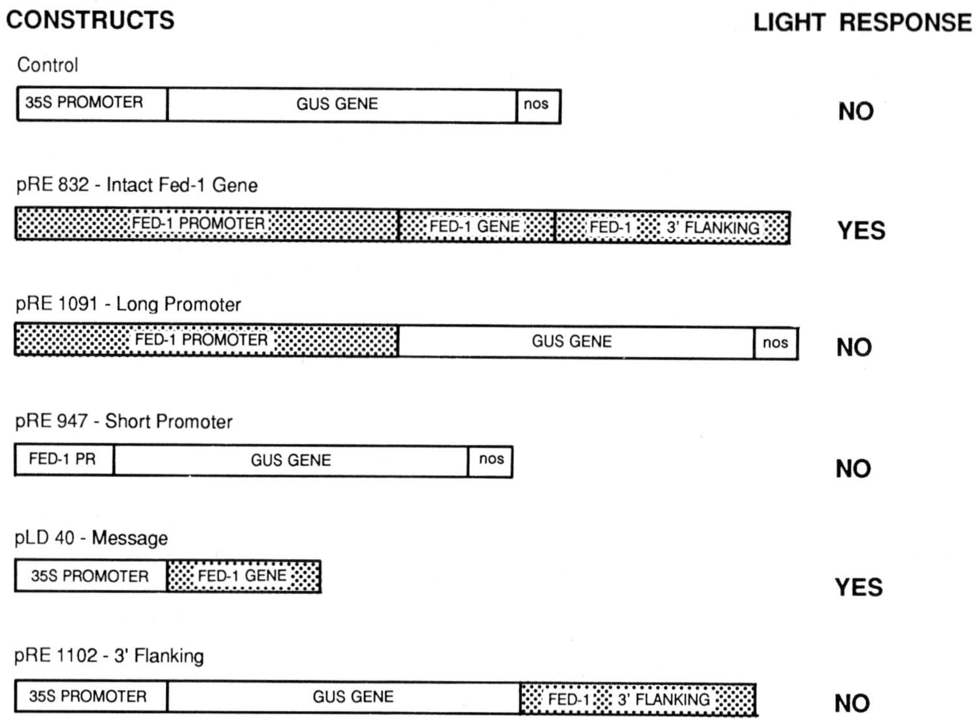

CONSTRUCTS	LIGHT RESPONSE
Control 35S PROMOTER | GUS GENE | nos	NO
pRE 832 - Intact Fed-1 Gene FED-1 PROMOTER | FED-1 GENE | FED-1 3' FLANKING	YES
pRE 1091 - Long Promoter FED-1 PROMOTER | GUS GENE | nos	NO
pRE 947 - Short Promoter FED-1 PR | GUS GENE | nos	NO
pLD 40 - Message 35S PROMOTER | FED-1 GENE	YES
pRE 1102 - 3' Flanking 35S PROMOTER | GUS GENE | FED-1 3' FLANKING	NO

Figure 5. Summary of additional constructs. Each construct listed has been tested in at least 8 independent transgenic plants using a protocol similar to that described in Figure 4. However, instead of simply comparing dark adapted and light grown plants, we now compare dark adapted plants to plants which have been dark adapted and then returned to the light for 6 hours.

light-grown pea plants. Since we observe this response in the absence of either 5′ or 3′ flanking sequences, it cannot be attributed to the *Fed-1* promoter or to other transcriptional regulatory sequences outside the transcribed portion of the gene. The *cis*-acting elements responsible for light effects on *Fed-1* mRNA must therefore be located within the transcribed portion of the gene. The promoter region, although transcriptionally active, is not light responsive. This arrangement is unprecedented in higher plant systems, and provides strong support for our earlier contention that light responses of different genes can differ in important ways.

We are trying to further localize the light response elements within the transcription unit by testing a variety of additional constructs. Unfortunately, in spite of a number of attempts, we have not yet been successful in achieving good light responses in a transient assay system. Thus each new construct must still be tested in regenerated transgenic plants. Figure 5 summarizes the present status of this work. The 3′ untranslated portion of the *Fed-1* transcription unit does not appear to confer, or to be essential for, light responsiveness. Neither the 5′ untranslated region alone or the coding region alone confer light responsiveness. However, a construct containing both the 5′ untranslated and the coding sequences is light responsive. Thus our current hypothesis is that essential *cis*-acting elements are present in both parts of the transcript, or that a single element overlaps the junction between these regions.

RESPONSES IN ETIOLATED SEEDLINGS AND THE ROLE OF TRANSCRIPTION

Our observation that the light response *cis* elements are located inside the *Fed-1* gene does not rule out the possibility of a transcriptional control mechanism, particularly one affecting elongation or termination rather than initiation. Although control of initiation by elements 5′ to the promoter is the most common form of transcriptional regulation in animal cells, internal elements affecting transcript elongation or termination have been reported in a number of systems (Bently and Groudine, 1988; Farnham and Means, 1990; Hurt *et al.*, 1989; Kerppola and Kane, 1988; Mösinger *et al.*, 1987; Rougvie and Lis, 1988; Weiss *et al.*, 1989; Wright and Bishop, 1989). As one approach to evaluating the relative importance of such elements in regulating *Fed-1* expression, we have carried out a number of *in vitro* 'run-on' transcription assays. In such assays, RNA polymerases in the process of transcription at the time of nuclear isolation are allowed to continue elongation *in vitro*. Although sometimes assumed to measure only the number of engaged polymerases, 'run-on' assay are also sensitive to factors affecting chain elongation (Bently and Groudine, 1988; Mösinger *et al.*, 1987; Rougvie and Lis, 1988). Thus by working with intact nuclei in the absence of chromatin dissociating agents we should be able to detect changes in elongation as well as in initiation, provided that the factors responsible for altering elongation rates are not lost during nuclear isolation.

Initial experiments comparing nuclei from dark-adapted pea leaves before and a few hours after returning them to light suggested a light effect of about 2-3 fold. However, this difference is unlikely to account for the light effects on mRNA levels, since similar experiments with transgenic tobacco plants carrying the 'message construct' driven by the 35S promoter showed little or no change in transcriptional activity. Large changes in mRNA abundance were observed in both cases, so if transcriptional changes account for these effects they must not be measurable with the 'run-on' assay.

Much larger effects on transcriptional activity were seen when nuclei from etiolated and light grown pea seedlings were compared. These results raise the possibility that transcriptional control is much more important for the *initial Fed-1* response in etiolated seedlings than for subsequent responses in green leaves. The many differences between the light responses of green plants and etiolated seedlings (e.g., Chory *et al.*, 1989; Fluhr and Chua, 1986) make it clear that we need not expect the same mechanisms to work in both situations.

In vivo data relevant to this question was obtained by comparing the light responses of various 'first generation' constructs in tobacco seedlings and mature plants. For the seedling experiments, large numbers of seedlings are sown on filter paper in a dilute mineral nutrient solution and grown for 6 days in the dark. (Some light is required for germination of tobacco seed. We satisfy this requirement by imbibing seed in laboratory light for 30 minutes before they are transferred to the dark room. Subsequent growth is in total darkness). Some are then transferred to white light, and both illuminated and control seedlings are harvested 24 hours later. Both the 'long' and 'short' promoter constructs are expressed in etiolated seedlings, and - in contrast to the situation in light grown leaves - the amount of GUS transcript in seedlings increases somewhat in response to light in most transformants. Together with the run-on transcription data mentioned above, these results indicate that a part of the *Fed-1* light response in etiolated material is controlled at the transcriptional level by *cis* elements in the promoter. However we believe that a large part of the response in etiolated material is still controlled by a mechanism involving internal *cis* elements since *Fed-1* light responses are much larger in seedlings containing the intact gene than in seedlings containing the promoter constructs.

This supposition can be tested by examining the expression of the 'message' construct, which lacks *Fed-1* promoter elements. If internal *cis*-acting elements are important in seedlings as well as in leaves, we would expect to see strong light effects on the expression of this construct in seedlings.

Initially we saw only very small light effects with the message construct, but we now believe this result was an artifact produced by using entire seedlings for RNA extraction. Since the 35S promoter does not provide the organ specificity of the *Fed-1* promoter, plants containing the message construct synthesize *Fed-1* mRNA in roots and other parts of the seedlings in which it would not normally be present. If cells in which the message is not normally expressed lack the ability to properly regulate its abundance we would not expect to see large light/dark differences in RNA pooled from entire seedlings. This hypothesis is supported by recent experiments in which we have observed strong light regulation of *Fed-1* mRNA levels in separately harvested cotyledons of seedlings containing the message construct. Thus we believe that *Fed-1* mRNA levels in seedling cotyledons are regulated in two ways. There is a response at the level of transcriptional initiation, as noted above. However, there is also a response which depends on gene-internal *cis*-acting elements in much the same way as the response we originally observed in light-grown leaves.

CONCLUSIONS

Much more needs to be learned before we can understand how the internal regulatory element(s) in the *Fed-1* gene exert their effect on mRNA abundance. We cannot yet exclude the possibility that such elements mediate a light effect on transcriptional elongation or termination, but we feel it is more likely that light acts post-transcriptionally to permit the accumulation of transcripts which would otherwise be rapidly degraded, or to stimulate mRNA degradation in the dark. Such a mechanism would explain the remarkable quantitative and qualitative similarity in the behavior of the intact gene transcribed from its own promoter and the message construct driven by the much stronger 35S promoter (Elliott *et al.*, 1989a). We are currently trying to measure mRNA stability *in vivo*, as well as to further define essential light response element(s) within the mRNA sequence.

Although we do not yet fully understand the mechanism for light induction of *Fed-1* mRNA, it is clear that it differs from that for RbcS and Cab transcripts. Thus our results with *Fed-1* further emphasize the diversity of signal transduction pathways and the need to appreciate differences as well as similarities in light responsive gene systems.

REFERENCES

Anderson LE (1986) Light/dark modulation of enzyme activity in plants. Advances in Botanical Research 12: 1-46

Benfey PN, Chua N-H (1989) Regulated gene expression in transgenic plants. Science 244:174-181

Bently DL, Groudine, M (1988) Sequence requirements for premature termination of transcription in the human c-*myc* gene. Cell 53:245-256

Bevan M (1984) Binary *Agrobacterium* vectors for plant transformation. Nucleic Acids Res 12:5711-8721

Buchanan BB (1980) Role of light in the regulation of chloroplast enzymes. Ann Rev Plant Physiol 31:371-374

Bustos MM, Guiltinan MJ, Jordano J, Begum D, Kalkan FA, Hall, TC (1989) Regulation of ß-glucuronidase expression in transgenic tobacco plants by an A/T-rich, *cis*-acting sequence found upstream of a French bean ß-phaseolin gene. Plant Cell 1: 839-852

Cammack R, Rao K, Hall DO (1985) Ferredoxins: Structure and function of a ubiquitous group of proteins. Physiol Veg 23:649-658

Chory J, Peto CA, Ashbaugh M, Saganich R, Pratt L, Ausubel F (1989) Different roles for phytochrome in etiolated and green plants deduced from characterization of *Arabidopsis thaliana* mutants. Plant Cell 1:867-880

Crawford NA, Droux M, Kosower NS, Buchanan BB (1989) Evidence for function of the ferredoxin/thioredoxin system in the reductive activation of target enzymes of isolated intact chloroplasts. Arch Biochem Biophys 271:223-239

Datta N, Cashmore AR (1989) Binding of a pea nuclear protein to promoters of certain photoregulated genes is modulated by phosphorylation. Plant Cell 1: 1069-1077

Dobres MS, Elliott RC, Watson JC, Thompson WF (1987) A phytochrome-regulated pea transcript encodes ferredoxin I. Plant Mol Biol 8:53-59

Dutton JE, Rogers LJ, Haslett BG, Takruri IAH, Gleaves JT, Boulter D (1980) Com parative studies of two ferredoxins from *Pisum sativum* L. J Exp Bot 31:379-391

Elliott RC, Dickey LF, White MJ, Thompson WF (1989a) *cis*-Acting elements for light regulation of pea ferredoxin I gene expression are located within transcribed sequences. The Plant Cell 1:691-698

Elliott RC, Pedersen TJ, Fristensky B, White MJ, Dickey LF, Thompson WF (1989b) Characterization of a single copy gene encoding ferredoxin I from pea. The Plant Cell 1:681-690

Farnham PJ, Means, AL (1990) Sequences downstream of the transcription initiation site modulate the activity of the murine dihydrofolate reductase promoter. Mol Cell Biol 10:1390-1398

Fluhr R, Chua N-H (1986) Developmental regulation of two genes encoding ribulose bisphosphate carboxylase small subunit in pea and transgenic petunia plants: Phytochrome response and blue light induction. Proc Natl Acad Sci USA 83:2358-2362

Fluhr R, Kuhlemeier c, Nagy f, Chua N-H (1986) Organ-specific and light-induced expression of plant genes. Science 232:1106-1112

Goodwin G, Bustin M (1988) The HMG proteins and their genes. Architecture of Eukaryotic Genes. Weinheim, VCH Verlagsgesellschaft

Green PJ, Kay SA, Chua N-H (1987) Sequence-specific interactions of a pea nuclear factor with light-responsive elements upstream of the rbcS-3A gene. EMBO J 6:2543-2549

Guiltinan MJ, Thomas JC, Nessler CL, Thomas TL (1989) Expression of DNA binding proteins in carrot somatic embryos that specifically interact with a *cis* regulatory element of the French bean phaseolin gene. Plant Mol Biol 13:605-610

Hoekema A, Hirsch PR, Hooykaas PJJ, Schilperoot RA (1983) A binary vector strategy based on separation of the *vir*- and T-region of the *Agrobacterium tumefasciens* Ti-plasmid. Nature 303:179-180

Holdsworth MJ, Laties GG (1989) Site-specific binding of a nuclear factor to the carrot extensin gene is influenced by both ethylene and wounding. Planta 179:17-23

Horsch RB, Fry FE, Hoffman NL, Eicholtz D, Rogers SG, Fraley RT (1985) A simple and general method for transferring genes into plants. Science 227:1229-1231

Hurt MM, Pandey NB, Marzluff WF (1989) A region in the coding sequence is required for high level expression of murine histone H3 gene. Proc Natl Acad Sci USA 86:4450-4454

Jacobsen K, Laursen NB, Jensen EO, Marcker A, Poulsen C, Marcker KA (1990) HMG I-like proteins from leaf and nodule nuclei interact with different AT motifs in soybean nodulin promoters. Plant Cell 2:85-94

Jensen EØ, Marker KA, Schell J, de Bruijn FJ (1988) Interaction of a nodule specific, *trans*-acting factor with distinct DNA elements in the soybean leghaemoglobin *Ibc3* 5′ upstream region. EMBO J 7:1265-1271

Jofuku KD, Okamuro JK, Goldberg RB (1987) Interaction of an embryo DNA binding protein with a soybean lectin gene upstream region. Nature 328:734-737

Kaufman LS, Briggs WR, Thompson WF (1985) Phytochrome control of specific mRNA levels in developing pea buds: the presence of both very low fluence and low fluence responses. Plant Physiol 78:388-393

Kaufman LS, Roberts LR, Briggs WR, Thompson WF (1986) Phytochrome control of specific mRNA levels in developing pea buds. Kinetics of accumulation, reciprocity, and escape kinetics of the low fluence response. Plant Physiol 81:1033-1038

Kerppola TK, Kane CM (1988) Intrinsic sites of transcriptional termination and pausing in the c-*myc* gene. Mol Cell Biol 8:4389-4394

Kimat Y, Hase T (1989) Localization of ferredoxin isoproteins in mesophyll and bundle sheath cells in maize leaf. Plant Physiol 89:1193-1197

Kuhlemeier CP, Green J, Chua N-H (1987) Regulation of gene expression in higher plants. Ann Rev Plant Physiol 38:221-257

Marcotte WRJ, Russell SH, Quatrano RS (1989) Abscisic acid-responsive sequences from the *Em* gene of wheat. Plant Cell 1:969-976

Mösinger E, Batschauer A, Vierstra R, Apel K, Schäfer E (1987) Comparison of the effects of exogenous native phytochrome and in vivo irradiation on in vitro transcription in isolated nuclei from barley (*Hordeum vulgare*) Planta 170:505-514

Rougvie AE, Lis JT (1988) The RNA polymerase II molecule at the 5′ end of the uninduced *hsp*70 gene of D. melanogaster is transcriptionally engaged. Cell 54:795-804

Russell DW, Smith M, Cox D, Williamson VM, Young ET (1983) DNA sequences of two yeast promoter-up mutants. Nature 304:652-654

Sagar AD, Briggs WR, Thompson WF (1988a) Nuclear-cytoplasmic partitioning of phytochrome-regulated transcripts in *Pisum sativum*. Plant Physiol 88:1397-1402

Sagar AD, Horwitz BA, Elliott RC, Thompson WF, Briggs WR (1988b) Light effects on several chloroplast components in norfluorazon-treated pea seedlings. Plant Physiol 88:340-347

Sakihama N, Shin M (1987) Evidence from high pressure liquid chromatography for the existence of two ferredoxins in plants. Arch Biochem Biophys 256:430-434

Simpson J, Herrera-Estrella L (1990) Light-Regulated Gene Expression. Critical Reviews in Plant Sciences 9:95-109

Smeekens S, Binsbergen JV, Weisbeek P (1985) The plant ferredoxin precursor: nucleotide sequence of a full length cDNA clone. Nucleic Acids Res 13:3179-3194

Solomon MJ, Strauss F, Varshavsky A (1986) A mammalian high mobility group protein recognizes any stretch of six A•T base pairs in duplex DNA. Proc Natl Acad Sci (USA) 83:1276-1280

Somers DE, Caspar T, Quail PH (1990) Isolation and characterization of a ferredoxin gene from *Arabidopsis thaliana*. Plant Physiol 93:572-577

Spiker S (1984) High mobility group chromosomal proteins of wheat. J Biol Chem 259:12007-12013

Spiker S (1988a) Histone variants and high mobility group non-histone chromosomal proteins of higher plants: Their potential for forming a chromatin structure that is either poised for transcription or transcriptionally inert. Physiol Plant 75:200-213

Spiker S (1988b) Histones and HMG proteins of higher plants. Architecture of Eukaryotic Genes. Weinhein, VCH Verlagsgesellschaft

Spiker S, Murray MG, Thompson WF (1983) DNase 1 sensitivity of transcriptionally active genes in intact nuclei and isolated chromatin of plants. Proc Natl Acad Sci (USA) 80:815-819

Struhl K (1985) Naturally occurring poly(dA-dT) sequences are upstream promoter elements for constituitive transcription on yeast. Proc Natl Acad Sci (USA) 82:8419--8423

Takahashi Y, Hase T, Wada K, Matsubara H (1983) Ferredoxins in developing spinach cotyledons: the presence of two molecular species. Plant Cell Physiol 24:189-198

Thompson WF (1988) Photoregulation: diverse gene responses in greening seedlings. Plant, Cell and Env 11:319-328

Thompson WF, Kaufman LS, Watson JC (1985) Induction of plant gene expression by light. Bioessays 3:153-159

Vorst O, v Dam F, Oosterhoff-Teertstra R, Smeekens S, Weisbeek P (1990) Tissue specific expression directed by an *Arabidopsis thaliana* pre-ferredoxin promoter in transgenic tobacco plants. Plant Mol Biol 14:491-499

Wada K, Onda M, Matsubara H (1989) Amino acid sequences of ferredoxin isoproteins from radish roots. J Biochem 105:619-625

Weiss EA, Michael A, Yuan D (1989) Role of transcriptional termination in the regulation of μ mRNA expression in B lymphocytes. J Immunol 143:1046-1052

Wright S, Bishop JM (1989) DNA sequences that mediate attenuation of transcription from the mouse protooncogene *myc*. Proc Natl Acad Sci USA 86:505-509

Section C.

PHYTOCHROME ACTION - DEVELOPMENTAL PHYSIOLOGY

DO THE MEMBERS OF THE PHYTOCHROME FAMILY HAVE DIFFERENT ROLES? PHYSIOLOGICAL EVIDENCE FROM WILD-TYPE, MUTANT AND TRANSGENIC PLANTS

H. Smith, G.C. Whitelam and A.C. McCormac
Department of Botany,
University of Leicester,
Leicester LE1 7RH,
UK.

INTRODUCTION

The realisation that phytochrome represents a family of photoreceptors, encoded by at least three, and probably more, nuclear genes (Sharrock and Quail, 1989), has led to the proposal that the members of the family may have differential roles in the regulation of development and metabolism (see Smith and Whitelam, 1990, for review). In principle, this idea simplifies our understanding of the complexity of photomorphogenesis, by providing a potential resolution of the many conflicts in the physiological data. Investigations over several decades have shown that a number of different response modes, separable on classical photobiological grounds, can be incontrovertibly attributed to phytochrome, but this multiplicity of roles has been extremely difficult to reconcile with the classical concept of phytochrome as a single photoreceptor. Such conflicts may be traced back at least two decades, to the proposal by Hillman (1967) that there must be at least two populations of phytochrome, one that is 'active' but present in low concentration, and one that is 'bulk', i.e., present in high concentration and responsible for the spectrophotometric observations but not active in the physiological responses. It is, therefore, tempting to speculate that different phytochrome-mediated response modes may be mediated by different phytochromes. This article considers this idea and some of the evidence that currently appears to support it. We conclude that the hypothesis is at least sustainable on present evidence, but the attractiveness of the concept must be weighed in balance with the implications it has for the central dogmas of photomorphogenesis.

NATO ASI Series, Vol. H 50
Phytochrome Properties and Biological Action
Edited by B. Thomas and C. B. Johnson

Once we accept the plurality of phytochromes, as we must do on molecular grounds, we are also obliged to accept the possibility for plurality in all other considerations relevant to phytochrome. At a minimum, we can no longer exclude the following:

1. Multiple primary mechanisms of action;
2. Multiple cellular locations;
3. Multiple controls on the expression of the different phytochromes;
4. Multiple physiological roles.

We here concentrate on the last of these considerations; namely, do the different members of the phytochrome family have different physiological roles? Evidence is presented that different physiological response modes may be identified. Following this, we describe some new data from studies on mutant and transgenic plants that indicate that arriving at conclusions on the question of which phytochrome is responsible for which response, will not be easy. The first priority, however, is to consider the criteria that may be used to attribute physiological response modes to the identifiable forms of phytochrome.

TERMINOLOGY

As discussed at this NATO Workshop, the terminology relating to the molecular forms of phytochrome is far from ideal, and is generally used loosely and inconsistently. In this article, we have attempted to follow the nomenclatural guidelines agreed at the Workshop. In particular, when referring to the phytochrome genes, we use the format *phyA*, *phyB*, *phyC* etc. In referring to the phytochrome species that can currently be recognised on biochemical and immunochemical grounds, we use the terms Type I and Type II. Finally, at the physiological level, we refer to light-labile phytochromes as 'phytochrome-L', and the phytochromes that are apparently stable in the light as 'phytochrome-S'.

For the purposes of this article, the distinctions and similarities between, on the one hand, phytochromes of Types I and II, and on the other hand phytochrome-L and phytochrome-S, are crucially important. In operational terms, the only distinction that can be tested physiologically is whether a particular response is mediated by a form of phytochrome that is light-labile (i.e., phytochrome-L), or one that is light-stable (i.e., phytochrome-S). On this basis, as will be seen, it is possible to attribute certain response modes to phytochrome-L or to phytochrome-S. It cannot be stated with certainty, however, that phytochrome Type I is synonymous with phytochrome-L, or that phytochrome Type II is synonymous with phytochrome-S.

PHYTOCHROME RESPONSE MODES

Table 1 shows the several different phytochrome response modes that can be recognised on classical photobiological grounds.

Very-Low-Fluence Responses. In the VLFR, saturation is reached at extremely low fluences and Pfr concentrations; consequently FR-reversibility is not possible since even FR establishes sufficient Pfr to induce the response. Because of the low fluences for saturation, no-one has yet attempted to determine whether reciprocity is valid. VLFRs have been observed for both the stimulation of the germination of some light-requiring seeds (Raven and Spruit, 1973), and for inhibition of seedling extension growth (Blaauw *et al.* 1968). VLFRs can only be seen in seeds or seedlings that have been imbibed or grown in total darkness.

Low-Fluence Responses. The LFR is the classical phytochrome response, as established and characterised by Hendricks and Borthwick and their co-workers. The LFR is saturated at moderate R fluences, reversible by FR, and normally exhibits full reciprocity. Escape from FR-reversibility is commonly rapid, often being complete within less than 2 hours. LFRs have been observed for seed germination, and for most of the phenomena associated with de-etiolation. However, seed germination is often induced by brief R in a manner that appears identical with the LFR, but in which FR reversibility is retained for many hours in darkness.

High-Irradiance Responses. The HIR requires prolonged irradiation treatment and, in the classical sense, is dependent on fluence rate. HIRs in seeds and etiolated seedlings commonly have an action maximum in the FR, and although R/FR reversibility is not possible, elegant bichromatic experiments have shown that FR-HIRs are indubitably mediated by phytochrome (Hartmann, 1966). The fluence rate range of FR-HIRs is relatively low, that examined by Hartmann, for example, being saturated at about 5 umol

m^{-2} s^{-1}. As seedlings de-etiolate in white light, the FR-HIR tends to be lost, but other HIRs, including responses with a R action maximum, appear or are retained (Beggs *et al.* 1980; Holmes *et al.* 1982).

Photoperiodism. In the fully de-etiolated plant, a number of other response modes may be recognised, which appear to be different from the responses modes of the etiolated plant. The earliest to be studied were the roles of phytochrome in night break and extended day responses of the photoperiodic induction of flowering. The involvement of phytochrome in photoperiodism is complicated by its interaction with underlying endogenous rhythms, and although in some cases pulse treatments may be FR-reversible and may show reciprocity, in others prolonged irradiation is required for maximum response. It is also frequently true that FR effects may be exerted even after very long periods in extended darkness following the end of the photoperiod.

End-of-Day Responses. Extension rate in light-grown plants is often very responsive to the state of phytochrome at the end of the daily light period (Downs *et al.* 1957). Seedlings given a short period of FR at the end-of-day commonly show marked enhancement of extension growth, which may be 'reversed' by R; repeated R/FR reversibility is usually observed and reciprocity is fulfilled. An important feature of end-of-day responses is that the FR is still active in enhancing extension rate for many hours in darkness after the white light has been switched off. This extended FR reversibility distinguishes end-of-day responses from the typical LFRs of etiolated seedlings.

R:FR Ratio Perception. Light-grown plants have the capacity to perceive the ratio of R and FR radiation in the incident light (Holmes and Smith, 1975; Smith and Holmes, 1977). This property allows the detection of neighbouring or shading vegetation, because chlorophyllous tissues reflect and transmit FR, and absorb most of the R. Proximity perception (Ballaré *et al.* 1987; Smith, Casal and Jackson, 1990) is ecologically important in that it confers the capacity for shade avoidance reactions, and thus contributes to the ability of individual plants to compete for the resource of light (Smith, 1982; Casal and Smith, 1989). Plants are often very responsive to daytime R:FR, and responses, such as extension growth, frequently express themselves in an inverse linear relation to the proportion of Pfr established by the actinic radiation (Morgan and Smith, 1976; 1978).

Table 1. The Identifiable Phytochrome Response Modes

Response Mode	Developmental Stage	Characteristics
VLFR	Imbibed Seeds Etiolated Seedlings	10^{-9} - 1^{-7} mol m^{-2} Not FR Reversible Reciprocity Unknown
LFR	Imbibed Seeds Etiolated Seedlings	1^{-6} - 10^{-3} mol m^{-2} R/FR Reversible Reciprocity Valid
HIR	Imbibed Seeds Etiolated Seedlings De-etiolated Plants	5 - 30 μmol m^{-2} s^{-1} Not R/FR Reversible Fluence Rate Dependent Reciprocity Invalid
End-of-Day	Light-Grown Plants	10^{-6} - 10^{-3} mol m^{-2} R/FR Reversible Reciprocity Valid
R:FR perception	Imbibed Seeds Light-Grown Plants	Linear with Pfr/P Fluence Rate Compensated
Photoperiod Perception	Light-Grown Plants	Night-Breaks R/FR-Reversible Coupled to Rhythm

ATTRIBUTION OF THE RESPONSE MODES

On physiological grounds, the only criterion that can be applied is based on whether a response is mediated by a stable or light-labile phytochrome. This is not easy, and certainly not always definitive. The protocol essentially is to test whether a particular R-induced response can be reversed by FR after a prolonged period of darkness. On the generally-accepted hypothesis that phytochrome operates solely through the activity of Pfr, such extended FR-reversibility can only occur if the Pfr mediating the response is stable during the prolonged dark period. Although this test is definitive for situations in which stable Pfr is implicated, it is not equally definitive in the reverse sense. Lack of extended FR-reversibility does not necessarily mean that a stable-Pfr is responsible for the response, since a response may escape from FR-reversibility for reasons other than Pfr destruction. Furthermore, the reversibility approach cannot be applied to responses requiring continuous irradiation. Consequently, only a restricted set of definitive statements can be made.

As mentioned above, extended FR-reversibility is evident for end-of-day elongation responses, and has been observed in photoperiodism and the germination of some seeds. The attributions of these responses to phytochrome-S can therefore be reasonably certain (Table 2). In no other cases can such an attribution yet be considered reliable, not even for R:FR ratio perception, which otherwise has many characteristics that might indicate a similarity with end-of-day responses.

To attribute responses to light-labile phytochrome is equally uncertain on physiological grounds. The only response that seems unequivocally to be mediated by phytochrome-L is the FR-HIR. FR-mediated HIRs are extremely difficult to understand on any basis other than light-labile Pfr, in which Pfr destruction is a function of Pfr concentration. The general hypothesis is response represents a balance between Pfr action and Pfr destruction, and that the FR action maximum is due to the maintenance by FR of a concentration of Pfr that is sufficient to induce the response, but insufficient to allow significant loss of the phytochrome responsible for the response. Thus, any response that has the characteristics of the FR-HIR may be concluded as being mediated by phytochrome-L.

Thus, all that can be said at present is that end-of-day responses, when characterised by the action of FR after extended dark periods, are mediated by phytochrome-S, and the FR-HIR is mediated by phytochrome-L. It is tempting to regard phytochrome-S as synonymous with Types II phytochromes, and phytochrome-L as synonymous with Types I phytochromes, and this might well prove to be the case; however, strictly speaking such a conclusion is still premature.

In recent years, the possibility of studying physiological responses in plants with altered levels of different phytochrome molecular species, either using mutant or transgenic plants has arisen. In the rest of this articles we describe some recent results from our laboratory bearing on the use of mutants and transgenic plants to elucidate the putative roles of the different molecular species of phytochrome.

Table 2. Attribution of photoresponses to phytochromes 1 and 2. (Parentheses indicate uncertainty of attribution)

Response	Responsible Phytochrome	
Seed Germination		
R-induced promotion	(L)	S
FR-induced inhibition	L	
De-etiolation		
Inhibition of axis elongation	(L)	
Thylakoid protein gene expression	(L)	
Anthocyanin synthesis	(L)	
Regulation of Type-I synthesis	L	(S)
End-of-Day Responses		
Stem Elongation		S
R:FR Perception		
Stem Elongation		(S)
Promotion of Flowering		(S)
Photoperiodism		
Night-break inhibition in SDP	(L)	
Night-break promotion in LDP	L	S

EVIDENCE FROM MUTANT PLANTS

Photoreceptor mutations, affecting the abundance or functional integrity of particular phytochrome species, would represent valuable tools in assigning specific roles to different phytochromes. In recent years, a number of putatively phytochrome-deficient mutations have been described, including the tomato *au* mutation, the *Arabidopsis hy* mutations and the cucumber *lh* mutation (see Adamse *et al.* 1988, for review).

At present, none of these 'photoreceptor' mutants has been shown to result from a lesion in a phytochrome structural gene. However, the *au* mutation appears not to affect the accumulation of all pools of phytochrome detectable in those seedlings. Etiolated tissues of the tomato *au* mutant fail to accumulate spectrophotometrically-, or immunochemically-detectable phytochrome (Parks *et al.* 1987; Sharrock *et al.* 1988; Oelmüller *et al.* 1989). However, the *au* mutation does not result from a lesion in the structural gene for the tomato *phyA* homologue, and normal levels of translatable mRNA for Type I phytochrome are produced (Sharrock *et al.* 1988). Thus, the *au* mutant is apparently deficient in phytochrome because of the instability of the phytochrome polypeptide. Thus phytochrome deficiency in *au* mutants appears to be restricted to the labile phytochrome pool that would normally accumulate inetiolated tissues (i.e., phytochrome-L), because phytochrome is spectrally-detectable in light-grown tissues of the mutant (Adamse, 1988). However, spectrally-measured phytochrome levels in light-grown *au* tissues are somewhat reduced compared to the isogenic wild type. This may mean that in wild type seedlings some phytochrome-L persists upon growth in the light, or that the *au* mutant is also partially deficient in a light-stable phytochrome species (i.e., phytochrome-S). In either event, the severe deficiency in one or more pools of phytochrome in the *au* mutant potentially allows the role of those phytochromes to be assessed.

The *au* mutant displays a phenotype that is reminiscent of an etiolated appearance, even after growth in continuous white light. Dark-grown seedlings of *au* show a reduced photoregulation of elongation growth, chlorophyll accumulation and expression of several normally light-responsive genes. For two responses, accumulation of cab transcripts, and the potentiation of chlorophyll accumulation, a VLFR detectable in wild type seedlings is absent in the *au* mutant (Sharrock *et al.* 1988; Ken-Dror and Horwitz, 1990). This finding is consistent with the view that the VLFR is mediated by light-labile phytochrome-L, which in the *au* mutant may be said to be indistinguishable from Type I phytochrome. Not all phytochrome responses are modified by the *au* mutation. Light-grown *au* plants respond normally to end-of-day FR, with respect to elongation growth (Adamse, 1988). Furthermore, Figure 1 shows that day-time supplementary FR leads to qualitatively similar promotion of elongation growth in wild type and *au* seedlings. These responses would appear, therefore, to be mediated by the light-stable phytochrome-S pool that is detectable in light-grown *au* tissues.

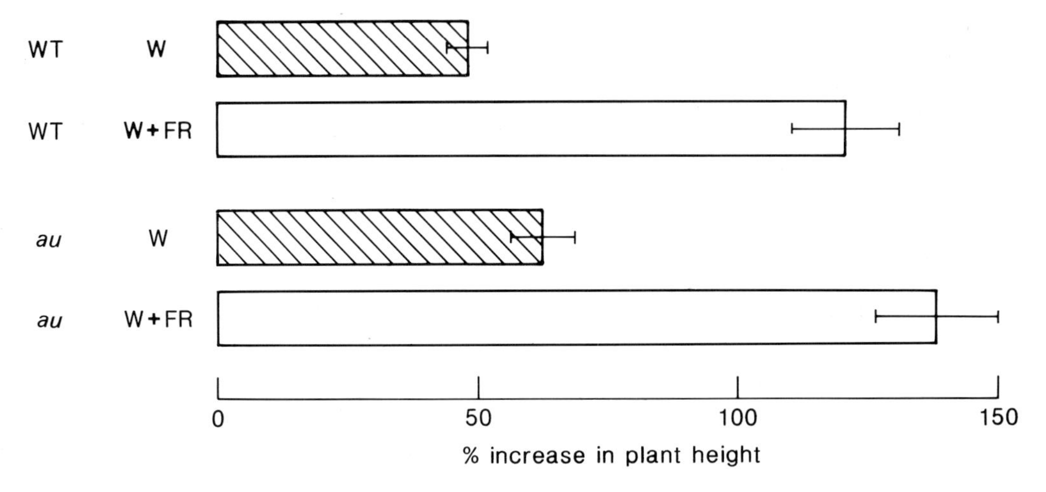

Figure 1. Elongation growth of wild-type (WT) and *aurea* mutant (*au*) tomato seedlings grown under continuous white light (W) (R:FR = 6.8) or white light with supplementary far-red (W+FR) (R:FR = 0.07). Seedlings were grown for 2 weeks in W prior to treatment for 7 days.

Since the *au* mutation only affects some phytochrome responses, and only some of the detectable phytochrome species, it offers the potential to elucidate, at least partially, which phytochrome species mediate which responses. This will require a more extensive analysis of the *au* mutant in order to identify the lesion leading to the mutation, and to determine which molecular species of phytochrome are present or absent in the mutant.

The long hypocotyl mutations *hy*-1, *hy*-2, and *hy*-6 of *Arabidopsis thaliana* fail to accumulate spectrally-detectable phytochrome in etiolated tissues. However, unlike the tomato *au* mutant, etiolated tissues of these *Arabidopsis* mutants apparently accumulate normal levels of immunochemically-detectable phytochrome (Parks *et al.* 1989; Chory *et al.* 1989). The molecular basis of these mutations is unknown, although lesions affecting chromophore synthesis/attachment are suspected, and it is not known whether all pools of phytochrome are deficient, or only phytochrome-L. It is evident, however, that *hy*-1, *hy*-2 and *hy*-6 all contain some functional phytochrome, since light-grown plants of all three mutant types are capable of R:FR ratio perception and show phytochrome--

mediated regulation of extension growth and flowering (G. C. Whitelam; unpublished data). Consequently, speculation on the roles of different phytochromes based on results from the *hy*-1, *hy*-2 and *hy*-6 mutants would be premature.

The *hy*-3 mutant of *Arabidopsis* is another long hypocotyl mutation, but in this case the mutation does not affect the levels of spectrally-detectable phytochrome in etiolated tissues. Unlike the *hy*-1, *hy*-2 and *hy*-6 mutations, etiolated *hy*-3 seedlings display a more-or-less normal FR-HIR inhibition of hypocotyl elongation. However, *hy*-3 seedlings do not respond well to continuous R. In mustard seedlings, inhibition of hypocotyl elongation by continuous R, unlike the FR-HIR, is retained after de-etiolation (Beggs *et al.* 1980), consistent with the idea outlined above that inhibition by continuous R may be mediated by the phytochrome-S pool. On this reasoning, it may be speculated that the *hy*-3 mutation affects the functioning (photoreceptor or transduction) of a light-stable phytochrome. This view is supported by the findings that seeds of *hy*-3 have a reduced content of spectrally-detectable phytochrome (Cone, 1985), and that light-grown *hy*-3 display reduced extension growth responses to R:FR ratio (G. C. Whitelam; unpublished observations). Further analyses of the phytochromes and photoresponses of *hy*-3 will be required in order to test this speculative hypothesis.

EVIDENCE FROM TRANSGENIC PLANTS

As the various gene sequences are isolated and characterised, it becomes feasible to develop transgenic plant procedures designed to elucidate the putative roles of the different molecular species of phytochrome. The first transgenic plants overexpressing a foreign introduced phytochrome gene used a cereal (oat or rice) *phyA* gene, with tobacco or tomato as the host (Keller *et al.* 1989; Boylan and Quail, 1989; Kay *et al.* 1989a). In the case of the plants transformed with the oat gene (Keller *et al.* 1989; Boylan and Quail, 1989) the resultant phenotype was characterised by marked dwarfism when the plants were grown in white light. Initially, tobacco plants transformed with the rice *phyA* gene, did not exhibit developmental differences from the wild-type, but subsequent transformations led to lines characterised again by marked dwarfism (Kay *et al.* 1989b). Overexpression of the cereal *phyA* genes led to substantial increases in the levels of Type I phytochrome in the transformed lines, with there apparently being no limitation imposed by the capacity of the host plant to synthesise sufficient chromophore and to assemble it correctly with the apoprotein. Although degradation of the oat Type I phytochrome in the tobacco transgenics occurs in the light (J. Cherry and R.D. Vierstra, personal communication), nevertheless a substantial amount of the introduced Type I phytochrome remains in plants grown in white light (Keller *et al.* 1989).

If it is true that Types I and II phytochromes have different physiological roles, then the physiological responses of transgenic plants that overexpress an introduced Type I phytochrome should be of considerable interest. In the experiments outlined here, transgenic tobacco plants that overexpress an introduced oat Type I gene were subjected to three physiological tests designed to determine whether the introduced Type I phytochrome operates in the expected manner in the tobacco hosts. The experiments tested whether (a), the presence of Type I phytochrome allowed the persistence of a FR-HIR in de-etiolated seedlings; (b), end-of-day FR treatments were affected by the presence of excess Type I phytochrome; and (c), typical R:FR ratio responses could occur in the presence of the additional Type I phytochrome. The transgenic plants used were derived from selfed progeny of the homozygous plants produced by Keller *et al* (1989). This article presents an outline summary of the results of these investigations; a fully-detailed treatment has been submitted for publication (McCormac, Vierstra, Hershey, Cherry and Smith, in press).

Phenotypic Differences Between Wild-type and Transgenics. Growth in white light confirmed the substantial phenotypic differences between seedlings of the wild-type and the transgenic overexpressers that had been previously reported (Keller *et al.* 1989). The transgenic seedlings were markedly dwarfed, the overall pattern indicating a ca 40% reduction in extension growth partially compensated by an increased production of leaf matter, and an enhanced level of chlorophyll per unit leaf area. The phytochrome overexpressers flowered more rapidly than the wild-types, the time for bud emergence being reduced by up to six weeks. Only ca 60% of the wild-types produced flower buds within the treatment period, compared to 95% of the transgenics.

Response of Seedlings to Continuous FR. Figure 2 shows the inhibition of extension growth by 48 h continuous FR in seedlings previously grown either in darkness or white

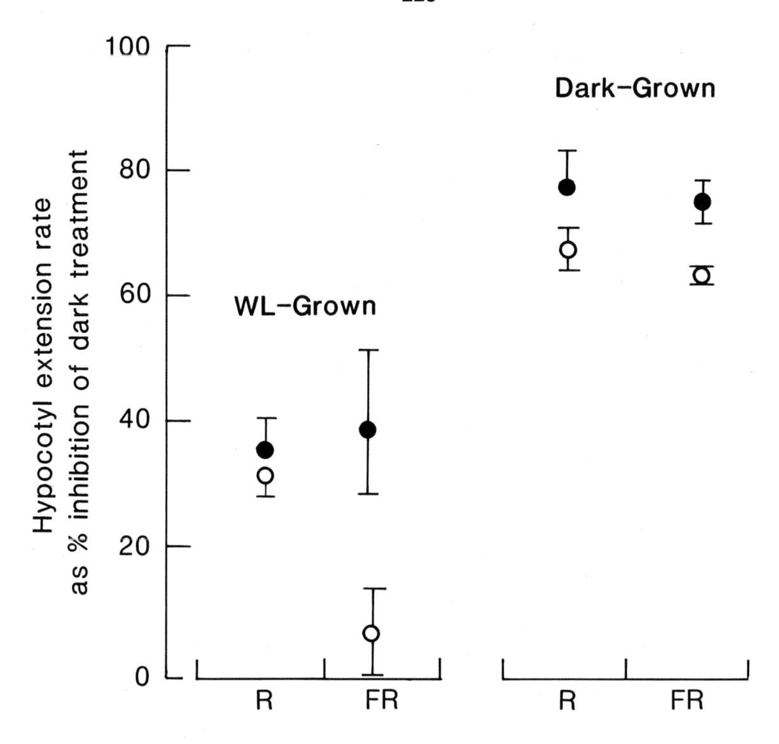

Figure 2. Growth of wild-type (open symbols) and phytochrome-overexpressing transgenic (solid symbols) tobacco cv Xanthi seedlings as affected by 48 h of continuous red (R) or far-red (FR) light. Prior to treatment, seedlings were either grown in the darkroom until hypocotyl length was at least 5 mm, or grown in continuous W (80 umol m^{-2} s^{-1}) for 6 days. Hypocotyl length was measured before and after treatment, and the ordinate values represent the percentage inhibition compared to growth in the dark for the 48 h treatment period.

light for 6 days. In the etiolated state, both wild-types and transgenics showed substantial inhibition under FR treatments, indicating the existence of a FR-HIR. De-etiolated wild-type seedlings lose the capacity for FR-mediated growth inhibition, but the transgenics still retain substantial inhibition. This indicates that the presence of relatively large amounts of Type I phytochrome in the light-grown transgenics allows the persistence of a FR-HIR, as predicted above.

Effect of End-of-Day FR Treatments. Plants were grown in a regime of approximately 12 h white light and 12 hours darkness, and some were given 30 min of FR at the end of the light period. The results of these experiments are as yet preliminary, but both wild-types and transgenics appear to exhibit very similar responses to end-of-day FR. This contrasts with the major disruption of R:FR ratio perception (next section) and is consistent with the conclusion above that end-of-day FR is mediated by a phytochrome with the properties of a Type II. On the other hand, such a conclusion cannot be definitively made because of the action of the introduced oat Type I phytochrome. Since the Type I clearly suppresses extension growth in the light-grown seedlings (see below), removal of Pfr by end-of-day FR would relieve the transgenic seedlings from such inhibition. In other words, Type I phytochrome can mediate end-of-day FR effects, although in wild-type plants it would probably not be present at sufficiently high levels.

Effect of Continuous Supplementary Far-Red. The typical growth responses of shade avoiding species to daytime supplementary FR include enhanced stem and petiole elongation, reduced leaf development, elevated epinastic curvature of petioles, decreased root:shoot ratio, changes in assimilate distribution towards stems and away from leaves and storage organs, and acceleration of flowering. The major growth parameters of the typical shade avoidance responses were analysed in the wild-type and transgenic phytochrome overexpressers.

In these experiments. seedlings were grown under fluorescent white light (WL) of 130 umol m^{-2} s^{-1} (R:FR = 6.8) up till the third true leaf-pair had expanded (approximately 5 weeks) after which they were transferred to cabinets providing R:FR ratios of either 6.8 or 0.07 at three different PAR levels (55, 75, and 160 umol m^{-2} s^{-1}). These light sources established approximately 0.8 Pfr/P and 0.1 Pfr/P respectively. The fluence rate values given relate to the 400-800 nm waveband only; in order to depress the R:FR to such a large extent, the amount of added FR in the WL+FR was up to ca 200 umol m^{-2} s^{-1} (700-800 nm) at the highest fluence rates.

Wild-type plants exhibited large growth responses to being exposed to a R:FR ratio of 0.07. A R:FR as low as this represents very dense vegetational shade, and substantial developmental effects are expected. In the case of tobacco, however, the extent of the responses were affected by the PAR under which the treatments were given.

Figure 3 shows plant height at 2 weeks after transfer. Extension growth at high R:FR is obviously less at higher, than at lower fluence rates. Similar effects of fluence rate were observed for internode length and petiole length. The reason for this effect of fluence rate is not yet known; it may represent a HIR of the type reported by Beggs *et*

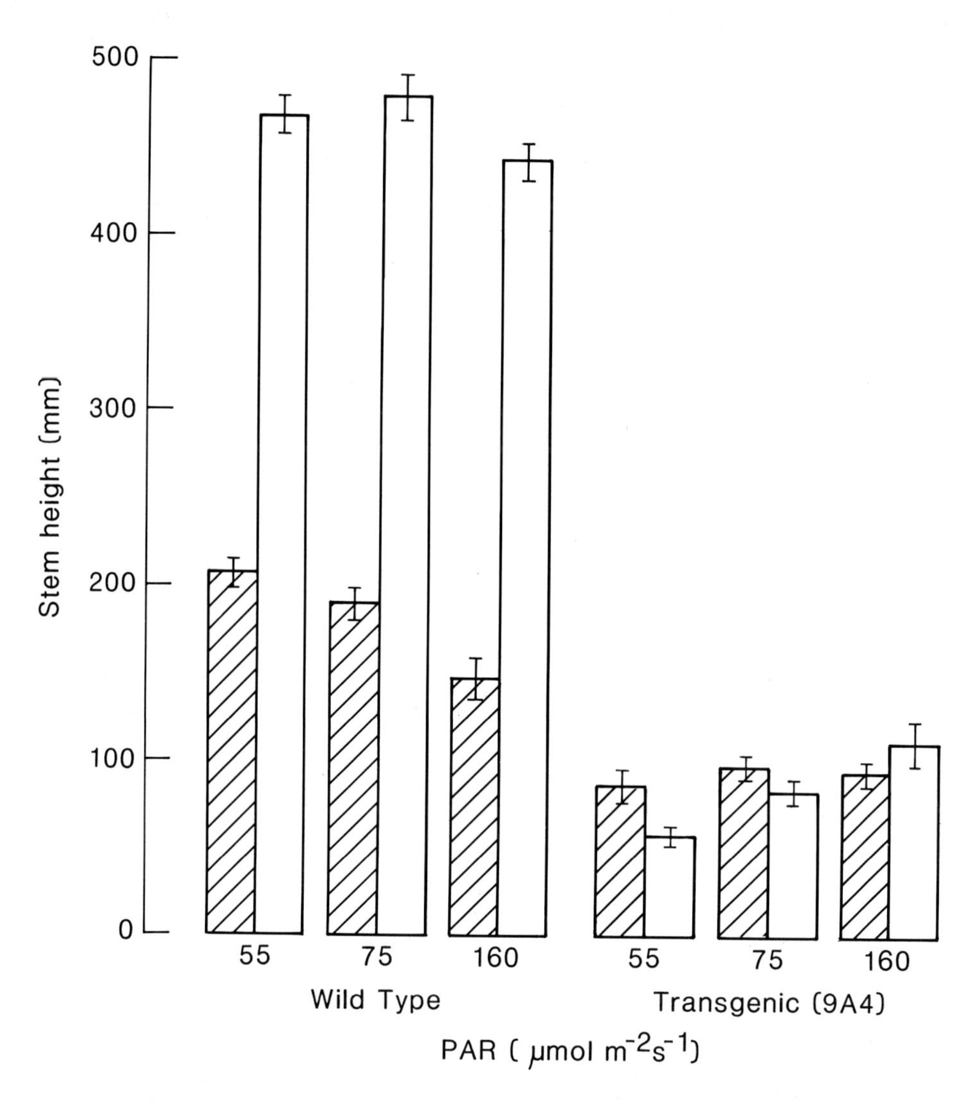

Figure 3. Elongation growth of wild-type and transgenic tobacco seedlings in continuous white light (hatched bars; R:FR = 6.8), and in continuous white plus far-red light (open bars; R:FR = 0.07). Fluence rate was varied, but R:FR held constant, by growing the seedlings at different heights within the growth chambers; the fluence rate values stated represent the 400-700 nm PAR levels. Prior to treatment the seedlings were grown in continuous W for 5 weeks, and the treatment period was a further 2 weeks.

al. (1980) in de-etiolated tissues, or it may involve a blue-light-mediated response. Growth under WL+FR (R:FR = 0.07) caused up to a three-fold increase in extension rate in the wild-type seedlings (Figure 3), a typical shade avoidance response. In the transgenics, however, the addition of FR did not result in a statistically significant increase in extension at any fluence rate; indeed, at the lower fluence rates, extension rate was actually *reduced* by additional FR. This effect can be seen for plant height but was even more marked for internode and petiole extension. In no plant yet studied, even shade tolerant species, has a depression of extension growth been reported as a result of growth under WL with supplementary FR. These results are, therefore, exceptionally surprising.

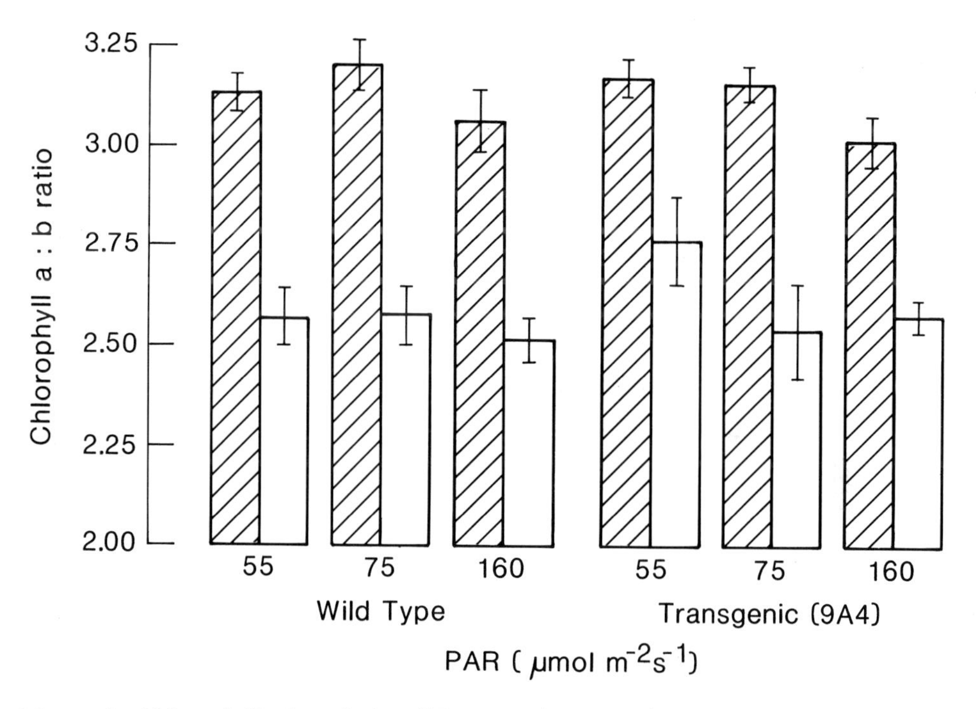

Figure 4. Chlorophyll a:b ratio in wild-type and transgenic tobacco seedlings. Experimental conditions and symbols as in the legend to Figure 3.

Although major differences between the wild-type and transgenic plants were observed for R:FR effects on extension growth parameters, no such differences were seen for leaf development parameters. For many shade avoiding species, the levels of photosynthetic pigments, chlorophyll a/b binding (cab) proteins, and chlorophyll a:b ratio are markedly affected by growth under WL+FR. This is clearly seen in Figure 4 for chlorophyll a:b ratio, but no differences were discernible between the wild-type and transgenic seedlings. Similarly, no differences were observed for the effects of R:FR ratio on total chlorophyll levels, or on cab protein levels, between the wild-types and the transgenics (data not shown). This is a significant point, since it demonstrates that the overexpression of a Type I phytochrome does not disrupt all shade avoidance responses.

A SPECULATIVE INTERPRETATION

It is clearly too early to attempt a full interpretation of the physiological effects of the over-expression of Type I phytochrome, as much more data will need to be gathered. Even so, some of the findings reported here are at first sight unexpected, and speculation as to the mechanism is an interesting intellectual exercise. The findings that overexpression of a Type I phytochrome in de-etiolated seedlings confers a typical FR-HIR, but does not substantially disrupt end-of-day responses to brief FR, are consistent with - but do not prove - the proposed roles of phytochrome-L and phytochrome-S proposed above. On the other hand, the major disruption of R:FR ratio perception, and especially the depression of extension by supplementary FR, are not easy to understand.

Are Types I and II Phytochromes Interchangeable? There are basically two principal ways in which the overexpression of Type I phytochrome could disrupt the typical shade avoidance reactions of the light-grown plant. These, in turn, depend upon two different scenarios for the action of phytochrome in the normal, wild-type plant.

In the first scenario, any type of phytochrome may be thought of as being able to mediate R:FR perception and thereby initiate shade avoidance reactions. Type I phytochrome accumulates in etiolated seedlings to very high levels, and the first exposure to light establishes a high level of Pfr, which inhibits axis elongation severely. Such severe inhibition of extension would not be compatible with growth of the mature plant in the natural light environment, and consequently it is important for the high levels of Type I phytochrome to be reduced to very low levels; this happens through the combination of Pfr destruction and autoregulation of Pr synthesis. Total phytochrome (i.e., both Types I and Types II) would then operate to detect R:FR ratio and mediate the shade avoidance responses via the action of Pfr.

On the second scenario, which is only subtly different, a Type II phytochrome (or a phytochrome-S) alone mediates R:FR ratio perception and shade avoidance. The presence of significant amounts of Type I in the light-grown plant would interfere with action of the responsible Type II because of the inhibitory effect of Type I Pfr on extension growth. Thus, Type I levels must be reduced upon de-etiolation to allow the much smaller amounts of Type II to operate effectively. Type I may still have a role in light-grown plants, as is evidenced by the demonstration of phenomena that are dependent on light-labile phytochrome (Smith and Whitelam, 1990).

On the first scenario, one would expect the existence of a large (ca 9-fold; R.D.Vierstra, personal communication) excess of Type I over Type II phytochrome in the transgenics to lead to depression of all responses to added FR, because the concentration of Pfr (i.e., Types I + Types II) would be very much higher than in the wild-types at all R:FR ratios. That this is not the case (e.g., chlorophyll a:b ratio is unaffected by the overexpression) suggests that Type I Pfr cannot do everything that Type II Pfr can do. This is probably an unsafe conclusion, but it is an interesting speculation. On the second scenario, the excess Type I phytochrome would operate as if the plant were still etiolated, and at all R:FR ratios enough Pfr would be present to cause severe extension growth inhibition. This means that the excess Type I overides the operation of Type II and in this way disrupts R:FR ratio perception and shade avoidance reactions. If it is assumed that regulation of chlorophyll a:b ratio is not a normal role for Type I phytochrome, then the typical response of this parameter to R:FR ratio in the transgenics may be understood.

How Does Fluence Rate Affect the Action of the Overexpressed Type I Phytochrome?. This issue is complicated by the effect of fluence rate on extension growth in the wild-types grown at high R:FR. Nevertheless, the *depression* of extension rate by added FR at low fluence rate in the transgenics is a striking phenomenon which is difficult to interpret. If the Type I effects are entirely due to the concentration of Pfr established, then it seems, on the face of the data, that there is more Pfr established at low R:FR and low fluence rate, than at high R:FR and low fluence rate. Also, for the transgenics grown at low R:FR, there appears to be more Pfr established at low, than at high, fluence rate.

A partial resolution of this apparent paradox might lie in a combination of the instability of Type I Pfr, and the accumulation of photoconversion intermediates at high fluence rates. Cherry *et al* (in press) have recently shown that oat Type I phytochrome in transgenic tobacco is degraded relatively slowly in the light, but that its synthesis is more-or-less unaffected by light. If degradation rate in continuous light is a function of Pfr concentration, total Type I phytochrome would be higher in those plants grown at low R:FR than at high. In the light sources used here, the WL establishes about 80% Pfr, whereas the WL+FR establishes about 10%. Furthermore, the plants were grown for 2 weeks under continuous illumination, conditions that might be expected to result in maximum degradation of the Type I oat phytochrome. On these grounds, the equilibrium level of total Type I phytochrome established between continued synthesis and degradation would be much higher at low R:FR than at high R:FR. Assuming this difference could be more than 10-fold, the paradox would be resolved. On this basis, however, it is difficult to understand why extension rate in the transgenics is higher at high than at low fluence rates. One may be forced back to the hypothesis (Smith, Jackson and Whitelam, 1988) that the accumulation of photoconversion intermediate levels at the high fluence rate causes a substantial reduction in effective Type I Pfr concentration, thus allowing a higher rate of extension.

CONCLUSIONS

Transgenic and mutant plants should, in principle, prove valuable in attempts to elucidate which phytochrome mediates which physiological response. However, we must expect very complex physiological results, and severe problems of interpretation; facile conclusions based on limited analyses of phenotypic characteristics will tend to be misleading. At the present time, it is not possible to conclude definitively that different phytochromes do, indeed, have different physiological roles. The balance of the evidence supports this attractive hypothesis, but its very attractiveness should make us extremely wary of premature conclusions. It is of crucial importance that effort is concentrated on the application of defined physiological tests, rather than reliance on a superficial observation of crude phenotypic changes. In other words, the proper analysis of transgenic and mutant plants will depend crucially upon the expert application of sophisticated photophysiological methods.

REFERENCES

Adamse P (1988) Mutants as an aid to the study of higher plant photomorphogenesis. PhD Thesis, Agricultural University, Wageningen, The Netherlands

Adamse P, Kendrick RE, Koornneef M (1988b) Photomorphogenetic mutants of higher plants. Photochemistry and Photobiology 48:833-841

Ballaré CL, Sânchez RA, Scopel AL, Casal JJ, Ghersa CM (1987) Early detection of neighbour plants by phytochrome perception of spectral changes in reflected sunlight. Plant Cell and Environment 10:551-557

Beggs CJ, Holmes MG, Jabben M, Schäfer E (1980) Action spectra for the inhibition of hypocotyl growth by continuous irradiation in light- and dark-grown Sinapis alba L. seedlings. Plant Physiology 66:615-618

Blaauw OH, Blaauw-Jensen G, Van Leeuwen WJ (1968) An irreversible red-light-induced growth response in *Avena*. Planta 82:87-104

Boylan MT, Quail PH (1989) Oat phytochrome is biologically active in transgenic tomatoes. The Plant Cell 1:765-773

Casal JJ, Smith H (1989) The function, action and adaptive significance of phytochrome in light-grown plants. Plant Cell and Environment 12:855-862

Chory J, Peto CA, Ashbaugh M, Saganich R, Pratt LH, Ausubel F (1989) Different roles for phytochrome in etiolated and green plants deduced from characterization of *Arabidopsis thaliana* mutants. The Plant Cell 1:867-880

Cone JW (1985) Photocontrol of seed germination of wild-type and long-hypocotyl mutants of *Arabidopsis thaliana*. PhD Thesis, Agricultural University, Wageningen, The Netherlands

Downs RJ, Hendricks SB, Borthwick HA (1957) Photoreversible control of elongation in Pinto beans and other plants under normal conditions of growth. Botanical Gazette 118:199-208

Hartmann KM (1966) A general hypothesis to interpret 'high energy phenomena' of photomorphogenesis on the basis of phytochrome. Photochemistry and Photobiology 5:349-366

Hillman WH (1967) The physiology of phytochrome. Annual Review of Plant Physiology 18:301-324

Holmes MG, Beggs CJ, Jabben M, Schäfer E (1982) Hypocotyl growth in *Sinapis alba* L: the roles of light quality and quantity. Plant Cell and Environment 5:45-51

Holmes MG, Smith H (1975) The function of phytochrome in plants growing in the natural environment. Nature 254:512-514

Kay SA, Keith B, Shinozaki K, Chye M-L, Chua N-H (1989a) The rice phytochrome gene: structure, autoregulated expression and binding of GT-1 to a conserved site in the 5' upstream region. The Plant Cell 1:351-360

Kay SA, Nagatani A, Keith B, Deak M, Furuya M, Chua N-H (1989b) Rice phytochrome is biologically active in transgenic tobacco. The Plant Cell 1:775-782

Keller J, Shanklin J, Vierstra RD, Hershey HP (1989) Expression of a functional monocotyledonous phytochrome in transgenic tobacco. EMBO Journal 8:1005-1012

Ken-Dror S, Horwitz BA (1990) Altered phytochrome regulation of greening in an *aurea* mutant of tomato. Plant Physiology 92:1004-1008

McCormac, AC, Vierstra RD, Hershey HH, Cherry J, Smith H (1991) Physiological analysis of the effects of the overexpression of an introduces oat phytochrome gene in transgenic tobacco seedlings. (in press)

Morgan DC, Smith H (1976) Linear relationship between phytochrome photoequilibrium and growth in plants under natural radiation. Nature 262:210-212

Morgan DC, Smith H (1978) The relationship between phytochrome photoequilibrium and development in light grown *Chenopodium album* L. Planta 142:187-193

Oelmüller R, Kendrick RE, Briggs WR (1989) Blue-light mediated accumulation of nuclear-encoded transcripts coding for proteins of the thylakoid membrane is absent in the phytochrome-deficient *aurea* mutant of tomato. Plant Molecular Biology 13:223-232

Parks BM, Jones AM, Adamse P, Koornneef M, Kendrick RE, Quail PH (1987) The *aurea* mutant of tomato is deficient in spectrophotometrically and immunochemically detectable phytochrome. Plant Molecular Biology 9:97-107

Parks BM, Shanklin J, Koornneef M, Kendrick RE, Quail PH (1989) Immunochemically detectable phytochrome is present at normal levels but is photochemically non-functional in the *hy1* and *hy2* long hypocotyl mutants of *Arabidopsis*. Plant Molecular Biology 12:425-437

Raven CE, Spruit CJP (1973) Induction of rapid chlorophyll accumulation in dark-grown seedlings. III. Transport model for phytochrome action. Acta botanica Neerlandica 22:135-143

Sharrock RA, Parks BM, Koornneef M, Quail PH (1988) Molecular analysis of the phytochrome deficiency in an *aurea* mutant of tomato. Molecular and General Genetics 213:9-14

Sharrock RA, Quail PH (1989) Novel phytochrome sequences in *Arabidopsis thaliana*: structure evolution and differential expression of a plant regulatory photoreceptor family. Genes and Development 3:1745-1757

Smith H (1982) Light quality photoperception and plant strategy. Annual Review of Plant Physiology 33:481-518

Smith H, Casal JJ, Jackson GM (1990) Reflection signals and the perception by phytochrome of the proximity of neighbouring vegetation. Plant Cell and Environment 13:73-78

Smith H, Holmes MG (1977) The function of phytochrome in the natural environment III. Measurement and calculation of phytochrome photoequilibrium. Photochemistry and Photobiology 25:547-550

Smith H, Jackson GM, Whitelam GC (1988) Photoprotection of phytochrome. Planta 175:471-477

Smith H, Whitelam GC (1990) Phytochrome, a family of photoreceptors with multiple physiological roles. Plant, Cell and Environment 13:695-707

THE PHYSIOLOGY OF PHOTOMORPHOGENETIC TOMATO MUTANTS

J.L. Peters[1], J.C. Wesselius[1], K.C. Georghiou[3], R.E. Kendrick[1], A. van Tuinen[2], and M. Koornneef[2]
Departments of [1]Plant Physiology Research and [2]Genetics,
Wageningen Agricultural University,
Generaal Foulkesweg 72,
NL-6703 BW Wageningen,
The Netherlands.

INTRODUCTION

Photomorphogenesis of higher plants is a complex process resulting from the co-action of at least 3 different photoreceptors: phytochrome, a blue light (B)/UV-A photoreceptor (cryptochrome) and a UV-B photoreceptor (Mohr, 1986). The existence of multiple photoreceptor types, *e.g.* type I (PI) or light-labile phytochrome and type II (PII) or light-stable phytochrome, adds to the complexity (Furuya, 1989; Tomizawa *et al.*, 1990). The assignment of specific functions to the distinct molecular species of the photoreceptor is therefore being studied with the aid of photomorphogenetic mutants in which certain parts of the morphogenetic pathway are eliminated or altered. The relevance of the changed part in the mutant is directly indicated by its difference in response compared to its isogenic wild type (Koornneef and Kendrick, 1986). Mutants can be found (isolated) from natural populations or varieties (cultivars) or more efficiently after mutagenic treatment: using *e.g.* chemicals, irrradiation; somaclonal variation; transposon insertion; transformation; introduction of antisense RNA.

Photomorphogenetic mutants can be divided into three groups: *photoreceptor* mutants, lacking the photoreceptor or containing a modified photoreceptor which is non-functional; *transduction chain* mutants and *response* mutants. The first two will be pleiotropic for all responses regulated by the photoreceptor while the latter are modified with respect to particular responses. Tomato (*Lycopersicon esculentum* Mill.) has several features which make it suitable for genetic analysis. It is widely studied since it is a crop species of economic importance and many mutants are available (Hille *et al.*, 1989; Tanksley and Mutschler, 1990). In addition it has a relatively small genome, is diploid

[3]Department of General Botany, University of Athens, 157-84 Athens, Greece.

NATO ASI Series, Vol. H 50
Phytochrome Properties and Biological Action
Edited by B. Thomas and C. B. Johnson

with 12 chromosomes, is self pollinating, individual plants producing a large number (≈2000) of relatively large seeds which result in seedlings suitable for physiological analysis and it is amenable for *Agrobacterium*-mediated transformation (McCormick *et al.*, 1986; Koornneef *et al.*, 1987). In this review mutants of tomato which are important for the study of photomorphogenesis are discussed.

SURVEY OF MUTANTS

Aurea mutants. During selection of gibberellin (GA)-responsive mutants, a mutant was isolated that required GA for germination, but in contrast to GA-deficient mutants was characterized by a long hypocotyl and a marked reduction in chlorophyll content when grown in white light (WL) (Koornneef *et al.*, 1981). A genetic analysis revealed that this recessive mutant was allelic with the previously described *aurea* (*au*) mutant. This gene is located on chromosome 1. Another mutant at the *au* locus was recently isolated in the progeny of tomato plants derived from tissue culture by Lippuci di Paola *et al.* (1988).

The *au* mutant is the best characterized example of a higher plant photomorphogenetic photoreceptor mutant and contains less than 5% of the spectrophotometrically (Koornneef *et al.*, 1985) and immunologically (Parks *et al.*, 1987; Oelmüller *et al.*, 1989) detectable phytochrome in etiolated tissues as compared to its isogenic wild type. However, light-grown tissues of the *au* mutant contain about 50% of the *in vivo* spectrophotometrically detectable (Adamse *et al.*, 1988b) and extracted (López-Juez *et al.*, 1990b) phytochrome compared to wild type, suggesting that the light-stable phytochrome pool accumulates (presumably PII).

Investigation of the nature of the *au* lesion was carried out by Parks *et al.* (1987) and Sharrock *et al.* (1988) and they have shown that phytochrome mRNA is produced to a similar extent in *au* as wild type and that this mRNA is functional in an *in vitro* translation system, yet *in vivo* the protein fails to accumulate. In addition, there appear to be multiple genes coding for phytochrome in tomato (Sharrock *et al.*, 1988; Hauser and Pratt, 1990). One gene has been mapped to chromosome 10, whereas the *au* locus is situated on chromosome 1 (Sharrock *et al.*, 1988). Therefore the deficiency of spectrophotometrically active phytochrome in the *au* mutant appears not to result from a mutation of the structural gene or the lack of PI gene expression.

One aspect of the *au* phenotype is its reduced germination in darkness (D) compared to wild type, which can be overcome by GA (Koornneef *et al.*, 1985; Table 1). Since GA influences seedling growth and development, where seedlings are required for study of their photomorphogenesis, conditions have been sought which result in high germination of *au* seed batches in D. Recently, improvement of germination by chilling, high temperature and/or nitrate treatment have been studied. The freshly harvested seeds, which are dormant, can be induced to germinate after treatment with a the combination of chilling and nitrate. Moreover, exposure to continuous red light (R) led to an increase in germination of *au* seed batches, while no inhibitory effect of continuous far-red light (FR) was observed, in contrast to wild type which exhibits a strong irradiance dependent inhibition of germination (Table 1). Lipucci di Paola *et al.* (1988) even found a promotion of seed germination by FR for *au* mutants and suggested that this is the consequence of the absence of a FR high irradiance reaction (HIR).

Table 1. Germination percentages ± SE of wild-type and *au* tomato seeds (cv. Moneymaker) harvested in different years and imbibed in either H_2O or in 10^{-5} *M* GA_{4+7}, under different light conditions: continuous red (cR); continuous far-red (cFR); darkness (D) at 25°C (K.C. Georghiou, unpublished data).

Harvest	Wild type		*au*			
year	D	cFR	D	cR	cFR	D
	H_2O	H_2O	H_2O	H_2O	H_2O	GA_{4+7}
1983	ND	ND	45±3	97±2	44±3	98±2
1984	ND	ND	13±1	60±2	12±1	95±2
1985	ND	ND	15±3	90±1	13±3	92±4
1988	96±2	7±2	22±2	96±1	28±3	95±3
1989	89±3	8±3	5±1	18±2	4±2	95±2

ND: not determined.

At the etiolated seedling stage, the *au* mutant is characterized by a reduction in: hypocotyl growth inhibition in WL (Koornneef *et al.*, 1985), R, B and UV-A (Adamse *et al.*, 1988b); chlorophyll and chloroplast development (Koornneef *et al.*, 1985; Ken-Dror and Horwitz, 1990); anthocyanin content (Adamse *et al.*, 1989; Peters *et al.*, 1989) and the photoregulation of the transcript levels of chlorophyll *a/b*-binding proteins of photosystem

I and II, plastocyanin and subunit II of photosystem I (Sharrock *et al.*, 1988; Oelmüller *et al.*, 1989). This pleiotropic phenotype, coupled with lack of phytochrome is precisely that predicted for a photoreceptor mutant.

Although the adult light-grown *au*-mutant plants are taller and have yellow leaves compared to the wild type (Table 2) they survive and produce fruits under daylight conditions. Apparently the phytochrome content, which is below detection limits in etiolated *au*-mutant seedlings, is sufficient for survival. Oelmüller and Kendrick (submitted) showed that the levels of 4 nuclear-encoded transcripts coding for plastidic proteins are strongly reduced in the *au* seedlings exposed to R, while B leads to wild-type levels of these transcripts in the *au* mutant. They concluded that B absorbed by the B/UV-A photoreceptor sensitizes the mutant to the low level of phytochrome present and that this allows normal gene expression and survival of the mutant under daylight conditions. Anthocyanin synthesis in seedlings is also strongly reduced in the *au* mutant and is discussed in detail below.

In adult light-grown plants both wild type and the *au* mutant exhibit a quantitatively similar elongation growth response to end-of-day FR (Downs *et al.*, 1957) treatment indicating the presence of functional phytochrome in light-grown *au*-mutant plants

Table 2. Total chlorophyll content and the effect of a 20 min far-red (FR) pulse at the end of 7 daily photoperiods of 14 h white fluorescent light (25 Wm^{-2}) on plant height and anthocyanin content of comparable leaf samples of wild-type (WT), *au*, *hp*, *pro* and *auhp* tomato plants (cv. Ailsa Craig) (J.L. Peters and J.C. Wesselius, unpublished data).

Experiment	Genotype	Total Chlorophyll (mg.g fr wt^{-1})	Increase in height (mm)		Anthocyanin content (A_{535}.g fr wt^{-1})	
			-FR	+FR	-FR	+FR
1	WT	3.10	76.3	146.3	1.79	0.57
	au	1.41	103.6	200.7	ND*	ND
	hp	3.01	39.0	90.0	1.67	0.43
	auhp	1.62	82.0	160.7	ND	ND
2	WT	ND	23.1	55.1	4.7**	1.27**
	pro	ND	93.9	161.1	0.52**	0.15**

*ND: not determined **Measurement after 20 days treatment

(Adamse *et al.*, 1988b; López-Juez *et al.*, 1990b and Table 2). Although the most plausible inference is to ascribe this response to PII which is predicted to accumulate in the mutant, the molecular nature of phytochrome detected in light-grown tissues (Adamse *et al.*, 1988b; López-Juez *et al.*, 1990b) has not yet been positively identified, since PI and PII specific antibodies are not yet available for tomato.

Other long-hypocotyl mutants. A recessive mutant which has a phenotype similar, but less extreme than that of *au* has been called *yellow green* (*yg-2*) and has been mapped to chromosome 12 (Koornneef *et al.*, 1985). An additional mutant allele at the same locus called *auroid (aud)* was later isolated. The *yg-6* mutant has also an *au* phenotype and was demonstrated to be an *au* allele. The location of *yg-6* on chromosome 11 can be explained by a translocation between chromosome 1 and 11 during the radiation induction (Koornneef *et al.*, 1986). The *yg-2* mutants, like the *au* mutants are characterized by elongated hypocotyls, reduced germination, and a reduced anthocyanin and chlorophyll content (Koornneef *et al.*, 1985; Buurmeijer *et al.*, 1987). Electron micrographs showed that the number of thylakoids and their degree of stacking in the *au* and the *yg-2* mutants is greatly reduced (Koornneef *et al.*, 1985). The *yg-2* mutant is less extreme than the *au* mutant and older etiolated seedlings have been shown to accumulate some spectrophotometrically detectable phytochrome (Koornneef *et al.*, 1985).

The phenotype of the *procera* (*pro*) mutant is remarkably similar to that of the wild type treated with GA (Jones 1987; Jupe *et al.*, 1988). However, this mutant does not have increased GA levels (Jones 1987). It was suggested by Adamse *et al.* (1988a) that the *pro* mutant could be similar to the cucumber *lh* mutant and is a possible candidate for a mutant deficient in the function of light-stable phytochrome. However, in contradiction to this possibility, *pro* exhibits an end-of-day FR elongation response (Table 2), which is absent in the *lh* mutant (López-Juez *et al.*, 1990a). The double mutant *aupro* survives and shows characteristics of both the *au* and *pro* mutant: the yellow-green leaves of the *au* mutant and a reduction in indentation of the main leaflets and suppression of the development of minor leaflets (juvenile appearance) of the *pro* mutant. In addition, the light-grown *aupro* double mutant is taller than either the *au* or *pro* mutant (Table 3). Crossing the *pro* mutant with a GA-deficient mutant (Koornneef *et al.*, in press), which is an extreme dwarf and requires spraying with GA to develop flowers and fruits, results

Table 3. The mean internode length ± SE of mature of wild-type (WT), *au*, *hp*, *pro*, *auhp*, *aupro* tomato plants (cv. Ailsa Craig) growing in a greenhouse during the summer.

Genotype	Internode length (cm)
WT	5.30± 0.27
au	6.96± 0.29
hp	5.23± 0.35
pro	7.38± 0.47
auhp	5.85± 0.28
aupro	7.69± 0.39

in a plant with *pro* characteristics which no longer requires GA spraying. In other words, the GA deficiency is compensated by the *pro* mutation. Perhaps the *pro* mutation influences the GA receptor site or a step between the site of perception of the GAs and the response, allowing the activation of GA-mediated responses, such as stimulation of elongation growth. Since GAs have been proposed to be involved in the end-of-day FR response (García-Martínez *et al.*, 1987), the *pro* mutant can not be saturated for the particular GA response involved, since it exhibits a typical end-of-day FR response (Table 2).

High pigment mutants. The monogenic recessive *hp* mutant shows characteristics opposite to the *au* phenotype: seedlings have an increased anthocyanin content (Adamse *et al.*, 1989; Peters *et al.*, 1989), a reduced hypocotyl length in R, B, UV-A (Peters *et al.*, 1989) and yellow light (Mochizuki and Kamimura, 1985) compared to wild type. Moreover, in light-grown plants the chlorophyll content is particularly high in immature fruit tissues (Sanders *et al.*, 1975) and mature fruits have a higher lycopene and carotenoid content than wild type (Kerr, 1965). Unlike wild type, the *hp* mutant does not require the activation of the B photoreceptor to exhibit high levels of anthocyanin synthesis and enables complete de-etiolation under R (Peters *et al.*, 1989). Since etiolated *au* and *auhp* seedlings, both deficient in PI, show either no or a small (3% compared to *hp*) R/FR reversible anthocyanin response after a B pretreatment, respectively (Adamse *et al.*, 1989), it has been concluded that it is the PI pool which regulates anthocyanin synthesis at the seedling stage and that the *hp* mutation does not result in the constitutive expression of genes involved in anthocyanin synthesis. Fluence-rate response curves for induction of anthocyanin under continuous B are shown in Figure 1 for wild type, *hp*, *au* and *auhp*. While the *au* mutation almost completely prevents anthocyanin

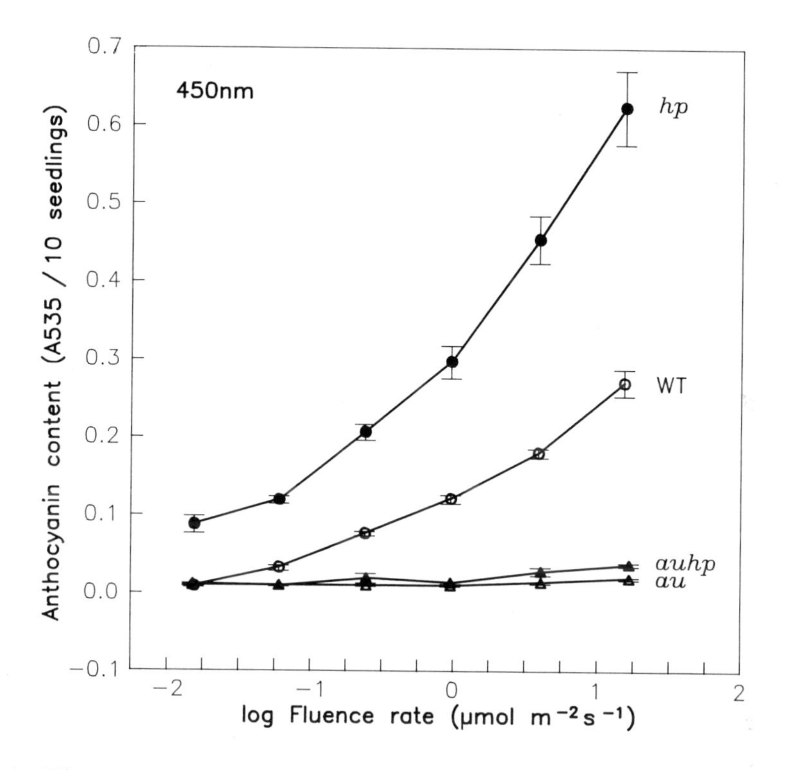

Figure 1. Fluence rate response curves for the accumulation of anthocyanin (A535) during a 24-h irradiation period with 450 nm blue light, in 3-day-old etiolated wild-type (WT), *au*, *hp* and *auhp* tomato seedlings (cv. Ailsa Craig) at 25° C. In control plants exposed to 24 h darkness, no anthocyanin could be detected. Results represent the mean ± SE for 4 replicate experiments (J.L. Peters unpublished data).

accumulation, there is a small yet significant accumulation at the highest fluence rates in the *auhp* double mutant. Photoregulation key enzyme genes in the flavonoid biosynthesis pathway (phenylalanine ammonia lyase and chalcone synthase) are currently under investigation in these mutants (R.P. Sharma, pers. comm.).

The nature of the processes influenced by the *hp* mutation is unknown. The phytochrome content of comparable samples of *hp* and wild-type etiolated seedlings is similar (Peters *et al.*, 1989). Therefore the difference observed can not be explained by a higher absolute level of the active FR-absorbing form of phytochrome (Pfr). Peters *et al.* (1989) proposed that the *hp* mutation increases the responsiveness to Pfr. Whether the *hp* mutant is: (i) hypersensitive to Pfr, (ii) only exhibits a higher response to a specific amount of Pfr or (iii) lacks inhibitors of Pfr action is currently under investigation.

Besides mutants with *hp* characteristics which are allelic with the *hp* mutation *e.g.* the dark-green mutant of Manapal tomato (Sanders *et al.*, 1975), $hp^{\underline{w}}$ (collection number WB3) (Peters *et al.*, 1989), there are also mutants which are similar in some aspects of the mutant phenotype of *hp*, but are non-allelic with *hp*, such as *hp-2* (Soressi, 1975), *atroviolatia* (*atv*) (Rick *et al.*, 1968) and intensive pigment (*Ip*) (Rick 1974). Mutants with a similar phenotype to the *hp* mutants, but which are caused by mutations in other genes are particularly interesting and will assist our understanding of photomorphogenesis.

Furthermore, plants with *hp* characteristics at the seedling stage (*i.e.* short hypocotyls with high anthocyanin levels) were obtained when high levels of an oat PI gene were expressed in tomato (Boylan and Quail, 1989). Light-grown transgenic plants classified as either null or as low-level expressors resembled wild type, while high-level expressors, in contrast to *hp*, were extreme dwarfs, with dark-green foliage and fruits. Regardless of whether the height of the adult plant was normal or dwarf, seedlings expressing oat phytochrome all had short hypocotyls at the seedling stage. The overproduction of oat PI results in the persistence of oat PI phytochrome in light-grown plants, demonstrating that PI can be biologically active in fully green tissue. This does not necessarily mean that tomato PI plays a dominant role in elongation growth of light-grown wild-type plants, since quantitative difference in the rate and the extent of degradation of the Pfr forms of tomato and oat phytochrome in the transgenic plants were observed. Boylan and Quail point out that the *au* mutant, wild type and transgenic PI overexpressors represent a continuum of phenotypic expression in response to increasing levels of PI. Although the *hp* mutant shows characteristics at the seedling stage similar to the transgenic PI overexpressors, it does not have an elevated PI level.

CONCLUDING REMARKS

Our studies of mutants enable us to suggest possible functions of PI and PII in tomato (Table 4). The lack of the large PI pool in the *au* mutant clearly correlates with its

Table 4. A tentative assignment of the physiological roles played by PI and PII phytochromes in tomato.

Response	PI	PII
Seed germination		
Promotion		+
Inhibition	+	
Depth of seed dormancy	+	
De-etiolation		
Gene expression	+	
Growth inhibition	+	
End-of-day far-red light		+

difficulty to de-etiolate. However, despite the reduced chlorophyll levels in light-grown *au*-mutant plants compared to wild type, presumably due to a continuing deficiency in PI, they exhibit a strong growth elongation response to end-of-day FR, which we propose is regulated by PII. Only further experiments will reveal if this hypothesis of discrete functions for the PI and PII pools is correct. To do this, specific probes will be necessary to determine the phytochrome species present at different stages during development, as well as the tissue specificity for the expression of the different phytochrome genes.

ACKNOWLEDGEMENTS

Supported by the Foundation for Fundamental Biological Research (BION), which is subsidized by The Netherlands Organization for the Advancement of Research (NWO). We are most grateful to Gerda H Heeringa, Mariëlle E L Schreuder, Corrie J Hanhart and Sebastiaan Verduin for valuable technical assistance and the greenhouse staff of the Genetics Department, Wageningen Agricultural University, for cultivation of the plants.

REFERENCES

Adamse P, Kendrick RE, Koornneef M (1988a) Photomorphogenic mutants of higher plants. Photochem Photobiol 48:833-841

Adamse P, Jaspers PAPM, Bakker JA, Wesselius JC, Heeringa GH, Kendrick RE, Koornneef M (1988b) Photoregulation of a tomato mutant deficient in labile phytochrome. J Plant Physiol 133:436-440

Adamse P, Peters JL, Jaspers PAPM, van Tuinen A, Koornneef M, Kendrick RE (1989) Photocontrol of anthocyanin synthesis in tomato seedlings: A genetic approach. Photochem Photobiol 50:107-111

Boylan MT, Quail PH (1989) Oat phytochrome is biologically active in transgenic tomatoes. The Plant Cell 1:765-773

Buurmeijer WF, Roelofs TA, Vredenberg WJ (1987) Some aspects of altered structure and function of the photosynthetic apparatus in phytochrome-less mutants of tomato. In: Biggins J (ed) Progress in Photosynthetic Research. Vol. II. Martinus Nijhoff Publ., Dordrecht pp.383

Downs RJ, Hendricks SB, Borthwick HA (1957) Photoreversible control of elongation in pinto beans and other plants under normal conditions of growth. Bot Gaz 118:199-208

Furuya M (1989) Molecular properties and biogenesis of phytochrome I and II. Adv Biophys 25:133-167

García-Martínez JL, Keith B, Bonner BA, Stafford AE, Rappaport L (1987) Phytochrome regulation of the response to exogenous gibberellins by epicotyls of *Vignor sinensis*. Plant Physiol 85:212-216

Hauser B, Pratt LH (1990) Initial characterization of tomato phytochrome genes. Plant Physiol (Suppl) 93:137

Hille J, Koornneef M, Ramanna MS, Zabel P (1989) Tomato: a crop species amenable to improvement by cellular and molecular methods. Euphytica 42:1-24

Jones, MG (1987) Gibberellins and the *procera* mutant of tomato. Planta 172:280-284

Jupe SC, Causton DR, Scott IM (1988) Cellular basis of the effects of gibberellin and the *pro* gene on stem growth in tomato. Planta 174:106-111

Ken-Dror S, Horwitz BA (1990) Altered phytochrome regulation of greening in an *aurea* mutant of tomato. Plant Physiol 92:1004-1008

Kerr EA (1965) Identification of high-pigment, *hp*, tomatoes in the seedling stage. Can J Plant Sci 45:104-105

Koornneef M, van de Veen JH, Spruit CJP, Karssen CM (1981) Isolation and use of mutants with an altered germination behaviour in *Arabidopsis thaliana* and tomato. In: Induced mutations - A Tool in Plant Breeding. International Atomic Energy Agency, Vienna, p227

Koornneef M, Cone JW, Dekens RG, O'Herne-Robers EG, Spruit CJP, Kendrick RE (1985) Photomorphogenic responses of long-hypocotyl mutants of tomato. J Plant Physiol 120:153-165

Koornneef M, Bosma TDG, de Jong M, Ramanna MS, van de Veen JH (1986) *Yellow-green* (*yg-6*) is an allele of *aurea* (*au*). Tomato Genet Coop Rpt 36:8-9

Koornneef M, Jongsma M, Weide R, Zabel P, Hille J (1987) Transformation of tomato. In: Nevins DJ, Jones RA (eds) Tomato Biotechnology. Alan R Riss Inc., New York, p169

Koornneef M, Kendrick RE (1986) A genetic approach to photomorphogenesis. In:

Kendrick RE, Kronenberg GHM (eds) Photomorphogenesis in Plants. Martinus Nijhoff Publ., Dordrecht, p521

Koornneef M, Bosma TDG, Hanhart CJ, van der Veen JH, Zeevart JAD (to be published) The isolation and characterization of gibberellin-deficient mutants in tomato. Theor Appl Genet

Lipucci di Paola M, Collina Grenci F, Caltavuturo L, Tognoni F, Lercari B (1988) A phytochrome mutant from tissue culture of tomato. Adv Hort Sci 2:30-32

López-Juez E, Buurmeijer WF, Heeringa GH, Kendrick RE, Wesselius JC, (1990a) Response of light-grown wild-type and long-hypocotyl mutant cucumber plants to end-of-day far-red light. Photochem Photobiol 52:143-150

López-Juez E, Nagatani A, Buurmeijer WF, Peters JL, Furuya M, Kendrick RE, Wesselius JC (1990b) Response of light-grown wild-type and *aurea*-mutant tomato plants to end-of-day far-red light. J Photochem Photobiol B Biology 4:391-405

McCormick S, Niedermeyer J, Fry J, Barnason A, Horsch R, Fraley R (1986) Leaf disc transformation of cultivated tomato (*L. esculentum*) using *Agrobacterium tumefaciens*. Plant Cell Reports 5:81-84

Mochizuki T, Kamimura S (1985) Photoselective method for selection of *hp* at the cotyledon stage. Tomato Genet Coop Rpt 35:12-13

Mohr H (1986) Coaction between pigment systems In: Kendrick RE, Kronenberg GHM (eds) Photomorphogenesis in Plants. Martinus Nijhoff Publ., Dordrecht, p547

Oelmüller R, Kendrick RE, Briggs WR (1989) Blue-light mediated accumulation of nuclear-encoded transcripts coding for proteins of the thylakoid membrane is absent in the phytochrome-deficient *aurea*-mutant of tomato. Plant Mol Biol 13:223-232

Oelmüller R, Kendrick RE (submitted) Blue light is required for survival of the tomato phytochrome-deficient *aurea* mutant and the expression of four nuclear genes coding for plastidic proteins. Plant Mol Biol.

Parks BM, Jones AM, Adamse P, Koornneef M, Kendrick RE, Quail PH (1987) The *aurea* mutant of tomato is deficient in spectrophotometrically and immunocytochemically detectable phytochrome. Plant Mol Biol 9:97-107

Peters JL, van Tuinen A, Adamse P, Kendrick RE, Koornneef M (1989) High pigment mutants of tomato exhibit high sensitivity for phytochrome action. J Plant Physiol 134:661-666

Rick CM (1974) High soluble-solids content in large-fruited tomato lines derived from wild green-fruited species. Higardia 42:493-510

Rick CM, Reeves AF, Zobel RW (1968) Inheritance and linkage relations of four new mutants. Tomato Genet Coop Rpt 18:34-35

Sanders DC, Pharr DM, Konsler TR (1975) Chlorophyll content of a dark-green mutant of 'Manapal' tomato. Hort Science 10: 262-264

Sharrock, RA, Parks BM, Koornneef M, Quail PH (1988) Molecular analysis of the phytochrome deficiency in an *aurea* mutant of tomato. Mol Gen Genet 213:9-145

Soressi GP (1975) New spontaneous or chemically-induced fruit-ripening tomato mutants. Tomato Genet Coop Rpt 25:21-22

Tanksley SD, Mutschler MA (1990) Linkage map of tomato (*Lycopersicon esculentum*) (2N = 24). In: O'Brian SJ (ed) Genetic Maps. Cold Spring Harbor Laboratory Press, p6.3

Tomizawa K, Nagatani A, Furuya M (1990) Phytochrome genes: Studies using the tools of molecular biology and photomorphogenetic mutants. Photochem Photobiol 52:265-275

IMMUNOLOCALIZATION OF PHYTOCHROME AND IMMUNODETECTION OF ACTIN IN *MOUGEOTIA*

C. Hanstein, U. Lange, H.A.W. Schneider-Poetsch[1], F. Grolig and G. Wagner
Botanisches Institut I,
Membran- und Bewegungsphysiologie,
Justus-Liebig-Universität,
Senckenbergstrasse 17,
D-6300 Giessen, Germany.

INTRODUCTION

The photomorphogenic and photomodulatory pigment phytochrome (Rüdiger, 1980) induces diverse photoresponses in plants such as germination and flowering (Furuya, 1987), but also chloroplast movement in *Mougeotia*, *Mesotaenium* and *Adiantum* (Wada and Kadota, 1989). Recently, two different species of phytochrome, phytochrome I and phytochrome II (Furuya, 1989), have been discovered with considerable difference in amino acid sequence in higher plants, and a family of possibly five phytochrome genes has been identified in *Arabidopsis thaliana* (Sharrock and Quail, 1989). At present, no information is available on the action mechanism(s) of phytochrome and little information on the evolution of its gene(s) in higher and lower plants, including algae (Cordonnier *et al.*, 1986; Furuya, 1987; Dring, 1988; Wada and Kadota, 1989; López-Figueroa *et al.*, 1989). The Conjugatophycean green algae *Mougeotia* and *Mesotaenium* clearly show phytochrome-mediated responses, and *Mesotaenium* phytochrome has been purified and preliminarily characterized (Kidd and Lagarias, 1990).

Phytochrome, predominant in governing chloroplast re-orientation in *Mougeotia* and *Mesotaenium*, provides the cell with proper information on the quality and direction of light relevant for photosynthesis. Directionality is imprinted in a particular pattern (tetrapolar gradient; Haupt, 1982) of activated photoreceptor molecules, rapidly changed by alteration of the light direction through ultrafast photoreactions, with slowly responding cytoskeletal components. The key question in algae alike *Mougeotia* and *Mesotaenium*, but also in moss, fern and possibly higher plants is the mechanism, by

[1]Botanisches Institut der Universität zu Köln, Gyrhofstrasse 15, D-5000 Köln 41, Germany.

NATO ASI Series, Vol. H 50
Phytochrome Properties and Biological Action
Edited by B. Thomas and C. B. Johnson

which photoconverted phytochrome (Pfr) interacts with the cytoskeleton (Grolig and Wagner, 1988; Dring, 1988; Wada and Kadota, 1989). In non-muscle cells, such proteins including α-actinin and spectrin have well been described (Pollard and Cooper, 1986).

Our analysis centres at the identification of phytochrome-regulated membrane anchorage sites to actin in higher and in lower plants; immunolocalization and -detection of phytochrome and actin, respectively, will mainly be shown here in the lower plant *Mougeotia*.

METHODS

Plant materials. Mougeotia scalaris (Hassal) was cultured as described before (Russ *et al.*, 1988). *Glycine max* (soy bean), *Phaseolus coccineus* (scarlet runner) or *Phaseolus vulgaris* (kidney bean) were germinated in hydroculture (tapwater) in the greenhouse for 6 days. The roots were harvested and homogenized for immunoblotting in a multimix blender (MX 32, Braun, Frankfurt, Germany) at 4°C in extraction buffer of 25 mM imidazole pH 7.5, 0.5 mM ATP, 0.1 mM $CaCl_2$, 1 mM ß-mercaptoethanol, 5 mM phenylmethyl - sulfonylfluoride (PMSF) and 0.02 g polyclar AT ml^{-1} extraction buffer. *Mougeotia* was homogenized in liquid nitrogen in a mortar with the frozen powder then resuspended in the following extraction buffer: 0.1 M Tris pH 7.5, 0.2 M Na-tetraborate, 20 mM Na-metabisulfite, 0.25 M ascorbic acid, 26 mM diethyldithio-carbamic acid (Na-salt), 0.1 M maleic acid, 10% (w/v) sucrose, 10% (v/v) glycerol, 0,1% (v/v) 2-mercaptoethanol.

Immunoblotting. The *Mougeotia* or root extracts were centrifuged for 5 min or 20 min at 10.000 x g at 4°C, and the root supernatants further separated by 100.000 x g for 1 h. Protein of the final root supernatant was precipitated on ice by addition of an equal volume of 20% (w/v) trichloroacetic acid (TCA) to inhibit proteolysis. The same was done with the *Mougeotia* supernatant after the first centrifugation. After the extracts had been allowed to precipitate for 20 min, the protein pellet was collected by centrifugation at 10,000 x g for 10 min. The pellet was washed three times with 200 μl of acetone to remove chlorophyll where present and residual TCA, using the same conditions of centrifugation as before. The final air-dried pellet was resuspended and dissolved in sodium dodecyl sulfate sample buffer (Laemmli, 1970) with 2 min discontinuous mechanical shaking and 5 min heating in boiling water, or 15 min stirring by magnetic bar and 5 min heating in boiling water followed by 5 min stirring. Insoluble residue was pelleted at 10.000 x g for 10 min, and an aliquot of supernatant was resolved by SDS-PAGE on a 10% or a 12.5% mini-gel. Proteins were transferred overnight to nitrocellulose at 6.7 V cm^{-1}. The nitrocellulose was briefly washed in distilled water and stained for 15 min in 4% (w/v) Ponceau S dissolved in 3% (w/v) TCA to monitor transfer efficiency of total protein. Blocking was for 45 min in 5% (w/v) skim milk powder in Tris-buffered saline (10 mM Tris, 150 mM NaCl, pH 7.5) at room temperature and for 15 min in TBS-Tween (TBS plus 0.05% (v/v) Tween 20) at 37°C. Monoclonal anti-actin (C4, courtesy of J. Lessard, University of Cincinnati, Ohio, U.S.A; Lessard, 1988) was diluted in 1% (w/v) bovine serum albumin (BSA) in TBS-Tween and applied (5 μg/ml) for 1 h. After thorough washing in TBS-Tween, bound antibody was detected by a sandwich assay made of goat anti-mouse (IgG, 2 μg ml^{-1} for 1 h) followed by alkaline phosphatase, coupled to rabbit anti-goat IgG with glutaraldehyde (1 μg ml^{-1} for 1 h). Blots were developed in 0.17 mg ml^{-1} 5-bromo-4-chloro-3-indolylphosphate p-toluidine salt (BCIP), 0.33 mg ml^{-1} nitroblue tetrazolium chloride (NBT), 100 mM NaCl, 50 mM $MgCl_2$, 100 mM Tris-HCl, pH 9.0. NBT was prepared as a 250-fold concentrated stock solution in 70% (v/v) dimethlyformamide (Grolig *et al.*, 1988).

Isolation of Mougeotia protoplasts and immunofluorescence. About 2 g wet weight of algae were suspended in 100 ml 0.4 M mannitol, 1 mM $CaCl_2$, 1% (w/v) Cellulase TC (Serva, Heidelberg, Germany) in culture medium (Stabenau, 1978) at pH 5.0, and shaken at 50 rpm at 16°C overnight. Protoplasts were filtered successively through 120- and 50- μm nylon mesh to remove debris and washed in the same medium without enzyme by sedimentation at 1 x g. *Mougeotia* protoplasts, suspended in 100 μl of culture medium were fixed for 15 min in 440 μl of 1.5% (w/v) formaldehyde plus 0.5% (v/v) glutardialdehyde, 0.4 M mannitol in microtubule-stabilizing buffer (MTB) of 100 mM PIPES, 10 mM EGTA, 5 mM $Mg_2SO_4.7H_2O$, pH 6.9 (Traas *et al.*, 1987) for 15 min, followed by washing in 0.4 M mannitol in MTB for 10 min. The fixed protoplasts were incubated for 1 h in 40 μl of monoclonal anti-phytochrome (Z-3B1, mouse IgG, Schneider-Poetsch *et al.*, 1988) diluted 1:5 in MTBX (MTB including 1% (w/v) BSA, 0.03% Triton X-100 and 0.4 M mannitol). After washing twice in 40μl MTBX each, the protoplasts were incubated for 1 h in 90 μl MTBX with fluoresceine-isothiocyanate

(FITC)-labelled secondary antibody (sheep anti mouse IgG, Amersham N 1031, diluted 1:50), followed by washing twice in 200 μl MTBX. The stained protoplasts were incubated in mowiol 4-88 (Hoechst, Germany) plus 0.1% (w/v) phenylenediamine (Johnson and De Nogueira Araujo, 1981) overnight in darkness and inspected in the fluorescence microscope. Controls were run either in absence of both species of antibodies to check autofluorescence or in presence of the FITC-labelled secondary antibody only to check for nonspecific binding. Epifluorescence microscopy on a Leitz Diaplan used a Leitz Ploemopak L3 and a Fluotar 100x, NA 1.32 as the main objective. Photographs were taken on Kodak TMax 400 at 1600 ASA.

Actin viscometric assay, and plant factor to depolymerize F-actin. Bovine heart from a local slaughterhouse was used as an actin source. Actin was prepared according to the isolation procedure of Pardee and Spudich (1982) with a slight modification: the KCl concentration to separate tropomyosin in the supernatant from F-actin in the pellet was increased to 0.8 M KCl, followed by a one step ultracentrifugation and dialysis as described. Isolated bovine heart actin was purified from contamination of an endogenous actin-depolymerizing factor by Sephadex G-150 gel filtration (Pharmacia LKB, Freiburg, Germany); column dimensions in cm were (height/radius = H/R) 100/1 with a flow rate of 25 ml h^{-1} of the buffer: 5 mM Tris, 0.5 mM DTT, 0.2 mM ATP, 0.2 mM $CaCl_2$, pH 8.0 at 4°C (MacLean-Fletcher and Pollard, 1980). For the viscosity assay, 100 μl glass capillaries (Brandt, Wertheim, Germany) were used and the F-actin viscosity (0.6 mg/ml) as a function of added plant factor was determined at 37°C by the falling ball technique (low shear viscosity) at an angle of inclination of 45°C (Schleicher *et al.*, 1984). The plant factors to be tested were protein fractions from roots of *Zea mays* (maize) or *Phaseolus coccineus* (scarlet runner), cultured and homogenized as described above. The homogenate was prepared for the viscometric assay in an extraction buffer of 50 mM Tris, 10 mM EDTA, 3 mM EGTA, 2 mM DTT, 30% (w/v) saccharose, 3 mM benzamidine, 5 mM thiourea, 0.02% (w/v) azide at pH 8.0 and 0.02 g polyclar AT g^{-1} fresh weight of plant material. The homogenate was centrifuged at 15.000 x g for 20 min, 4°C, and the supernatant submitted to 110.000 x g for 2 h, 4°C, to pellet particles.

For anion exchange chromatography, DEAE Sepharose Cl-6B (Pharmacia-LKB, Freiburg, Germany) was used (column H/R 12/1.3; flow rate 150 ml h^{-1}). A volume of

300 ml extract was eluted in a buffer of 10 mM Tris, 1 mM EGTA, 1 mM DTT, 1 mM benzamidine at pH 7.5, by steps of 0 to 100 mM NaCl, 100 to 200 mM NaCl, 200 to 350 mM NaCl and 350 to 500 mM NaCl. The flow through fraction was precipitated in $(NH_4)_2SO_4$ of 85% (w/v) saturation.

In some of the experiments, the DEAE flow through fraction was submitted to cation exchange chromatography, using phosphocellulose Servacel P 80 (Serva, Heidelberg, Germany; column H/R 12/0.55, flow rate 40 ml h^{-1}) with a continuous buffer gradient of 80 ml each of the two components: Buffer A was 10 mM Tris, 1 mM DTT, 1mM EGTA, 0.002% azide at pH 7.5; Buffer B was buffer A plus 0.5 M NaCl. Alternatively, this fraction was submitted to hydroxylapatite chromatography (Biogel HT, BioRad, München, Germany; column H/R 8/1.1; flow rate 40 ml h^{-1}) with a continuous buffer gradient made of 100 ml each of the two components: Buffer A was 10 mM potassium phosphate at pH 7.2; Buffer B was 500 mM potassium phosphate at pH 7.2.

RESULTS

Immunoblotting. The monoclonal anti-actin (Mab C4) immunostained plant proteins of 45,000 M_r of *Glycine max*, *Phaseolus coccineus* and *Phaseolus vulgaris* (Fig. 1a) and of *Mougeotia* (Fig. 1b), co-migrating with the actin of bovine heart. Compared to actin of bovine heart, the data show two comigrating bands at 38,000 M_r in the cases of *Glycine max* and of *Phaseolus coccineus* which is not seen for *Phaseolus vulgaris* (Fig. 1a, lanes 2 and 3 *versus* lane 4). The band of 38,000 M_r presumably represents a degradation product of the actin from *Glycine max* and *Phaseolus coccineus*, as mentioned also by Pardee and Spudich (1982). The detection of a band with the same molecular weight of about 66,000 in both *Phaseolus coccineus* and *Glycine max* (Fig. 1a, lanes 2 and 3) may indicate another protein with a similar/same epitope (Williamson *et al.*, 1986, 1987) as that of actin probed with Mab C4. Whether the presentation of this epitope depends on the immunoblot procedure has not been tested.

Even though actin was clearly identified in *Glycine max*, *Phaseolus coccineus* and *Phaseolus vulgaris*, a regulatory factor of actin polymerization or depolymerization could not be clearly shown here. After fractionation by DEAE ion-exchange chromatography, a factor from *Zea mays*, as well as from *Phaseolus coccineus*, decreased the state of actin polymerization by 80%, both in absence and in presence of Ca^{2+}. The factor, however,

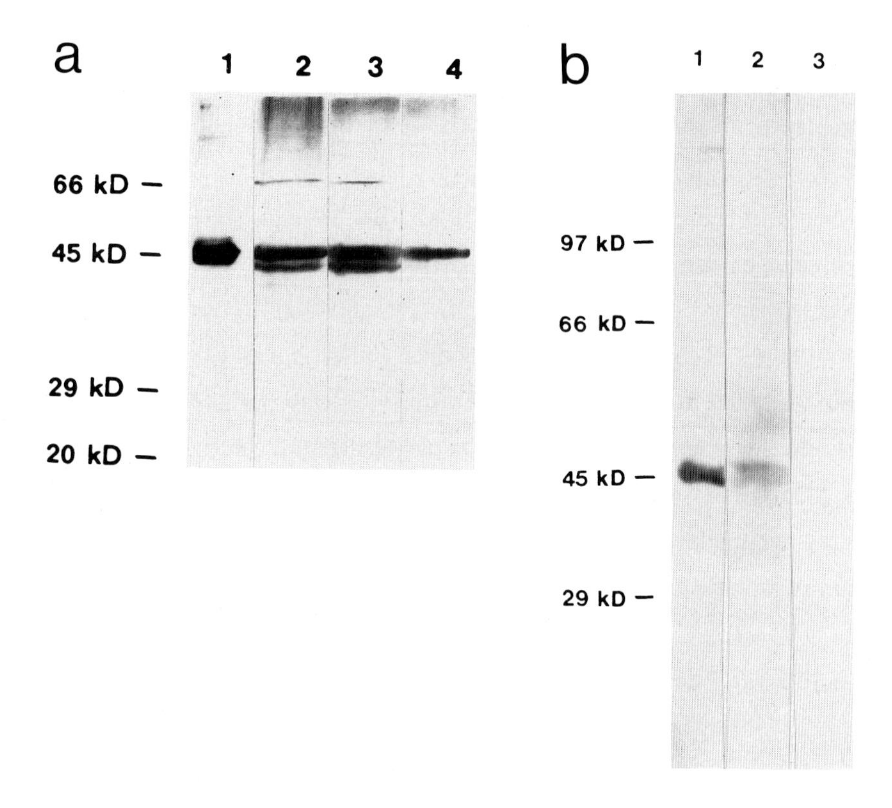

Figure 1. Identification of actin in different plant species by means of immunoblot analysis, in reference to bovine heart actin. The lanes in (a) are: 1 = Purified actin from bovine heart (reference); enriched fractions from: 2 = *Phaseolus coccineus*, 3 = *Glycine max* , 4 = *Phaseolus vulgaris.* Controls were done with secondary antibody only (not shown). The lanes in (b) are: 1 = Purified actin from bovine heart (reference); *Mougeotia* homogenate as probed with 2 = primary and secondary antibody and 3 = with secondary antibody only (control).

vanished after extended purification by cation exchange chromatography (phosphocellulose) or by hydroxylapatite chromatography. Hence, the protein nature of this tested plant factor of actin depolymerization is not validated so far.

Immunofluorescence. The anti-phytochrome Mab Z-3B1 strongly stained areas close to the protoplast circumference of *Mougeotia* (Fig. 2). At the first glance, this observation does not appear surprising and could be expected from the conclusion of Haupt (1982) on dichroic plasmalemma-bound phytochrome, based on physiological evidence.

The anti-phytochrome Z-3B1 (Schneider-Poetsch *et al.*, 1988) was raised against a

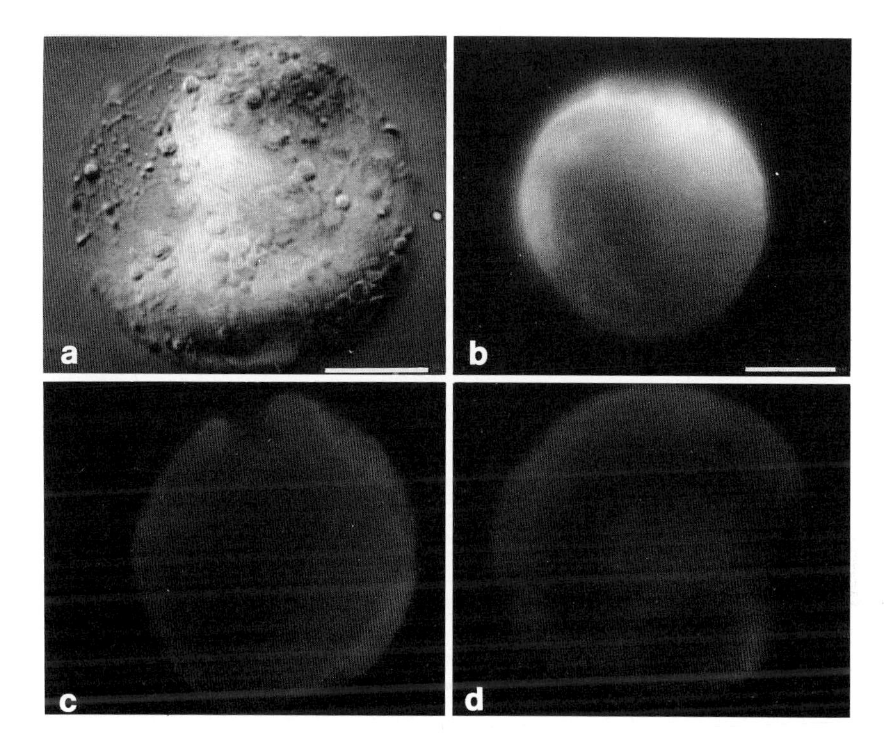

Figure 2. Micrographs of the video-enhanced Nomarski-view of a live *Mougeotia*-protoplast (a), as compared to the fluorescent view of a fixed protoplast as such (d) and after immunostaining with (b) or without primary monoclonal antibody (c) against phytochrome.

soluble phytochrome from *Zea mays*, and proved to be immunoreactive against a wide variety of very distantly related organisms. The epitope for this antibody has been retested and is located in the region of the amino acid residues 740-770 (Schendel *et al.*, in press). From our findings in immunofluorescence, it may be concluded, that *Mougeotia* contains a protein with an epitope similar/identical to an epitope of soluble phytochrome I in higher plants (monocots/dicots). This conclusion would match the findings of Kidd and Lagarias (1990) of soluble type-I phytochrome in *Mesotaenium*.

Although Z-3B1 was raised against purified native *Zea*-phytochrome (Schneider-Poetsch *et al.*, 1988), it has been used so far with high reactivity only in immunoblots (e.g. López-Figueroa *et al.*, 1989). In contrast to these findings, we succeeded in *Mougeotia* here to use this antibody in immunofluorescence (Fig. 2), but without convincing success in the immunoblot (not shown). The latter observation may be explained in part by protein-chemistry-inhibiting phenolics in *Mougeotia* (Grolig and Wagner, 1989), which probably affect various epitopes differentially. Thus, *Mougeotia* actin could be readily detected here in the immunoblot (Fig. 1), while *Mougeotia* phytochrome which has been detected previously in an immunoblot by the antibody Pea-25 (Cordonnier *et al.*, 1986), was not detected here by the antibody Z-3B1. The fact that Z-3B1 recognizes the epitope in the immunofluorescence procedure, presumably depends on the differential treatment of the protein. While in the procedure for the immunoblot a high concentration of protoplasts is needed for making an extract, the crucial phenolics are readily diluted in the procedure for immunofluorescence, resulting overall in the effects of the phenolics on immunolocalization being less than on immunodetection.

In the Nomarski transmission view of the spherical *Mougeotia* protoplast (Fig. 2a), the folded chloroplast is decorated by abundant calcium-binding vesicles (Grolig and Wagner, 1989). In the protoplast periphery, numerous organelles are seen (in agitational movement).

In the fluorescent view of fixed *Mougeotia* protoplasts after staining by primary and secondary antibody (Fig. 2b-2d), the label appears distributed unevenly with most of the fluorescence originating from the cell periphery. Occasional 'caps' in the spherical chloroplast reflect cytoplasmic flaps of different size within the sphere, which can also be visualized in living cells by the fluorescent dye Indo-1 for cytoplasmic free calcium, or *in situ* by rhodamine phalloidin for actin filaments (Russ *et al.*, in press; Grolig *et al.*, 1990).

ACKNOWLEDGEMENTS

This work has been supported by a "Sachbeihilfe" of the "Deutsche Forschungsgemeinschaft" (Wa 265/10-1). Technical assistance by Mrs. Andrea Quanz is thankfully acknowledged.

REFERENCES

Cordonnier M-M, Greppin H, Pratt LH (1986) Identification of a highly conserved domain on phytochrome from angiosperms to algae. Plant Physiol 80:982-987

Dring MJ (1988) Photocontrol of development in algae. Ann Rev Plant Physiol Plant Mol Biol 39:157-174

Furuya M (1987) Phytochrome and photoregulation in plants. Academic Press, Tokyo

Furuya M (1989) Molecular properties and biogenesis of phytochrome I and II. Adv

Biophys 25:133-167

Grolig F, Wagner G (1988) Light-dependent chloroplast reorientation in *Mougeotia* and *Mesotaenium*: Biased by pigment-regulated plasmalemma anchorage sites to actin filaments? Botanica Acta 101:2-6

Grolig F, Williamson RE, Parke J, Miller Ch, Anderton, BH (1988) Myosin and Ca^{2+}-sensitive streaming in the alga *Chara*: detection of two polypeptides reacting with monoclonal anti-myosin and their localization in the streaming endoplasm. Eur J Cell Biol 47:22-31

Grolig F, Wagner G (1989) Characterization of the isolated calcium-binding vesicles from the green alga *Mougeotia scalaris*, and their relevance to chloroplast movement. Planta 177:169-177

Grolig F, Weigang-Köhler K, Wagner G (1990) Different extent of F-actin bundling in walled cells and in protoplasts of *Mougeotia scalaris*. Protoplasma 157:225-230

Haupt W (1982) Light-mediated movements of chloroplasts. Ann Rev Plant Physiol 33:205-233

Johnson GD, De Nogueira Araujo GMC (1981) A simple method of reducing the fading of immunofluorescence during microscopy. J Immunol Meth 43:349-350

Kidd DG, Lagarias JC (1990) Phytochrome from the green alga *Mesotaenium caldariorum*: Purification and preliminary characterization. J Biol Chem 265:7029-7035

Kropf DL (1989) Calcium and early development in fucoid algae. Biol Bull, 176;2:5-8

Laemmli UK (1970) Cleavage of structural proteins during the assembly of the head of bacteriophage T4. Nature 227:680-685

Lessard JL (1988) Two monoclonal antibodys to actin: one muscle selective and one generally reactive. Cell Motil Cytoskel 10:349-362

López-Figueroa F, Lindemann P, Braslavsky SE, Schaffner K, Schneider-Poetsch HAW, Rüdiger W (1989) Action of some conserved domains in phytochrome - like proteins from algae. J Plant Physiol 136:484-447

MacLean-Fletcher S, Pollard TD (1980) Identification of a factor in conventional muscle actin preparations which inhibits actin filament self association. Biochem Biophys Res Comm 96;1:18-27

Pardee JD, Spudich JA (1982) Purification of muscle actin. Meth Enzymol 85:164-182

Pollard TD, Cooper JA (1986) Actin and actin-binding proteins: A critical evaluation of mechanisms and functions. Ann Rev Biochem 55:987-1035

Rüdiger W (1980) Phytochrome, a light receptor of plant photomorphogenesis. In Structure and Bonding (ed. Dunitz, JD *et al.*) 40:101-140 Springer Berlin, Heidelberg, New York

Russ U, Grolig F, Wagner G (1988) Differentially adsorbed vital dyes inhibit chloroplast movement in *Mougeotia scalaris*. Protoplasma (Suppl 1): 180-184

Schendel R, Schneider-Poetsch HAW, Rüdiger W (1991) Structural studies on the photoreceptor, phytochrome: Re-evaluation of the epitope of the monoclonal antibody Z-3B1. Botanica Acta in press

Schleicher M, Gerisch G, Isenberg G (1984) New actin-binding proteins from *Dictyostelium discoideum*. EMBO J 3:2095-2100

Schneider-Poetsch HAW, Schwarz H, Grimm R, Rüdiger W (1988) Cross reactivity of monoclonal antibodies against phytochrome from *Zea* and *Avena*. Planta 173:61-72

Sharrock RA, Quail PH (1989) Novel phytochrome sequences in *Arabidopsis thaliana*:

structure, evolution, and differential expression of a plant regulatory photoreceptor family. Genes Devel 3:1745-1757

Stabenau H (1978) Wachstum von *Mougeotia* in der Durchlüftungskultur. Ber Deutsch Bot Ges 91:251-255

Traas JA, Doonan JH, Rawlins DJ, Shaw PJ, Watts J, Lloyd CW (1987) An actin network is present in the cytoplasm throughout the cell cycle of carrot cells and associates with the dividing nucleus. J Cell Biol 105:387-395

Wada M, Kadota A (1989) Photomorphogenesis in lower green plants. Ann Rev Plant Physiol Plant Mol Biol 40:169-191

Williamson RE, McCurdy DW, Hurley UA, Perkin JL (1987) Actin of *Chara* giant internodal cells. Plant Physiol 85:268-272

Williamson RE, Perkin JL, McCurdy DW (1986) Production and use of monoclonal antibodies to study the cytoskeleton and other components of the cortical cytoplasm of *Chara*. Eur J Cell Biol 41:1-8

INTEGRATION OF PHYTOCHROME RESPONSE

H. Mohr
Albert-Ludwigs-Universität,
Institut für Biologie II/Botanik,
Schänzlestrasse 1,
D-7800 Freiburg,
Germany.

THE PROBLEM

Plant growth and development is controlled by light *via* the sensor pigment phytochrome (Shropshire and Mohr, 1983). Phytochrome-mediated responses in any system require the presence of active phytochrome (Pfr, the far-red absorbing form of phytochrome) and the 'competence' of the system to respond to Pfr. This competence varies endogenously in the course of development (Mohr, 1983). As a representative example, I remind you of the classical observation, spatial pattern and time course of competence in phytochrome-mediated anthocyanin synthesis (Figs. 1,2).

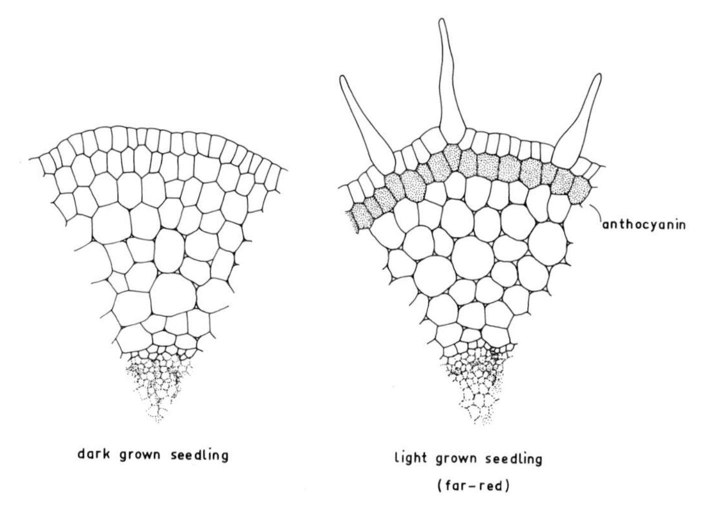

Figure 1. Spatial pattern of competence. Segments of cross-section through the hypocotyl of mustard (*Sinapis alba* L.) seedlings grown in the dark or in light (standard far-red light, operating exclusively via phytochrome). One sees that phytochrome causes certain cells of the epidermis (trichoblasts) to form hairs and all the cells of the subepidermal layer - but no other cells - to form anthocyanin. Obviously, the *specificity* of the response - hair formation vs. anthocyanin synthesis - depends on the specific competence of the determined cells rather than on phytochrome. Data from Mohr (1983).

NATO ASI Series, Vol. H 50
Phytochrome Properties and Biological Action
Edited by B. Thomas and C. B. Johnson

Efforts to define those factors which determine the spatial pattern of competence and the time course of responsiveness (Mohr, 1983) have not yet reached the molecular level - what is the chromatin structure of competent genes? - even though competence as a problem has now been accepted in molecular studies, including work with transgenic plants (Simpson *et al.*, 1986).

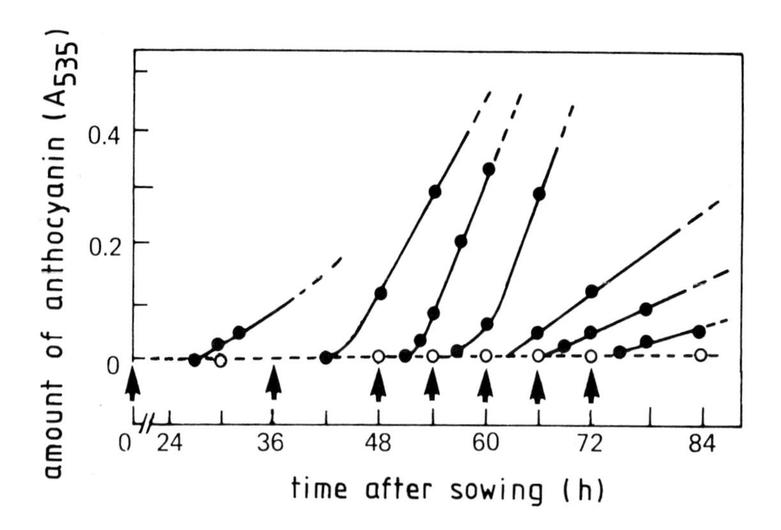

Figure 2. The time course of responsiveness toward phytochrome in the anthocyanin response of mustard seedlings. Dark-grown seedlings were transferred to continuous far-red light at various times after sowing (arrows). ○, dark control; ●, far-red light. Data from Schopfer (1984).

Moreover, even within the range of competence phytochrome action may not be conceived a matter of linear causality - as suggested by Fig. 3 - since phytochrome action appears to be integrated always in a *network* of regulator factors.

To illustrate this general principle, some recent case studies will now be discussed.

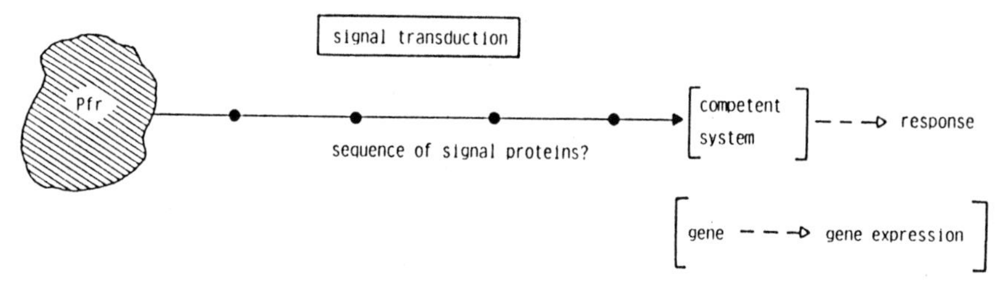

Figure 3. Scheme to illustrate the discussion about competence and signal transduction. Data from Mohr (1987).

CASE STUDIES

(1) Obligatory dependency of phytochrome-controlled nuclear gene expression on a 'plastid factor'. Using different plant materials and different approaches, it has been shown that the expression of nuclear genes encoding proteins which function in the chloroplast - such as the small sub-unit of ribulose-1.5-bisphoshate carboxylase (SSU) and the light-harvesting chlorophyll a/b-binding protein of photosystem II (LHCP) - depends on the integrity of the plastids (see Oelmüller, 1988, for review). It was inferred from these findings that a plastid-derived factor is involved in the transcriptional control of nuclear genes coding for proteins destined for the chloroplast (Fig. 4). The result of damage

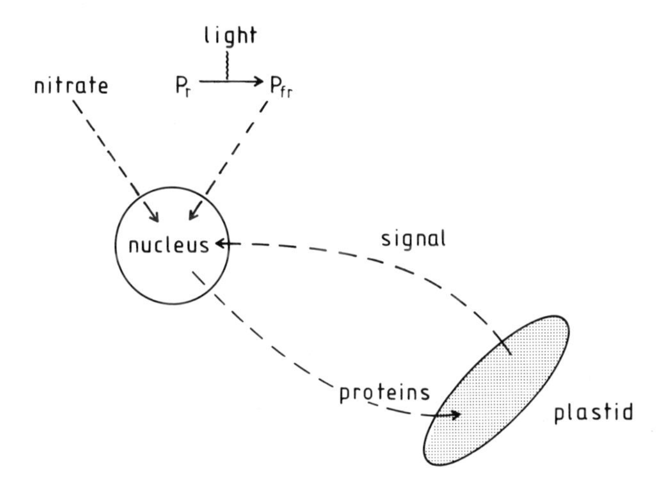

Figure 4. Scheme to illustrate the significance of a plastidic factor for expression of nuclear genes coding for proteins destined for the chloroplast.

done to the plastid, e.g. photooxidation of the plastid, would be to destroy the ability of the organelle to send off this signal. The nature of the postulated plastidic signal has so far remained elusive, but it is probably not a plastidic protein (Oelmüller *et al.*, 1986).

Cytosolic enzyme levels are not impaired by the photooxidative inactivation of the plastid, i.e. by the lack of plastidic factor, and morphogenesis of the seedling is normal (Reiss *et al.*, 1983). Deficiencies following a photooxidative treatment of the plastids appear to be confined solely to the plastid and peroxisomal compartments (Bajracharya *et al.*, 1987). The only exception found so far is nitrate reductase (NR), a cytosolic enzyme encoded in the nuclear DNA.

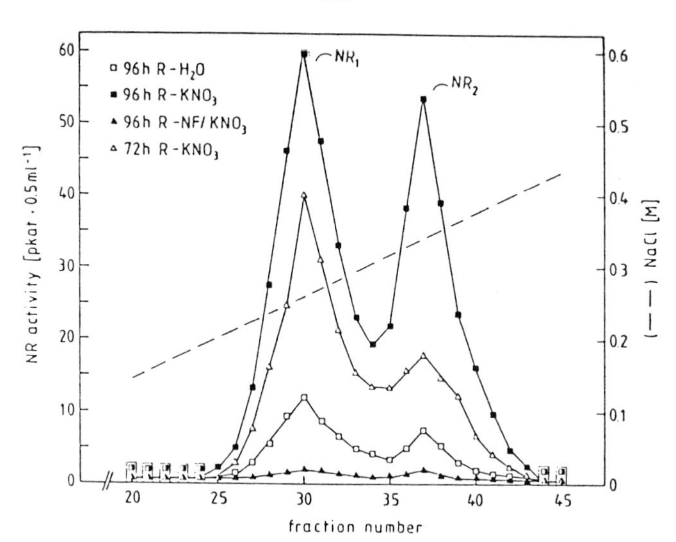

Figure 5. Elution profile of nitrate reductase (NR) from an extract of mustard cotyledons. The seedlings were grown on water, nitrate (15 mM) or nitrate plus Norflurazon (NF, $1 \cdot 10^{-5}$ M) in continuous red light (R) for 3 or 4 days until assay. It was shown that R operates only via phytochrome. Norflurazon was given to inhibit synthesis of coloured carotenoids and thus allows photo-oxidation of plastids. Data from Schuster *et al.* (1989).

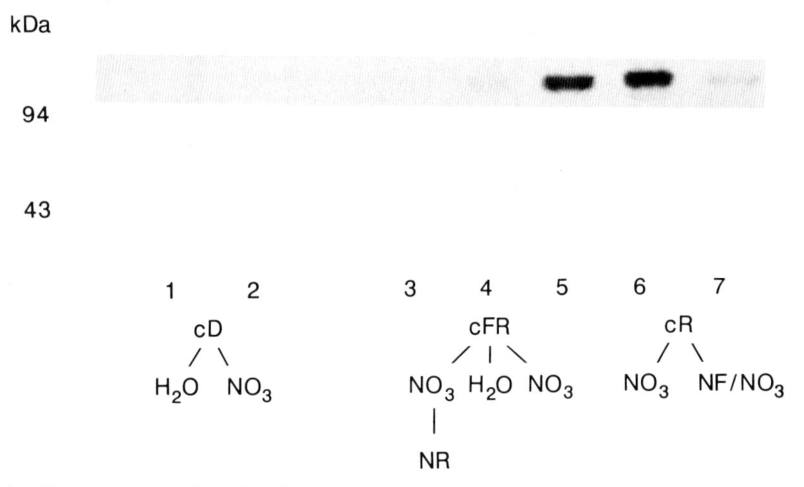

Figure 6. Fluorograms showing immunoprecipitation with NR antiserum of the *in vitro* translation products of total RNA isolated from cotyledons of 60 h old mustard seedlings. The seedlings were kept on continuous darkness (cD) (lanes 1,2), in continuous far-red light (cFR) (lanes 3,4,5), or in continuous red light (cR) (lanes 6,7). *Lane 7*, the seedlings were grown, in addition to nitrate, in the presence of Norflurazon (NF; $1 \cdot 10^{-5}$ M) from sowing onwards. *Lane 3*, the immunoprecipitation was carried out in the presence of 400 μl of enriched NR prepared from mustard seedling cotyledons, using anion-exchange chromatography (fast protein liquid chromatography, Mono Q column). Immunoprecipitation of the antigen-antibody complexes was facilitated by adsorption to Protein A-Sepharose. Data from Schuster and Mohr (1990a).

If the plastids are damaged by photooxidation, the action of nitrate and phytochrome on the appearance of cytosolic NR is abolished (Fig. 5), and the level of NR mRNA is strongly reduced (Fig. 6). Since NR is exclusively cytosolic, the level of NR is not directly affected by photooxidation of the plastids (Fig. 7). Rather, after the onset of a photooxidative treatment with red light, the NR level continues to increase unimpaired for several hours. The NR level decreases only once the mRNA level becomes limiting for the synthesis of the enzyme. NR has a considerable turnover with a half-life of the order of a few hours.

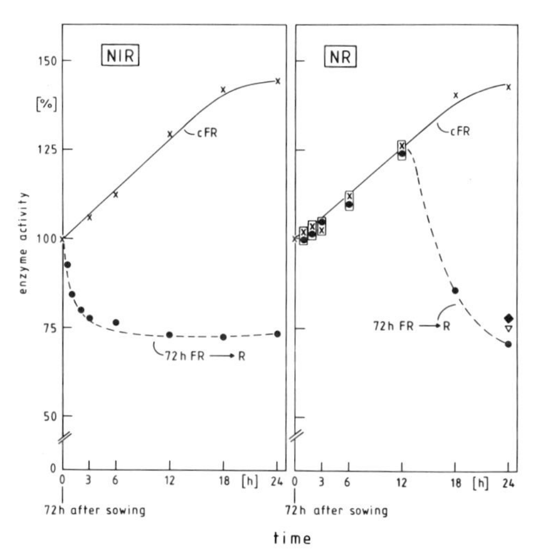

Figure 7. Time courses of nitrite reductase (NIR) and nitrate reductase (NR) isolated from the cotyledons of NF-treated mustard seedlings between 72 and 96 h after sowing. Seedlings were kept on KNO_3 (15 mM) from the time of sowing. *x*——*x*, seedlings kept in continuous far-red light (cFR) where no photo-oxidation takes place; O—O, seedlings transferred to photo-oxidative red light (R) at 72 h after sowing; ♦, seedlings received a photo-oxidative R treatment between 72 and 84 h after sowing and were then transferred back to FR; ▿, seedlings received a photo-odixative R treatment between 72 h and 84 h after sowing and were transferred back to darkness. Data from Oelmüller *et al.* (1988).

(2) Coaction of phytochrome and nitrate in plastidic nitrite reductase (NIR_2) gene expression in mustard (Sinapis alba L.). The second enzyme in the nitrate-assimilation pathway, NIR, catalyzes the reduction of nitrite to ammonia. The enzyme is generally considered to be plastidic, but nuclear-encoded.

In mustard (*Sinapis alba* L.) cotyledons we observed two isoforms of NIR (NIR_1 and NIR_2) using a chromatofocusing technique (Schuster and Mohr, 1990b). Only one of them (NIR_2) disappeared when the plastids were damaged by photooxidation in the

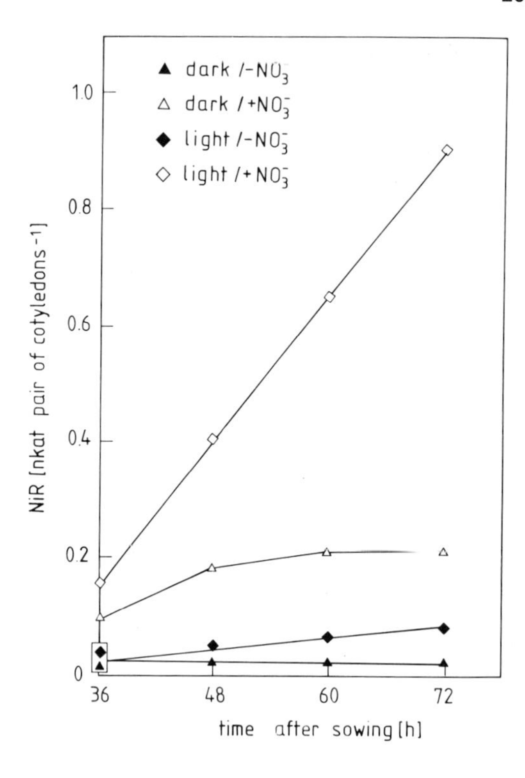

Figure 8. Time course of nitrite reductase (NIR) isolated from the cotyledons of mustard seedlings between 36 and 72 h after sowing. One sees that significant enzyme synthesis takes place only if nitrate is provided. However, *full** gene expression of NIR requires the coaction of nitrate and light. Data from Schuster and Mohr (1990b).

**Full* gene expression means the appearance of a final direct gene product - a protein - active at its physiological site of action (Lamb and Lawton 1983).

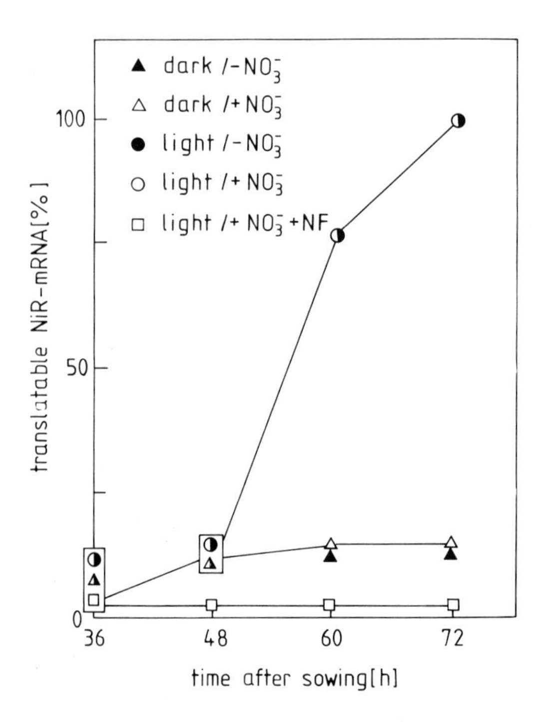

Figure 9. Time course of the amounts of translatable NIR_2 mRNA in the cotyledons of mustard seedlings between 36 and 72 h after sowing. One sees that nitrate has no effect on the NIR_2 mRNA level even though nitrate is required to allow synthesis of the NIR protein. □, in the presence of Norflurazon, i.e. when chloroplast becomes photodamaged, the NIR_2-mRNA remains below detectability. Data from Schuster and Mohr (1990b).

presence of Norflurazon (see Fig. 7). It was concluded that NIR_2 is plastidic while NIR_1 is extraplastidic and not affected by photooxidation of the plastids. Both isoforms appear to have the same molecular weight (60 kDa). Two distinct translation products which could be immunoprecipitated with NIR antiserum produced against total NIR from mustard were observed which differed slightly in molecular weight (60 versus 63 kDa). The 63 kDa polypeptide was considered to be the precursor of plastidic NIR_2.

While synthesis of NIR protein depended largely on nitrate (Fig. 8), the levels of *in vitro*-translatable NIR mRNAs were found to be either independent of nitrate and light (NIR_1) or controlled by phytochrome only (NIR_2) (Fig. 9). It appears that phytochrome strongly stimulates the level of NIR_2 mRNA while significant enzyme synthesis takes place only in the presence of relatively large amounts of nitrate. Since an increased enzyme level was strictly correlated with an increase of immunoresponsive NIR protein it is improbable that activation of a precursor plays a role. Rather, it is concluded that, *in situ*, nitrate controls translation.

As expected, a photooxidative treatment of the plastids (Fig. 9, R-NF/KNO_3, □) decreases the level of NIR_2, i.e. in the absence of the plastidic factor gene expression is blocked, and phytochrome is totally ineffective.

Moreover, a sequential action of phytochrome (Pfr) and nitrate in bringing about NIR_2 synthesis is a major character of the control. Expressed in the previously introduced terminology (Schuster *et al.*, 1988), coarse control of the appearance of translatable NIR_2 mRNA operates through phytochrome while fine tuning of gene expression, i.e. the actual appearance of the NIR_2 protein, is controlled by nitrate.

In a previously investigated case (Schuster *et al.*, 1988), control by phytochrome of the appearance of RuBPCase and the mRNA for its small subunit, a two-step control of gene expression (coarse control versus fine tuning) was also observed but both steps were found to be controlled by phytochrome. In the case of the appearance of the apoprotein of LCHII (Schmidt *et al.*, 1987) it was found that in continuous far-red light, operating through phytochrome, the appearance of mRNA was fully turned on while the final gene product, the apoprotein, was not detectable in the absence of chlorophyll. Thus, it appears, that in the case of plastidic proteins phytochrome is always involved in the coarse control of transcription while different mechanisms operate at the level of fine tuning.

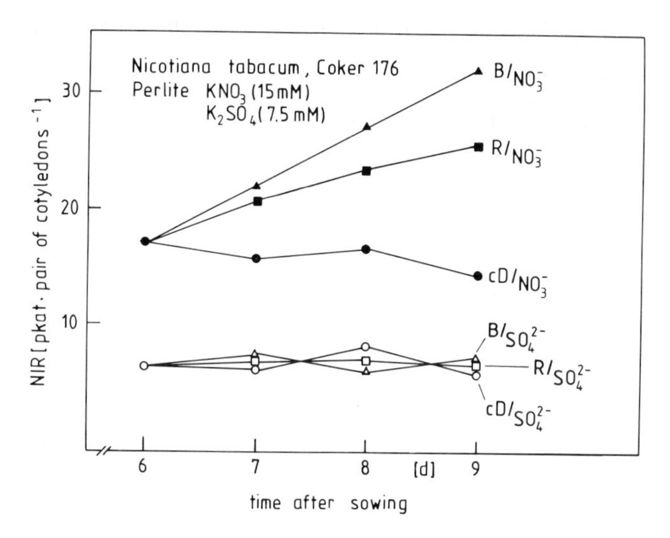

Figure 10. Time course of nitrite reductase (NIR) in the cotyledons of tobacco seedlings, cv. Coker 176, between 6 and 9 days after sowing. One sees that light has no effect in the absence of exogenous nitrate. However, as in mustard (see Fig. 8) full gene expression of NIR requires the coaction of nitrate and light. Data kindly provided by A. Neininger. *B* = blue light (10 Wm^{-2}); *R* = red light (6.8 Wm^{-2}).

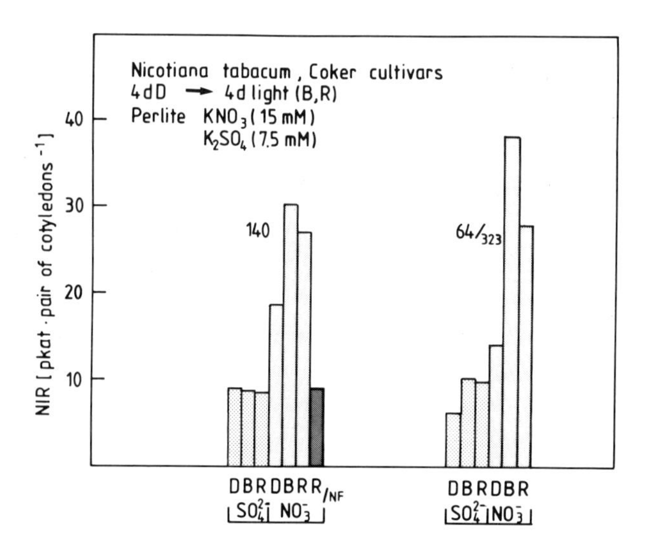

Figure 11. Levels of nitrite reductase (NIR) in the cotyledons of tobacco seedlings, 8 days after sowing. One sees that under identical experimental conditions the two cultivars differ conspicuously with regard to their responsiveness to light. Data kindly provided by A. Neininger. *B* = blue light (10 Wm^{-2}); *R* = red light (6.8 Wm^{-2}).

3. Action of nitrate and light in NIR synthesis in tobacco (Nicotiana tabacum L., *cv. Coker).* In the absence of nitrate light has no effect on NIR appearance in the tabacco cotyledons (Fig. 10), whereas a strong light effect is observed in the presence of nitrate. In fact, the situation in tobacco is much the same as in mustard except that the constitutive enzymelevel is higher, and blue light (applied at the same quanta fluence rate as red light) is considerably more effective than red light. The operation of cryptochrome (blue/UV-A photoreceptor) in addition to phytochrome is clearly indicated.

Six cultivars of Coker tobacco have been investigated with principally the same results. However, the cultivars differ conspicuously with regard to their light responsiveness. As an example, Coker 140 does not respond to light in the absence of nitrate, whereas Coker 64/329 responds considerably (Fig. 11). In case of the 64/323 cultivar responsiveness to blue light is much stronger in the presence of nitrate compared to Coker 140 whereas responsiveness to nitrate alone (D/NO_3^-) is low.

A major question had to be solved: does blue light control NIR synthesis independently of phytochrome? In the course of experimentation it turned out that the Scots pine seedling is far better suited to answer this question than the tobacco seedling.

(4) Obligatory blue light requirement in phytochrome-mediated Fd-GOGAT (= ferredoxin-dependent glutamate synthase) gene expression in Scots pine (Pinus sylvestris L.*).* In this case study we consider control by light of appearance of ferredoxin-dependent glutamate synthase (Fd-GOGAT) in the first whorl, the cotyledonary whorl, of Scots pine seedlings. Under all experimental circumstances we can detect only a single form of the enzyme (Elmlinger and Mohr, 1990). Since the enzyme plays a central role in nitrate/ammonium assimilation it was expected that nitrate plays a major part in gene expression. But this is not the case (Fig. 12, Table 1). Appearance of Fd-GOGAT is mainly controlled by light. The conspicuous feature is that up to 10 d after sowing the light effect on Fd-GOGAT appearance can be fully attributed to the operation of phytochrome whereas between 10 and 12 days after sowing, synthesis breaks down and degradation begins if no blue light is provided.

The question to be answered was whether phytochrome is still operating in B grown seedlings beyond 10 d after sowing, or is phytochrome action turned off and replaced by cryptochrome action as appearance suggests (Fig. 12).

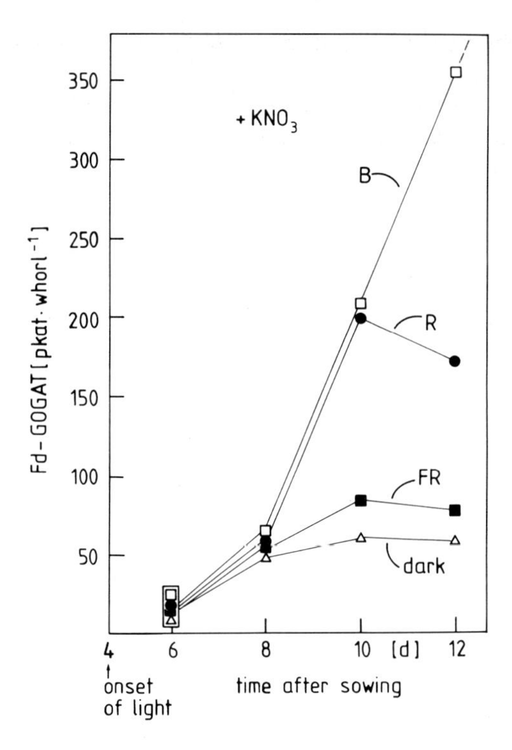

Figure 12. Time course of Fd-GOGAT (= ferredoxin-dependent glutamate synthase) in the first (cotyledonary) whorl of Scots pine seedlings. Seedlings were grown in perlite on 15 mM KNO_3 and kept in continuous red (R, 6.8 Wm^{-2}), blue (B, 10 Wm^{-2}) or far-red (FR, 3.5 Wm^{-2}) light. Data from Elmlinger and Mohr (1990).

Crucial experiments were performed with dichromatic light, i.e. the plant material was treated with two light beams simultaneously (Hartmann 1967). In addition to B, high-irradiance RG9-light (ψ_{RG9} < 0.01) or R (ψ_R = 0.8) were given to determine the Pfr/Ptot ratio (ψ) and thus the Pfr level during the B treatment. The major results are shown in Table 1.

For any increase of the enzyme level beyond 10 d light is indispensable (lines 1,2). In seedlings pretreated with B (6 d B) a similarly strong increase of enzyme level is observed in R and B (lines 3,4).

Table. 1. Action of dichromatic irradiation on appearance of Fd-GOGAT (= Fd-dependent glutamate synthase) in the cotyledonary whorl of Scots pine seedlings (*Pinus sylvestris* L.). Seedlings were grown in perlite without nitrate and pretreated with continuous blue light (B) between 4 and 10 days after sowing (see Fig. 12). Dichromatic treatment was then applied for 2 days. Enzyme assay was at day 12. Data from Elmlinger and Mohr (1990).

	Light treatment	**Fd-GOGAT [pkat • whorl^{-1}]**
(1)	6 d B	156 $\pm$ 6
(2)	6 d B + 2 d dark	130 $\pm$ 2
(3)	6 d B + 2 d B	251 $\pm$ 3
(4)	6 d B + 2 d R	259 $\pm$ 3
(5)	6 d B + 2 d [B + R]	254 $\pm$ 3
(6)	6 d B + 2 d [B + RG9]	155 $\pm$ 4
(7)	6 d B + 2 d RG9	147 $\pm$ 2
(8)	6 d R	152 $\pm$ 5
(9)	6 d R + 2 d R	144 $\pm$ 6
(10)	12 d dark	32 $\pm$ 2

B, blue light (10 Wm^{-2}); **R,** red light (20 Wm^{-2}); **RG9,** long wavelength far-red light (20 Wm^{-2}).

When the level of Pfr was kept very low in B - with simultaneously given RG9-light (ψ_{sim} < 0.01) - no inductive effect of B was seen (line 6). When the level of Pfr was kept high in B - with simultaneously given R (ψ_{sim} < 0.7) - the same high rate of enzyme synthesis was observed as in B and R alone (line 5).

These results show that B does not affect enzyme synthesis if the Pfr level is low. On the other hand, in a seedling which has never seen B, phytochrome is not capable of maintaining enzyme synthesis beyond 10 d after sowing. It is concluded that B/UV-A light has no direct effect on Fd-GOTAT gene expression. Rather, its action appears to be restricted to establishing responsiveness towards Pfr. This type of coaction between phytochrome and B/UV-A light appears to be wide-spread since it was observed previously in quite a number of photoresponses in angiosperms (see Mohr, 1986, for review).

CONCLUSION

The case studies discussed confirm the notion that phytochrome (Pfr) action (on gene expression) is always integrated in a network of regulatory factors. This is (almost) trivial when we include those (unknown) factors which determine the spatial and temporal patterns of competence towards Pfr. However, the dependency of phytochrome action on the plastidic factor or on blue light as well as the coaction of phytochrome and nitrate in bringing about *full* gene expression in some crucial enzymes indicate that even within the range of competence phytochrome action on gene expression can probably never be understood adequately in terms of linear causality as suggested by Fig. 3.

REFERENCES

Bajracharya D, Bergfeld R, Hatzfeld W-D, Klein S, Schopfer P (1987) Regulatory involvement of plastids in the development of peroxisomal enzymes in the cotyledons of mustard (*Sinapis alba* L.) seedlings. J Plant Physiol 126:421-436

Elmlinger MW, Mohr H (1990) Coaction of blue/ultraviolet-A light and light absorbed by phytochrome in controlling appearance of ferredoxin-dependent glutamate synthase (= Fd-GOGAT) in the Scots pine (*Pinus sylvestris* L.) seedling. Planta (submitted)

Lamb CJ, Lawton, MA (1983) Photocontrol of gene expression. In: Shropshire W, Mohr H (eds) Encyclopedia of Plant Physiology, Vol 16A. Springer, Heidelberg New York, pp.213-257

Levitan I, Bergfeld R, Rajasekhar VK, Mohr H (1986) Expression of nuclear genes as affected by treatments acting on the plastids. Planta 168:482-492

Mohr H (1983) Pattern specification and realization in photomorphogenesis. In: Shropshire W, Mohr H (eds) Encyclopedia of Plant Physiology, Vol 16A. Springer, Heidelberg New York, pp.336-357

Mohr H (1986) Coaction between pigment systems. In: Kendrick RE, Kronenberg GHM (eds) Photomorphogenesis in Plants. Nijhoff/Junk, Dordrecht, pp.547-564

Mohr H (1987) Future strategy in photomorphogenesis research. In: Furuya M (ed) Phytochrome and Photoregulation in Plants. Academic Press, Tokyo, pp.333-348

Oelmüller R (1989) Photo-odixative destruction of chloroplasts and its effect on nuclear gene expression and extraplastidic enzyme levels. Photochem Photobiol 49:229-239

Oelmüller R, Levitan I, Bergfeld R, Rajasekhar VK, Mohr H (1986) Expression of nuclear genes as affected by treatments acting on the plastids. Planta 168:482-492

Oelmüller R, Schuster C, Mohr H (1988) Physiological characterization of a plastidic signal required for nitrate-induced appearance of nitrate and nitrite reductases. Planta 174:75-83

Reiss T, Bergfeld R, Link G, Thien W, Mohr H (1983) Photo-oxidative destruction of chloroplasts and its consequences for cytosolic enzyme levels and plant development. Planta 159:518-528

Schmidt S, Drumm-Herrel H, Oelmüller R, Mohr H (1987) Time course of competence in phytochrome-controlled appearance of nuclear-encoded plastidic proteins and messenger RNAs. Planta 170:400-407

Schopfer P (1984) Photomorphogenesis. In: Wilkins M B (ed) Advanced Plant Physiology. Pitman, London

Schuster C, Mohr H (1990a) Photo-oxidative damage to plastids affects the abundance of nitrate-reductase mRNA in mustard cotyledons. Planta 181:125-128

Schuster C, Mohr H (1990b) Appearance of nitrite-reductase mRNA in mustard seedling cotyledons is regulated by phytochrome. Planta 181:327-334

Schuster C, Oelmüller R, Mohr H (1988) Control by phytochrome of the appearance of ribulose-1,5-bisphosphate carboxylase and the mRNA for its small subunit. Planta 174:426-432

Schuster C, Schmidt S, Mohr H (1989) Effect of nitrate, ammonium, light and a plastidic factor on the appearance of multiple forms of nitrate reductase in mustard (*Sinapis alba* L.) cotyledons. Planta 177:74-83

Shropshire W, Mohr H (eds) (1983) Encyclopedia of Plant Physiology, Vol 16A,B. "Photomorphogenesis". Springer, Heidelberg New York

Simpson J, Montagu van H, Herrera-Estrella L (1986) Photosynthesis-associated families: differences in response to tissue-specific and environmental factors. Science 2233:34-38

A QUANTITATIVE APPROACH TO THE MOLECULAR BIOLOGY OF PHYTOCHROME ACTION

C.B. Johnson, S.M. Allsebrook, H. Carr-Smith and B. Thomas[1]
Department of Botany,
University of Reading,
Whiteknights,
Reading RG6 2AS,
UK.

INTRODUCTION

Recently there have been enormous advances in our understanding of two major aspects of photomorphogenesis. Firstly, the photoreceptor itself: detail is beginning to emerge of at least three molecular forms of phytochrome and this has already led to considerable speculation as to the physiological roles of these different molecular species. Secondly, our understanding of the mechanism of light-mediated control of gene transcription is advancing rapidly, with the identification of the regulatory *cis*-acting elements involved in photocontrol of transcription and the isolation of protein factors which interact with them. Most of this work has necessarily been qualitative in nature.

The aim of this paper is two-fold: first, to assess the extent to which photocontrol of transcription can be regarded as the *end* of the phytochrome signal transduction chain and second, to examine the extent to which it is possible or prudent to postulate specific functions for the different molecular forms of phytochrome which have been identified. Examination of both these problems requires an approach which is quantitative rather than qualitative and this has characterized the work described here. We have examined two phenomena as models; the phytochrome control of nitrate reductase in *Sinapis alba*, and the photocontrol of flowering in wheat (*Triticum aestivum* L.).

An attractive idea, in view of the difficulties in identifying a single primary mechanism of phytochrome action (see Johnson, 1990) is the notion that the primary action of the stable form of phytochrome, which predominates in green light-grown

---

[1]Department of Molecular Biology, Horticulture Research International, Worthing Road, Littlehampton, BN17 6LP, UK.

NATO ASI Series, Vol. H 50
Phytochrome Properties and Biological Action
Edited by B. Thomas and C. B. Johnson

plants, might differ from that of the rapidly synthesised but labile form of Pfr predominant in seedlings. There is already considerable evidence for the idea of multigene families of enzymes in higher plants, with different members of the family having different functions (Smith, 1990). If such a notion is applicable to different members of a phytochrome "family" (see Smith, this volume) then the kinetic properties of these phytochrome species might be expected to have counterparts in the kinetic characteristics of the responses they promote. Thus, for example, the high irradiance response (HIR) of phytochrome has been shown to follow in general the time- and fluence-rate- course of Pfr in seedlings (Wall and Johnson, 1983), leading to the expectation that a response controlled by the labile phytochrome which predominates in such seedlings should show strong fluence rate dependence and a wavelength optimum that shifts progressively from the red to the far red with increasing periods of irradiation, as is in fact observed for most HIRs in seedlings. In contrast, the stable phytochrome which predominates in plants maintained in the light for prolonged periods is consistent with the observed lack of irradiance dependence and the absence of a far-red action maximum in the responses of mature light-grown plants to light.

These two response types, highly characteristic of seedlings and mature plants respectively, offer the best chance of distinguishing between the effects of stable (or "green plant") phytochrome and the labile "seedling" phytochrome. If the mechanisms of action of the two (or more than two, see later for a discussion of this) phytochrome species controlling these responses differ, one might expect;

(a) that specific phytochrome responses would be controlled by one rather than both species of phytochrome, and that as a consequence;

(a) the range of responses in seedlings compared with mature plants should be very different.

The two model systems we have investigated offer differing responses to these expectations.

PHYTOCHROME AND FLOWERING IN GREEN LIGHT-GROWN WHEAT

The use of action spectroscopy provides a good starting point for the investigation of light-controlled events in plant development. Action spectra can yield information regarding the nature of a photoreceptor and help to elucidate its mode of action (Schäfer *et al.,*. 1983). indeed, it was largely through the use of this technique that phytochrome was shown to be a major photoreceptor involved in the morphological responses of plants to light (see Schäfer *et al.,* 1983).

Action spectroscopy has also been widely used to investigate the identity of the photoreceptor(s) that mediate the photoperiodic control of flowering. In particular, similarities in the action spectra for the night-break inhibition of flowering in short-day plants (SDP) and for the promotion of flowering in long-day plants (LDP) have indicated that the same photoreceptor is involved in both cases (Parker *et al.,* 1946; Borthwick *et al.,* 1948; Parker *et al.,* 1950). Subsequent experiments in which action spectroscopy was combined with studies of photoreversibility confirmed that phytochrome is the photoreceptor for the night-break effect in both of these photoperiodic response groups (Downs, 1956).

However, when longer night-breaks were employed to control flowering, several workers observed that broad-band red light sources were often considerably less effective in promoting flowering in LDP than were broad-band sources containing both red and far-red wavelengths (Vince *et al.,* 1964), suggesting some similarity with the HIRs of seedlings. In order to investigate this possibility further we constructed an action spectrum for the photocontrol of flowering in *Triticum aestivum*, using apex length as a measure of floral stage (Carr-Smith *et al.,* 1989).

For construction of the action spectrum, four fluence rates were used (0.002-0.6 W m^{-2} at each of seven wavelengths from 640nm to 730nm and three fluence-rates of white light for the white-light controls. The plants were grown to the 4-leaf stage in the greenhouse with a 8h day as described above. At this stage, four 16h day-extensions of monochromatic or white light were given to the plants; they were then returned to the greenhouse and short days for two weeks before being harvested, dissected and apex length measured.

Percentage response (%R) was calculated from the equation:

$$\%R = (T - SDC)/(WLC-SDC).100$$

Where T=treatment mean, SDC=short-day control and WLC=white-light control. Within the range of fluence rates investigated there is a log-linear relationship between percentage response and fluence rate (Fig. 1). The fluence rate for a 50% response was calculated and the action spectrum plotted from these values; it has action maxima at

660nm and 716nm (see Fig. 2). The regression lines for the fluence rate-response curves (Fig. 1) are all nearly parallel, so although the 50% response level was chosen, in this case any other percentage response would give a very similar action spectrum.

This is the first action spectrum for light-grown, green plants that has dual action maxima in the red and far-red regions of the spectrum. Action spectra for HIRs of etiolated plants on the other hand, usually have these dual peaks. The action spectrum for inhibition of hypocotyl growth in etiolated *Sinapis alba* is typical in most ways, it has maxima in the blue region (which will not be considered here), at 655nm and at 716nm (Beggs *et al.,* 1980). This similarity between the action spectrum reported here and those for the HIR of etiolated plants is unexpected.

Current explanations of the HIR in etiolated plants are based on type-1 phytochrome. They account for the far-red maximum at least in part by the maintenance of higher Pfr levels in the far-red than in the red (Wall and Johnson, 1983). It is possible that a similar mechanism could account for the occurrence of a peak response in the far-red in green, light-grown plants. This would indicate that type-1, labile phytochrome must still be operative in mature green plants, whereas most of the responses of such plants seem to suggest the involvement of stable (or type-2) phytochrome. There are many examples where, after prolonged periods of growth in white light, day-extension promotion of flowering is maximal in the far-red (e.g. using broad-band light sources; Friend *et al.,* 1959; Vince *et al.,* 1964; using monochromatic light sources: Schneider *et al.,* 1967; Friend, 1968; Jacques and Jacques, 1969; Blondon and Jacques, 1970; Imhoff *et al.,* 1979). In all of these studies only one fluence rate at each wavelength was used. Wavelength-response curves can, however, change quite dramatically at different fluence rates (Holmes and Schäfer, 1981), so care has to be taken when drawing conclusions from these results. The results reported here show that when a full action spectrum is constructed for the floral response in LDP a peak still occurs in the far-red.

To investigate the possible involvement of type-1 phytochrome in this response we used monoclonal antibodies specific for wheat type-1 phytochrome to investigate the extent and kinetics of this molecular species of the pigment in light-grown wheat. The preparation and selection of these antibodies is discussed elsewhere (Carr-Smith *et al.,* 1991). The antibody finally used (MAC 56) shows more than 95% specificity for type-1 phytochrome. This antibody was used in an enzyme-linked immunosorbent assay

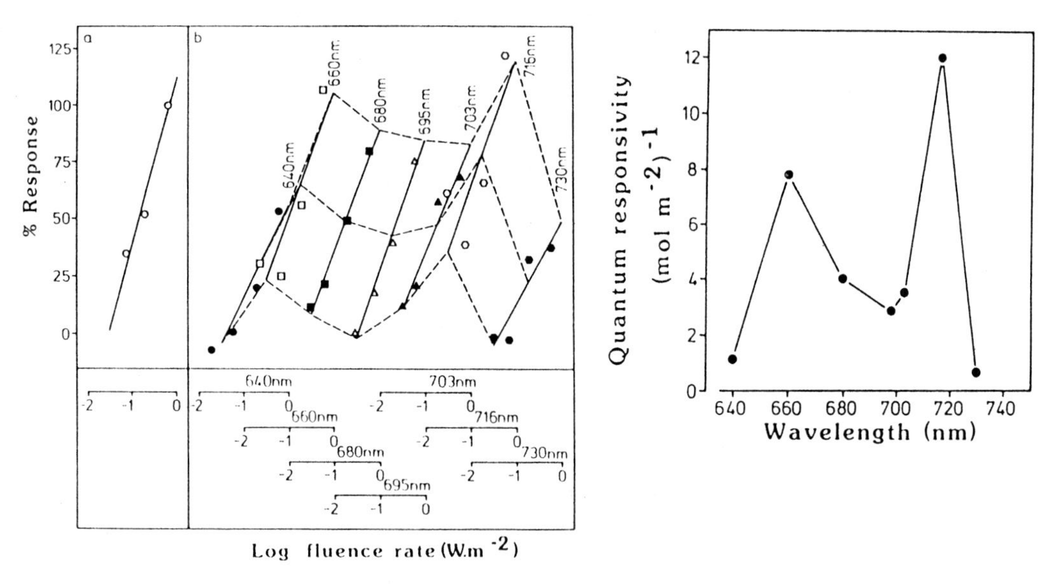

Figure 1. (left) Fluence rate response curves for promotion of floral development in spring wheat (*Triticum aestivum* L. cv.'Alexandria'). Before and after experimental treatments the plants were grown under short days in a greenhouse (max. temp. 30°C, min. temp. 15°C). From 08.00h to 16.00h the plants were allowed natural sunlight; for the remaining 16h of the 24h cycle they were kept in darkness by the use of covers made from thick black plastic. For the action spectrum, the plants were given four 16h day-extensions of monochromatic light or white light for the white-light controls.

Fluence rate-response curves for promotion of floral development in spring wheat by day-extensions of (a) white light and (b) monochromatic light. For clarity each wavelength in (b) has a separate x-axis, which, starting on the left with 640nm, is displaced one unit to the right for each wavelength in ascending order. The *broken lines* join points of equal fluence rates. Plants grown in short days were given four day-extensions of monochromatic or white light of different fluence rates, returned to short days for a period of growth and the apices dissected. ○=white light. •=640nm, □=660nm, ■=680nm, △=695nm, ▲=703nm, ○=716nm, •=730nm. Regression lines were calculated and found to be significant in all but two cases where they are only just not significant. White light: t=7.086, P<0.1 (1 d.f.). 640nm: t=4.854, P<0.05; 660nm: t=2.893, P>0.11 680nm: t=12.303, P<0.01; 695nm: t=13.430, P<0.01; 703nm: t=6.903, P<0.05; 716nm: t=2.208, P>0.1; 730nm: t=3.243, P<0.1 (all 2 d.f.). Standard errors are under 5% in every case.

Figure 2. (right) Action spectrum for promotion of floral development of spring wheat by day-extension. Photon fluence for 50% response was calculated from the regression coefficients of the fluence rate-response curves shown in Figure 1b and the reciprocal plotted against wavelength.

(ELISA) to measure the kinetics of type-1 phytochrome in light grown wheat plants of similar age to those used for the action spectroscopy.

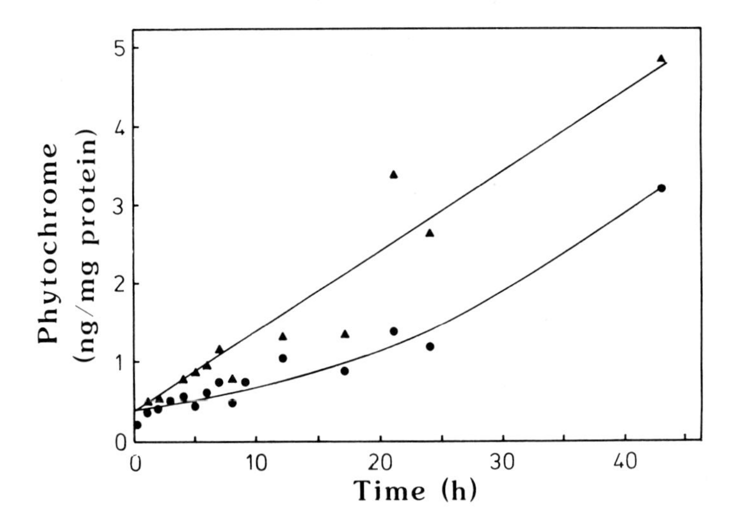

Figure 3. Comparison of the rates of type-1 phytochrome accumulation of green light-grown wheat plants transferred to darkness (zero time) following 8h in white light from fluorescent lamps with either 5min red (•) or 10min far-red (▲) end of day light pulse.

The kinetics of phytochrome accumulation in darkness after an 8h light period and of the loss of phytochrome in red light after a 16h dark period were investigated in separate experiments. The 8h light period and 16h dark period were used so that comparison could be made with the action spectrum for floral development. After an 8h day under fluorescent lamps in the growth cabinets, the plants were given saturating pulses of either red light (which leaves approx. 80% of phytochrome as Pfr) or far-red light (which leaves less than 3% of phytochrome as Pfr) and then transferred to darkness. Subsequently, samples were taken at regular intervals and assayed by ELISA. This revealed that only low levels of type-1 phytochrome remain after the 8h light period, and as predicted from phytochrome kinetics in etiolated seedlings, the initial rate of accumulation of phytochrome is more rapid in those plants given the far-red treatment compared with those given the red-light treatment (Fig. 3). After a 24h period in darkness the phytochrome content of the far-red-treated plants is approximately twelve times the level in the plants after 8h white light. This difference is due to a lag in phytochrome accumulation in the red-light-treated plants of about 15h. From the date presented here it is difficult to say exactly what causes this but the possibilities can be narrowed down. It occurs in darkness so could not be caused by the destructions of Pfr at high (Pfr); therefore, the lag in phytochrome accumulation must be due either to inhibition of Pr synthesis by Pfr with a long half-life or to Pr destruction. Results from either the far-red-treated plants or from the linear portion of the curve from the red-treated plants are commensurate with zero-order synthesis of Pr at a rate of 0.79ng phytochrome h^{-1} (g FW)$^{-1}$.

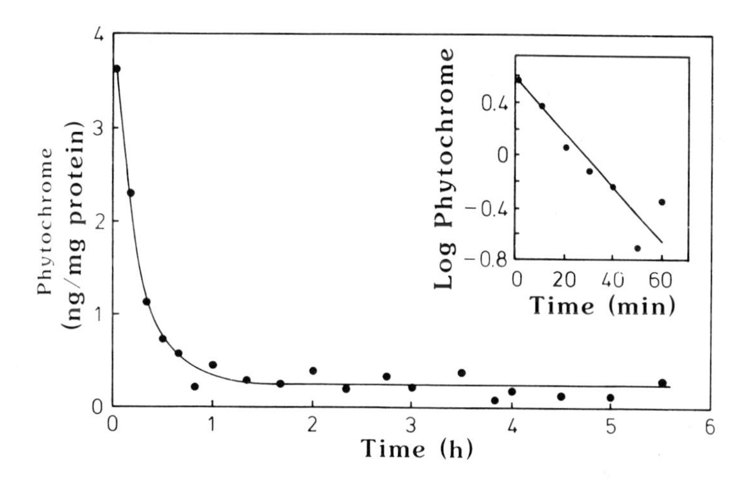

Figure 4. Time course of Pfr destruction in green light-grown wheat plants in continuous red light following 16h of darkness. Inset; semi-logarithmic plot of the first 60min of the time course.

Light treatments in experiments to investigate the destruction of phytochrome in red light were given after 16h of darkness. After this length of time in darkness substantial amounts of phytochrome have built up. On placing such plants in continuous red light the levels of type-1 phytochrome fall rapidly. From Figure 4 it can be seen that during the first 60min there is a rapid first-order loss of Pfr. In this case, Pfr has a half-life of about 15min, somewhat shorter than that of etiolated wheat where it is approximately 90min (Carr-Smith *et al.,* 1991). However, it appears that once the Pfr has fallen below a certain level (reached at about 100min) there is no further loss, suggesting that destruction may have ceased. It is possible that either Pfr destruction does not occur below a particular threshold level of Pfr or Pfr/Ptot, or that destruction is eventually balanced by synthesis. However, the lag in synthesis when plants are transferred from white light to darkness with a terminal red-light treatment (Fig. 4) suggests that synthesis, on the contrary, is inhibited by Pfr in red light. The former possibility is therefore more likely. This notion is further supported by data presented in Figure 5, which shows a comparison of the effect of continuous far-red light and continuing darkness on plants that have previously been in darkness for 16h. The far-red light produces approx. 3% Pfr

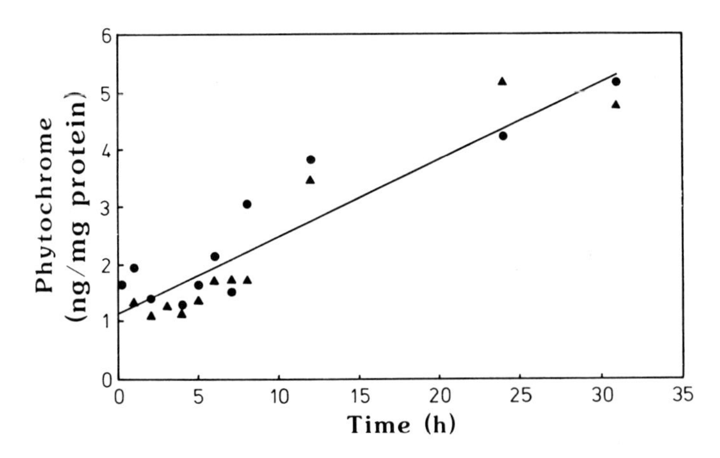

Figure 5. Comparison of the effects of continuous far-red light (▲) and continuing darkness (•) on phytochrome (type-1) levels of green, light-grown wheat plants that have previously been in darkness for 16h.

and might be expected to lower the level of phytochrome through destruction, as has been reported for etiolated plants of other species (Schäfer *et al.*, 1972). However, as can be seen, far-red light has no effect on phytochrome levels when compared with plants that have been left in darkness, even when the Ptot has increased five-fold, suggesting an absence of or reduction in Pfr destruction at low Pfr levels.

In summary, these experiments show clearly that there is a measurable pool of type-1 phytochrome in green, light-grown wheat. This pool of phytochrome has kinetic characteristics typical of type-1 phytochrome in etiolated seedlings, with zero-order synthesis of stable Pr and rapid first-order destruction of Pfr. Unusually, there appears to be a threshold level of Pfr or Pfr/Ptot below which Pfr destruction no longer occurs.

It also appears that Pr synthesis is controlled by light. For example, a lag of 15h in type-1 phytochrome accumulation in plants given an end-of-day red-light treatment (Fig. 3) would not be expected if this lag were solely due to destruction of type-1 Pfr because type-1 Pfr in green, light-grown wheat has a half-life of only 15min (Fig. 4). There must be inhibition of Pr synthesis by a pool of Pfr with a longer half-life. This could be type-2 phytochrome but the possibility of a sub-pool of type-1 phytochrome with a longer Pfr half-life cannot be completely excluded. Taken together with the action spectroscopy the kinetics data strongly support the idea that type-1 phytochrome is not restricted in activity to de-etiolating seedlings but can continue to be effective in mature light-grown plants.

PHOTOCONTROL OF NITRATE REDUCTION

The results shown above are not necessarily inconsistent with the idea of different members of the phytochrome 'family' controlling different responses. However, such specificity of action does not seem to be consistent with the characteristics of photocontrol of nitrate reductase (NR). Unlike many enzymes subject to phytochrome control, NR in *Sinapis alba* is responsive to phytochrome throughout the life of the plant, from 15h after sowing through to flowering and beyond. In the dark-grown mustard seedling the photocontrol of NR shows a typical HIR, whether NR activity is measured, or NR protein estimated by enzyme-linked immunosorbent assay (ELISA; Fig. 6). In contrast, we have shown previously that mature green plants of mustard and other species, as well as in cauliflower curd, NR activity shows a response approximately linearly related to phytochrome photo-equilibrium and irradiance independent (Whitelam *et al.,* 1979; Johnson and Whitelam, 1982, Whitelam and Johnson, 1982). In *Sinapis* this response too can be shown to involve control of enzyme synthesis as indicated by ELISA (Fig. 7).

Thus, if the kinetics of the response are a valid indication of the phytochrome species involved, the data pertaining to photocontrol of flowering in wheat clearly shows that a response to labile phytochrome can occur in mature plant tissue grown in the light for many weeks, whereas the example of nitrate reductase shows that a response apparently controlled by seedling phytochrome can at a later stage of development become under control of the stable (type-2) phytochrome, seemingly eliminating the simple idea that the response in any tissue necessarily changes in kinetic character as the predominant form of phytochrome changes from the labile (seedling) form to the stable (green plant) form.

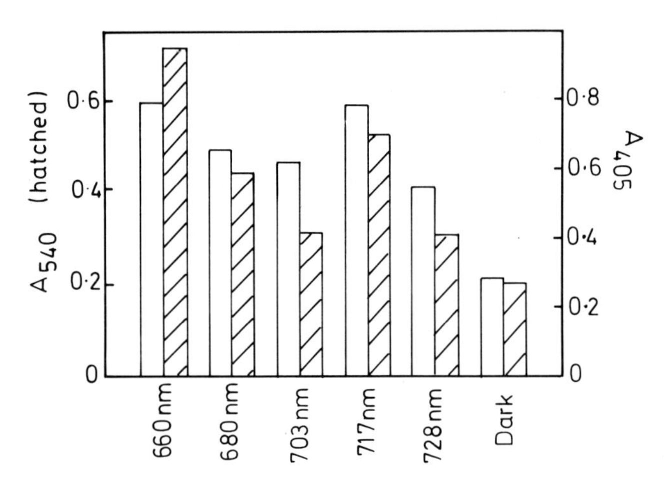

Figure 6. Effect of 12h irradiation of 48h-old seedlings of *Sinapis alba* with light of the wavelengths indicated on nitrate reductase activity (A_{540nm}) and NR protein measured by ELISA (A_{405nm}).

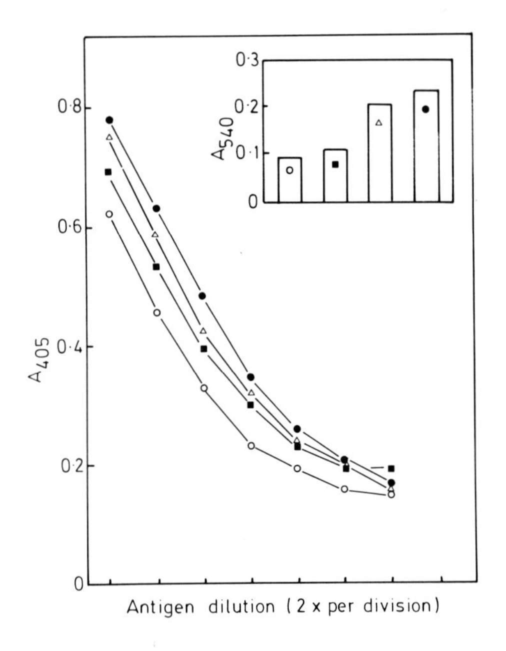

Figure 7. Effect of light quality on NR activity (inset; A_{540nm}) and NR protein measured by ELISA (A_{405nm}) in green light-grown mustard plants. Light treatments, 16h in each case were of constant photosynthetically active radiation (20μmol m^{-2} s^{-1}) combined with supplementary far-red light to yield calculated phytochrome photoequilibria of 0.25 (○); 0.39 (■); 0.54 (△) and 0.67 (•).

It is possible that day-length detection is insensitive to type-2 phytochrome whereas photomorphogenetic responses are sensitive to both types. Such data that have emerged from experiments involving expression of foreign type-1 phytochrome in transgenic tobacco and tomato are consistent with this idea, suggesting that type-1 phytochrome may be active whenever it is present (Boylan and Quail, 1989; Kay *et al.*, 1989; Keller *et al.*, 1989). Unfortunately the complementary experiment; transfer of type-2 phytochrome and its expression in seedlings has not yet been tested.

PHYTOCHROME CONTROL OF GENE EXPRESSION EXTENDS BEYOND CONTROL OF TRANSCRIPTION

The control of many enzymes by phytochrome has been shown to involve regulation of gene transcription; several chapters in this volume deal with this. One enzyme demonstrated to be subject to photocontrol of transcription is nitrate reductase (NR). For example, Deng *et al.* (1990) have shown that nitrate reductase mRNA and protein were induced simultaneously in dark-pretreated tobacco plants within 2h of transfer to white light. On the other hand, in plants given a light-dark cycle (16/8h), although both NR mRNA and NR protein showed transient increases during the light period the two phenomena were out of phase and differed quantitatively. Previously, Galancou *et al.* (1988) had shown that there were significant differences in the timing and magnitude of responses of NR mRNA, NR protein and NR activity in both tomato and tobacco growing in a light-dark cycle, suggesting that photocontrol of NR may involve regulation post-transcriptionally and post-translationally. Similar results have been obtained using maize shoots (Bowsher *et al.*, 1991). In none of these reports was the involvement of phytochrome directly proven. Phytochrome control of NR mRNA, protein and activity has been demonstrated in squash cotyledons both in response to pulses of red light and under continuous far-red light (Rajasekhar *et al.*, 1988).

Using seedlings of *Sinapis alba* as experimental material we have shown that NR is under phytochrome control and that phytochrome regulates at least one post-translational process. In these experiments, red light pulses were given as pretreatments to seedlings grown in the absence of nitrate. Upon addition of nitrate the induction of NR is rapid in the pretreated seedlings compared with the dark controls (Fig. 8a). The time course of induction of NR protein (measured by ELISA) and NR activity is shown in Figure 8b.

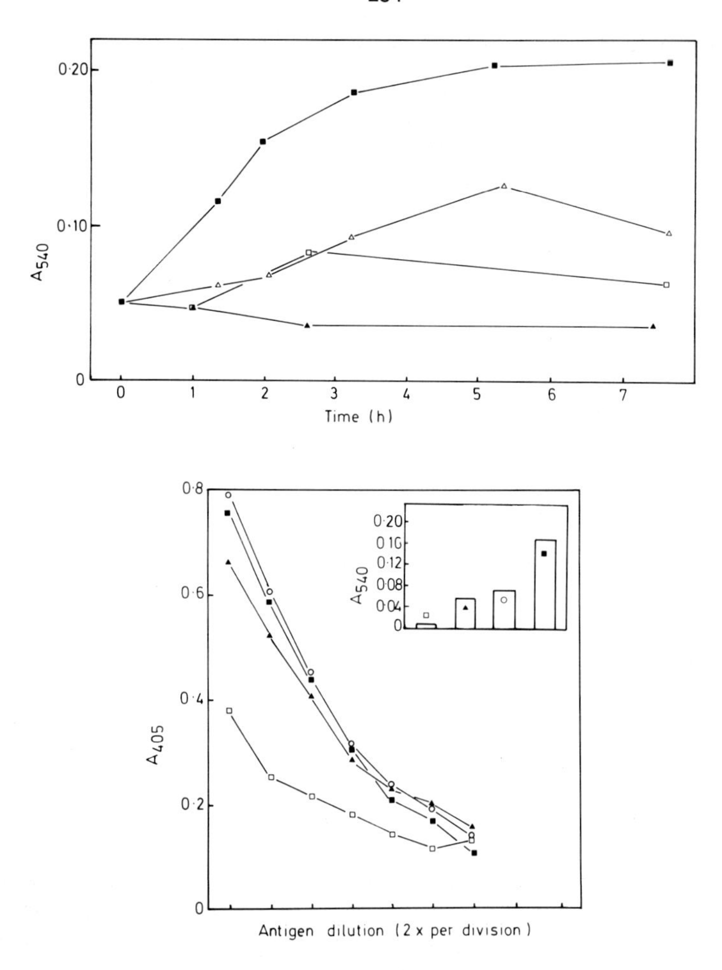

Figure 8. (top) Time course of nitrate reductase activity in seedlings of *Sinapis alba* grown for 66h on 1% agar in the dark. Three 10min pulses were given at 50h, 58h and 66h (treatments ■ and □). At 66h treatments ■ and △ were given 10ml of 100mM KNO_3. (bottom) Time course of induction following additions of KNO_3. Main figure, NR protein measured by ELISA (A_{405nm}); inset NR activity (A_{540nm}). Times 0h (□); 1h (▲); 2h (○); 3h(■).

It can be seen that there is a substantial lag between the induction of the enzyme protein and the appearance of enzyme activity. The involvement of two phytochrome-controlled processes is revealed when the pulses of red light are followed by far-red pulses: these completely reverse the induction of enzyme activity but fail to reverse the production of enzyme protein (Fig. 9a). Using much shorter red-light pulses followed by high fluence rate pulses of long wavelength (750nm) far-red light, it is possible to reverse the phytochrome effect on NR protein also (Fig.9b). Thus, the post-translational step in this

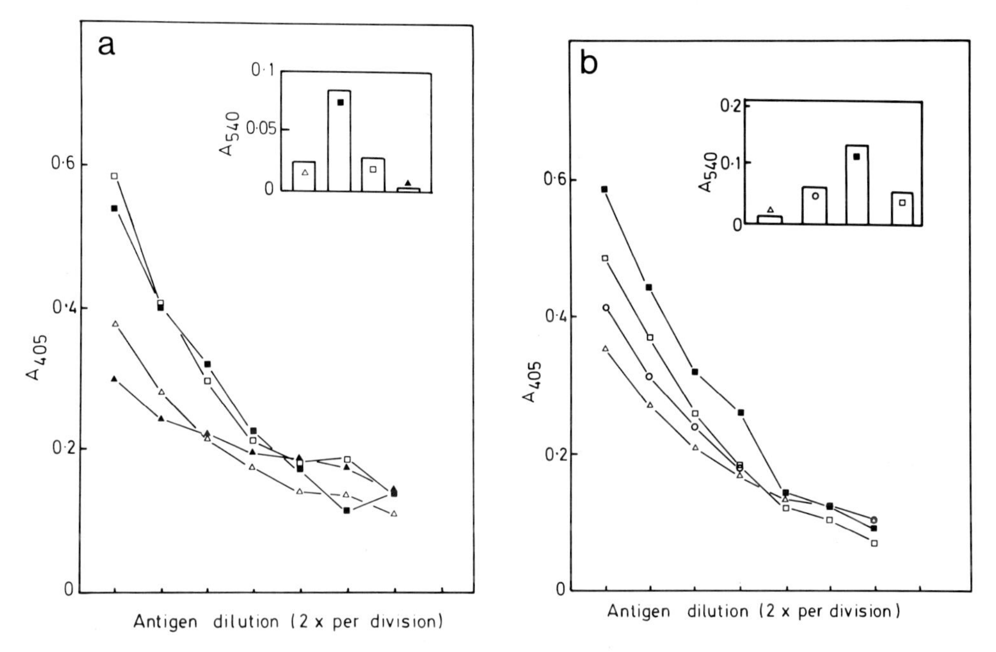

Figure 9. (a) NR and ELISA activity 3h after addition of KNO_3. Where indicated, far-red light (10min) was given immediately after each 10min red pulse. Symbols; Dark control (▵); red (■); red:far-red (□); dark [no nitrate] (▴). (b) As above, except that the red and far-red pulses were given at a high fluence rate for 2min and 5min (750nm) respectively. Symbols: dark control (▵); far-red (○); red (■); red:far-red (□).

photoresponse can be separated from the previous part of the induction process because it has a much shorter escape time. We have also been able to show differences between kinetics of induction of NR protein and activity in seedlings in a light-dark cycle (S. M. Allsebrook, unpublished). It is not yet clear whether it is the same molecular form of phytochrome that controls the two or more steps in the process leading to the appearance of NR activity.

CONCLUSIONS

The discovery of multiple forms of phytochrome (evidence is presented elsewhere in this volume for a third molecular form; see the chapter of Pratt) has understandably led to attempts to attribute specific roles or mechanisms of action for each species. However, the data presented here shows that at least one phytochrome response, the photocontrol of nitrate reductase, can be controlled both by seedling and green plant phytochrome. In contrast, the persistence of type-1 phytochrome as the effector of the day-extension response in mature light-grown wheat, long after this form of the photoreceptor has become the minority component suggests that type-2 phytochrome may not, for some reason, be able to participate in this response. There are other examples too where one form of the pigment seems to be implicated at a time when it is not the predominant form, the photopromotion of germination for example (Bartley and Frankland, 1982) and the control by phytochrome of its own synthesis (Hilton and Thomas, 1985, 1987). On the other hand, inhibition of elongation growth, like nitrate reductase, appears to be under the control of both seedling and green plant phytochrome, according to which form predominates at the time (Johnson, 1981). It would therefore that caution might be wisest at present in assigning specific roles to different members of the phytochrome 'family'.

ACKNOWLEDGEMENTS

The financial assistance of the Science and Engineering Research Council (H C-S) and the Agriculture and Food Research Council (CBJ) is gratefully acknowledged.

REFERENCES

Bartley MR, Frankland B (1982) Analysis of the dual role of phytochrome in the photoinhibition of seed germination. Nature 300:750-752

Beggs CJ, Holmes MG, Jabben, Schäfer E (1980) Action spectra for the inhibition of hypocotyl growth by continuous irradiation in light and dark grown *Sinapis alba* L. seedlings. Plant Physiol 66:615-618

Blondon F, Jacques R (1970) Action de la lumière sur l'initiation florale du *Lolium temulentum* L.: spectre d'action et rôle du phytochrome. C R Acad Sci (Paris) 270:947-950

Borthwick HA, Hendricks SB, Parker MW (1948) Action spectrum for photoperiodic control of floral initiation of a long-day plant, Wintex barley (*Hordeum vulgare*). Bot Gaz 110:103-118

Bowsher CG, Long DM, Oaks A, Rothstein SJ (1991) Effect of light/dark cycles on expression of nitrate assimilatory genes in maize shoots and roots. Plant Physiol 95:281-285

Boylan MT, Quail, PH (1989) Oat phytochrome is biologically active in transgenic tomatoes. The Plant Cell 1:765-773

Carr-Smith H, Johnson CB, Thomas B (1989) Action spectrum for the affect of day-extensions on flowering and apex elongation in green, light-grown wheat (*Triticum aestivum* L.). Planta 179:428-432

Carr-Smith H, Thomas B, Johnson CB, Plumpton C, Butcher GW (1991) The kinetics of type-1 phytochrome in green, light-grown wheat (*Triticum aestivum* L.) Planta (in press)

Deng M-D, Moureaux T, Leydecker M-T, Caboche M (1990) Nitrate-reductase expression is under the control of a circadian rhythm and is light inducible in *Nicotiana tabacum* leaves. Planta 180:257-261

Downs RJ (1956) Photoreversibility of flower initiation. Plant Physiol 31:279-284

Friend DCJ (1968) Spectral requirements for flower initiation in two long-day plants, rape (*Brassica campestris* cv. ceres) and spring wheat (*Triticum aestivum*). Physiol Plant 21:1185-1195

Friend DCJ, Helson VA, Fisher JE (1959) The relative effectiveness of standard cool fluoresecent and incandescent light in the photoperiodic response of Marquis wheat, Garnet wheat and Wintex barley. Can J Plant Sci 39:229-240

Galangau F, Daniele-Vedele F, Moureaux T, Dorbe M-F, Leydecker M-T, Caboche M (1988) Expression of leaf nitrate reductase genes from tomato and tobacco in relation to light-dark regimes and nitrate supply. Plant Physiol 88:383-388

Hilton JR, Thomas B (1985) A comparison of seed and seedling phytochrome in *Avena sativa* L. using monoclonal antibodies. J Exp Bot 36:1937-1946

Hilton JR, Thomas B (1987) Photoregulation of phytochrome synthesis in germinating embryos of *Avena sativa* L. J Exp Bot 38:1704-1712

Holmes MG, Schäfer E (1981) Action spectra for changes in the 'high irradiance reaction'in hypocotyls of *Sinapis alba* L. Planta 153:267-272

Imhoff Ch, Lecharny A, Jacques R, Brulfert J ((1979) Two phytochrome-dependent processes in *Anagallis arvensis* L.: flowering and stem elongation. Plant Cell Environ 2:67-72

Jacques M, Jacques R (1969) Spectre d'action de l'induction florale de deux Chenopodiacées de jour long. C R Acad Sci (Paris) 269:2107-2109

Johnson CB, Whitelam GC (1982) Phytochrome action in light-grown plants: the control of nitrate reduction as a model response. Photochem Photobiol 35:251-254

Johnson CB (1981) How does phytochrome perceive light quality? In: Plants and the daylight spectrum, H. Smith (ed) pp481-497. Academic Press London

Johnson CB (1990) Signal transduction mechanisms in phytochrome action. In: Ranjeva R, Boudet AM (eds) Signal Perception and transduction in higher plants. Springer-Verlag Berlin Heidelberg

Kay SA, Nagatani A, Keith B, Deak M, Furuya M, Chua N-H (1989) Rice phytochrome is biologically active in transgenic tobacco. The Plant Cell 1:775-782

Keller JM, Shanklin J, Vierstra RD, Hershey HP (1989) Expression of functional monocotyledonous phytochrome in transgenic tobacco. EMBO J 8:1005-1012

Parker MW, Hendricks SB, Borthwick HA, Scully NI (1946) Action spectrum for the photoperiodic control of floral initiation in short-day plants. Bot Gaz 108:1-26

Parker MW, Hendricks SB, Borthwick HA (1950) Action spectrum for the photoperiodic control of floral initiation of the long-day plant, *Hyoscyamus niger*. Bot Gaz 111:242-252

Rajasekhar VK, Gowri G, Campbell WH (1988) Phytochrome-mediated light regulation of nitrate reductase expression in squash cotyledons. Plant Physiol 88:242-244

Schäfer E, Marmé D, Marchal B (1972) *In vivo* measurements of the phytochrome photostationary state in far-red light. Photochem Photobiol 15:457-464

Schäfer E, Fukshansky L, Shropshire W Jr (1983) Action spectroscopy of photoreversible pigment systems. In: Encyclopedia of Plant Physiology NS, vol 16: Photomorphogenesis, pp39-68, Shropshre W Jr, Mohr H (eds) Springer Berlin Heidelberg New York

Smith H (1990) Signal perception, differential expression within multigene families and the molecular basis of phenotypic plasticity. Plant Cell Environ 13:585-594

Wall JK, Johnson CB (1983) An analysis of phytochrome action in the 'high-irradiance response'. Planta 159:387-397

Whitelam GC, Johnson CB, Smith H (1979) The control by phytochrome of nitrate reductase in the curd of light-grown cauliflower. Photochem Photobiol 30:589-594

Whitelam GC, Johnson CB (1982) Photomorphogenesis in *Impatiens parviflora* and other plant species under simulated natural canopy radiation. New Phytol 90:611-618

THE TRANSDUCTION OF LIGHT SIGNALS IN PLANTS: RESPONSES TO BLUE LIGHT

W.R. Briggs and T.W. Short
Department of Plant Biology,
Carnegie Institution of Washington,
290 Panama St., Stanford,
California, USA.

Although the majority of articles in this volume are devoted to studies of phytochrome, there has been considerable progress recently in understanding plant responses to blue light - mediated not by the blue light-absorbing bands of phytochrome, but rather by specific blue light photoreceptors. These inroads are being made at the cellular, biochemical, and molecular levels, and hold considerable promise of leading to an understanding of at least portions of the transduction chains set in action by blue light. While the systems under study do not involve phytochrome, they provide models for photoreceptor action that could apply to phytochrome, or phytochrome - blue light photoreceptor interaction.

There has developed over the years a large literature on the many effects of blue light on plants and fungi (Senger, 1987). Though significant progress has been made in understanding blue light-mediated photomorphogenetic phenomena, answers to two questions have remained elusive. First, what are some of the early steps in the respective transduction chains? Second, what are the chemical identities of the photoreceptors involved? These questions are clearly not independent of each other. For example, if a photoreceptor were a flavoprotein, then one might look for early steps in the signal transduction chain that involved light-induced oxidation/reduction changes. On the other hand, if a photoreceptor were a carotenoprotein, one might instead look for a cis-trans isomerization of the carotenoid moiety, perhaps leading to some conformational change in the underlying protein.

Because there is evidence that more than one blue light photoreceptor may be involved in blue light-regulated photomorphogenesis (Briggs and Iino, 1983; Iino, 1988b; Palit *et al.*, 1989) it would not be surprising to find more than one mechanism existent even in a single plant species. Even a superficial review of the phenomenology of blue

NATO ASI Series, Vol. H 50
Phytochrome Properties and Biological Action
Edited by B. Thomas and C. B. Johnson

light responses supports the suggestion that there may be a multiplicity of transduction chains. For example, pea seedlings grown in continuous red light exhibit two distinct responses to brief (e.g. 30 s) blue light irradiation - phototropism (Baskin, 1986) and the rapid inhibition of stem elongation (Laskowski and Briggs, 1988). These two responses differ significantly in their phenomenology: phototropism is induced by very low fluences of blue light, has a lag period of about 10 minutes, and may persist for a matter of hours (Baskin, 1986). By contrast, the rapid inhibition of stem elongation requires over an order of magnitude more blue light than phototropism, has a lag period of between 2 and 3 minutes, and persists only for an hour or less (Laskowski and Briggs, 1988). Against this context, the aim of this paper is to review briefly some promising systems currently being studied in several laboratories, and follow with a report on progress in our own laboratory in elucidating what may be an early step in the transduction chain for phototropism.

It is well-known that blue light plays an important role in inducing the swelling of stomatal guard cells (Zeiger, 1983). Elegant patch-clamp experiments with guard cell protoplasts (Assmann *et al.*, 1985; Shimizaki *et al.*, 1986) show that blue light, independent of its effect on photosynthesis, stimulates a rapid ATP-dependent efflux of cations - almost certainly protons (Shimizaki *et al.*, 1986) across the plasma membrane. A model for the role of this current in stomatal opening is that light activates a membrane proton ATPase, leading to proton extrusion. The resulting hyperpolarization opens voltage-dependent, inwardly rectifying potassium channels, leading to an influx of potassium. There is a consequent increase in the osmotic potential of the guard cell, and a flux of water following the osmotic gradient, leading to the increase in turgor that drives stomatal opening (Schröder, 1988). An increase in cytosolic calcium is thought to block these channels, leading to stomatal closing (Schröder and Hasegawa, 1989). Thus, in the stomatal response, it is thought to be ATPase activation that is the initial step in the transduction chain for blue light-induced opening, although there is, as yet, no evidence as to how blue light activates the ATPase.

Another well-known response to blue light in the plasma membrane is the flavin-mediated blue light-induced reduction of a b-type cytochrome (for a review, see Rubinstein and Stern, in press). Under the right conditions, blue light in the general fluence range of that for rapid inhibition of growth induces a prompt but relatively slow reduction of this plasma membrane cytochrome (Brain *et al.*, 1977; Goldsmith *et al.*, 1980). LIACs, as they have been named (for Light-Induced-Absorbance-Change, because they are monitored by observing the spectral shift of the cytochrome's Soret band on reduction) have been observed in membrane preparations from a wide range of fungi and higher plants as well in several intact fungi (see Rubinstein and Stern, in press), and the one LIAC action spectrum available clearly implicates a flavoprotein (Muñoz and Butler, 1975). The electron donor *in vivo* is unknown, and EDTA seems to be an effective electron donor *in vitro*.

For many years, LIACs have been phenomena in search of a physiological role. Recently, however, work in Rubinstein's laboratory is rekindling interest in the system. Dharmawardhane *et al.* (1989) have been studying electron transport across the plasma membranes of oat leaf mesophyll cells. They first expose the cells by removing the epidermis, and then measure ferricyanide reduction in the solution bathing them. With this technique, they have demonstrated that a class of lipids known as sphingoid bases will stimulate the light-induced flow of electrons across these plasma membranes by as much as a factor of four. The sphingoid bases have either no effect or are slightly inhibitory to such electron transport in the dark. Sphingoid bases are known to inhibit the electron transport across the plasma membrane leading to the oxidative burst caused in neutrophils by a range of agonists (Wilson *et al.*, 1986). They are known to inhibit protein kinase C, and are reported both to inhibit and to stimulate other protein kinases (see Hannun & Bell, 1989 for review).

Dharmawardhane *et al.* (1989) hypothesize that in oat cells it is the dephosphorylated state that is required for blue light activation of electron transport across the plasma membrane. In the presence of the the sphingoid bases, phosphorylation is prevented, and the system shows light sensitivity. These workers used broad bandpass filters to demonstrate that they were dealing with an effect of blue light, and not of red. The lag period for this phenomenon (10-15 seconds) is similar to that reported by Spalding and Cosgrove (1988) for a blue light-induced membrane depolarization of pea stem membranes as measured by external electrodes. However, the relationship between this depolarization, LIACs, the blue light-altered electron transport studied by Dharmawardhane *et al.*, and blue light-induced proton by guard cell protoplasts, is unclear. The depolarization recovers during continuous blue light and the LIACs, guard cell proton extrusion, and ferricyanide reduction do not.

Shinkle and Jones (1988) recently showed that the inhibition of elongation of cucumber hypocotyl sections by blue light could be prevented by addition of ascorbate, and that ascorbate also inhibited the action of extracted cell wall peroxidases, hence implicating peroxidase action in the blue light response. Rubinstein and Stern (in press) suggest that light-mediated electron transport across the plasma membrane and blue light-induced growth inhibition may in fact be coupled. They propose that electrons transported perhaps *via* the much investigated LIAC components, reduce oxygen to H_2O_2. Peroxidase then mediates formation of disulfide bridges in some wall component to bring about growth inhibition. While there are still large gaps in this hypothesis (for example, do plants contain a substance analogous to the sphingoid bases, and if so do they play a role in light-mediated trans-plasma membrane electron transport?) it has the appeal that most of its elements can be readily tested.

The three preceding examples all involve relatively rapid blue light-induced changes in transport at the plasma membrane. Kaufman's laboratory has recently been investigating longer term changes induced by blue light. Warpeha and Kaufman (1989) first characterized a growth response of pea stem elongation that, in contrast to that studied by Laskowski and Briggs (1989), required several hours before it could be detected. The fluence-response curve for this effect covered five decades of fluences, with a threshold below 10^{-1} μmol m^{-2} and was bell-shaped, showing saturation at 10^4 μmol m^{-2}. They then presented evidence that two separate blue light effects were involved - a low fluence effect that leads to growth inhibition, and a high fluence effect that somehow cancelled the inhibition. Both responses were present in seedlings grown under continuous red light (Warpeha and Kaufman, 1989), but only the low fluence response was observed in completely etiolated seedlings (Warpeha and Kaufman, 1990).

In studies analogous to those done over the past few years with phytochrome (see Schäfer and Briggs, 1986, Thompson, 1988), Kaufman's group then began investigating blue light-induced changes at the molecular level. They first chose four transcripts, two showing positive and two showing negative regulation by blue light, and investigated both their fluence requirements and the time courses for the light-induced changes in red light-grown pea seedlings (Warpeha *et al.*, 1989). Transcript abundance was first measured 24 h after a brief blue light irradiation. The two positively regulated transcripts

(Cab and pEA215) showed bell-shaped fluence-response curves similar to those for the long term growth inhibition (see Warpeha and Kaufman, 1989). However, the two negatively regulated transcripts showed only a decrease in abundance with increase in fluence. Furthermore, one of them (pEA207) was three orders of magnitude less sensitive than the other (pEA25), showing only a high-fluence or a low-fluence response, respectively. There were differences in kinetics as well, with both positively regulated transcripts showing a prompt response, and the negatively regulated transcripts showing no change until after 3 h. Reciprocity appeared to hold for all of these light responses.

Finally, Marrs and Kaufman (1989) used run-on transcription by isolated nuclei to investigate the possible role of blue light induced changes in transcription in the changes in transcript level previously observed. The changes they found were consistent with blue light-induced transcriptional regulation. For those transcripts that increased, there was a significant increase in measurable transcription. For those mRNAs that decreased, there was a concomitant decrease in measurable transcription. The kinetics of these transcriptional changes were also consistent with the changes in transcript abundance: where a transcript showed a prompt response, a large change in transcriptional activity was clearly evident after 3 h. Where a transcript showed a delayed response, only one of the two transcripts tested, pEA 207, showed a significant response after 3 h.

The molecular studies just discussed bear no obvious relationship to the other three systems already mentioned. They are orders of magnitude slower, and if they share part of a common transduction pathway with any of the three more rapid systems, there must be many steps in between. However, two other studies address possible earlier events in blue light-activated transduction chains. Both involve systems that can be activated by irradiation of isolated and purified membrane fractions.

In the first of these, Warpeha *et al.* (1990 and personal communication) have recently shown that blue light irradiation of sucrose gradient-purified plasma membrane from etiolated pea buds brings about a dramatic increase in the capacity of these membranes to hydrolyze GTP. The reaction is very sensitive to light, with a threshold near 10^{-1} μmol m^{-2}. It is also relatively rapid, showing a measurable change within 30 s and going to completion within 3-4 minutes. It is specific for GTP, and strongly inhibited by the GTP analogs guanosine thiotriphosphate (GTP-cS) and guanylyl imidodiphosphate

(GppNHp). Red light is completely ineffective, and the system shows no ATPase activity. Blue light irradiation of the plasma membranes strongly enhances the binding of GTPcS. Warpeha and Kaufman detected a 40 kDa protein on sodium dodecylsulfate (SDS) polyacrylamide gels that was both recognized by antibodies raised aginst the visual system G protein, transducin, and that bound an azido derivative of GTP when present with the membranes during irradiation. Finally, the 40 kDa protein could be ADP-ribosylated by both pertussis and cholera toxins. In the case of cholera toxin, full ribosylation required prior blue light treatment. Antibodies raised against the b and c subunits of transducin also recognized polypeptides in SDS gels of the plasma membranes. The work shows that the components of a blue light-activated G-protein system are present in pea, and makes it likely that G proteins are involved in the transduction of blue light signals in this plant.

While the response described in the last paragraph is very rapid, that which we have been studying, a blue light-induced change in the phosphorylation state of a plasma membrane protein, is even faster. We will begin by reviewing already published information on this system, and then consider some very recent work bearing on the nature of the events brought about by blue light.

About a year and a half ago, Gallagher *et al.* (1988) first reported a dramatic light-induced change in detectable phosphorylation of a membrane protein of molecular weight 120 kDa from elongating regions of etiolated pea stems. The response was detected by irradiating intact stem tissue, preparing membrane fractions, and then treating them with ^{32}P-labelled ATP prior to SDS polyacrylamide electrophoresis. The phosphorylation pattern thus obtained was compared with that of membranes from unirradiated tissue. Whereas strong phosphorylation of a protein near 120 kDa was detected in membranes from dark controls, phosphorylation at 120 kDa was vanishingly small or undetectable if the tissue had been irradiated (with several hours of white light in the earliest experiments). This first paper established three points: First, the response is mediated by a blue light photoreceptor rather than phytochrome. Second, the protein is located in the plasma membrane and is not found in any of the endomembranes. And, third, the association of the protein with the membrane is strongly lipophilic.

Recently we have examined in detail some of the photophysiological properties of the system (Short and Briggs, 1990), and have established a number of other properties for the system. First, the reaction is extremely rapid. Stem sections already in extraction medium in a chilled mortar on ice can be irradiated for as little as 0.3 s (provided the total fluence is sufficient to saturate the response), and ground immediately thereafter, and their membranes still show the typical light effect - an absence of phosphorylation at 120 kDa. Whatever the reactions leading to loss in capacity for phosphorylation, they go to completion during the brief time period required for irradiation and subsequent grinding (about 45 s), despite the low temperature. Second, the system can show dark recovery. If sections are incubated at room temperature for various times in darkness following a saturating irradiation and prior to membrane extraction, the capacity for phosphorylation at 120 kDa gradually recovers over about an hour, after a lag period of about 10 min. This dark recovery is not unlike that for phototropic sensitivity of maize coleoptiles following a saturating phototropic stimulus (Briggs, 1960). Third, the strongest phosphorylation is detected in membranes from the uppermost elongating stem tissue (on a unit protein basis) declining basipetally. However, wherever phosphorylation can be detected at 120 kDa in membranes from dark controls, it is lost in membranes from irradiated tissues. Curiously, the apical pea bud membranes showed no phosphorylation under either condition at that molecular weight. Fourth, at least within the limits of fluence rate and duration of irradiation tested, the system obeyed the reciprocity law. Fifth, the range of effective fluences, from threshold to saturation for the response, closely matched the range of fluences, from threshold to saturation, for phototropism. Hence we have concluded that the change in phosphorylation that we are seeing probably represents an early step in the phototropic transduction chain. It is sufficiently rapid, shows appropriate dark recovery, is found in the right place, and matches the fluence range precisely.

Short and Briggs (1990) published two additional intriguing observations about the system. First, when isolated membranes are irradiated instead of tissues, the capacity of the 120 kDa protein for phosphorylation by exogenous radiolabelled ATP is greatly increased, rather than being decreased as when the tissues are irradiated. Second, Coomassie-stained gels of membranes from dark-control stem sections show a clear band at 120 kDa. This band is missing on gels of membranes from blue light-treated seedlings. This light effect shares the same tissue distribution as the phosphorylation change, has the same fluence requirements, and shows the same dark recovery kinetics. Also, the protein is undetectable by similar staining of gels of membranes from buds - in parallel with the phosphorylation reaction, which is also undetectable in buds.

These two observations raise two immediate questions: First, why is the consequence of irradiation *in vitro* the opposite of the consequence *in vivo*? Second, what accounts for the disappearance of the protein band on gels from membranes of irradiated seedlings? The simplest hypothesis to answer the first question is the following: light in some manner induces a conformational change in the 120 kDa protein, exposing sites for phosphorylation. When irradiation is *in vivo*, these sites are immediately phosphorylated by endogenous ATP, and hence are not available for subsequent *in vitro* phosphorylation. When irradiation is *in vitro*, however, endogenous ATP is no longer available, and hence the exposed sites become rapidly phosphorylated by the exogenously provided radiolabelled ATP. This explanation can not completely account for the differences, however, since irradiation *in vivo* brings about decreased mobility on SDS gels whereas irradiation *in vitro* does not (see below).

There are at least three possible answers to the second question. First, the protein could become separated from the membranes on irradiation, though this result seems unlikely given its hydrophobic nature (Gallagher *et al.*, 1988). Second, the mobility of the protein could have been changed as a result of degradation or covalent modification, or as a direct result of phosphorylation. And, third, the relative affinity of the protein for the Coomassie stain may have been decreased by phosphorylation. We have recently obtained polyclonal antibodies against this protein and they have allowed us to address this question directly. Western blots of membrane proteins from dark control and light-treated stem sections clearly show that the protein remains with the plasma membrane fraction. Irradiation does not chase it either into another membrane fraction or into the cytoplasm. Second, irradiation *in vivo* induces both a mobility decrease and sufficient heterogeneity in electrophoresis that the protein band becomes considerably broader and more diffuse on the gel. Hence the apparent disappearance from the gel is probably both because the protein is no longer located in a single tight band, and because phosphorylation somehow reduces its capacity to bind Coomassie stain - a change with precedent in other proteins that undergo extensive phosphorylation (T. J. Guilfoyle, personal communication; see Beebe and Corbin, 1986 and Krust *et al.*, 1984 for references to mobility shifts).

The light-induced change in capacity for phosphorylation requires at minimum three components: the 120 kDa substrate protein, appropriate kinase activity, and the photoreceptor. At this time we are uncertain whether the system involves only a single polypeptide (e.g. a 120 kDa photoreceptor protein that has the capacity for auto-phosphorylation, a capacity that is somehow only expressed as a consequence of photoexcitation) or whether it involves two or even three polypeptides (e.g. in the extreme case, separate photoreceptor, kinase, and 120 kDa substrate polypeptides). However, the systems are now available to address this problem, and some progress has been made. First, the phosphorylation increase can be obtained by irradiating plasma membrane fractions purified by the two-phase technique of Widell and Larssen (1987). Hence, all three components are located in the plasma membrane. Second, the increase can be obtained by irradiation of membranes first solubilized either by Triton X-100 or by CHAPS -non-ionic and zwitterionic detergents, respectively. Furthermore, the quantum efficiency for the reaction is unchanged by solubilization. Hence all three solubilized components must not only reside within the same micelles but their physical state must be relatively unaltered from that in the intact membrane by solubilization. The results support the hypothesis that a single polypeptide is involved, but more rigorous tests must be applied before one can answer the question with certainty. We feel that by use of immobilized antibody and other techniques we are in a good position both to resolve this point and to isolate and characterize the photoreceptor itself.

We have also recently addressed a couple of other questions. First, what is the nature of the photoreceptor moiety itself? The controversy as to whether the photoreceptor for phototropism is a flavin or carotenoid has raged for decades (see Galston, 1959), and more recently Palit *et al.* (1989) and Schmidt *et al.* (1990) have raised the possibility that pterins may also be viable candidates. At least in our system a direct test for flavin involvement is possible. Both azide ions and phenylacetic acid can block flavins from carrying out their normal photochemistry. The former agent depopulates flavin excited states (Heelis *et al.*, 1978), while the latter reduces the flavin, undergoing at the same time an oxidative decarboxylation and forming a covalent bond with the flavin (Hemmerich *et al.*, 1967). Both of these agents selectively block the light-induced increase in potential for phosphorylation of the 120 kDa protein at concentrations that otherwise do not affect overall phosphorylation of the other membrane proteins that become phosphorylated during the assay. It is possible, although unlikely, that pterins might also be affected by azide and phenylacetic acid in a similar manner. Hence we conclude that the photoreceptor in this system is probably a flavin, and is certainly not a carotenoid.

Finally, we have asked whether the system we are studying is unique to pea seedlings or whether it may be ubiquitous in higher plants. If it really represents an early step in the phototropic transduction chain, one might expect it to be found in a wide range of plant species. If it were conserved, one might even be able to detect cross reactivity with the polyclonal antibody raised against the pea protein. Western blots were made from SDS polyacrylamide gels of crude membrane fractions from carrot cells and from zucchini, sunflower, *Arabidopsis*, maize, tomato, and fava bean seedlings and were probed with the pea antibody. In all cases, proteins were detected that cross-reacted with the antibody. Even more impressive, irradiation of membranes from all of these species in every case brought about a measurable and in some cases very strong enhancement of the capacity for phosphorylation. Though the proteins in the different species cover a range of molecular weights, between 116 and 130 kDa, and hence have undoubtedly diverged considerably during evolution, they share common epitopes and are similarly affected by irradiation of the isolated membranes in which they reside.

We hope soon to have in hand information from microsequencing of the 120 kDa protein and are in the process of constructing cDNA expression libraries to probe in the hopes of isolating and characterizing its gene. The overall task of understanding the system will of course become more complicated if, for example, three proteins are involved in the response rather than just one. We will then face carrying out a molecular attack on not one but three genes.

In conclusion, solid progress in understanding blue light effects at the cellular and molecular levels is emerging on several fronts. The results to date are still fragmentary, however. The relationships between the six blue light-induced systems discussed above - proton extrusion by guard cells, cytochrome reduction, electron transport across the plasma membrane, changes in gene transcription, effects on G proteins, and effects on phosphorylation, are simply not clear. There could certainly be a causal relationship between phosphorylation changes or G-protein activation and transcriptional changes. The possible link between LIACs and trans-plasma membrane electron transport has already been mentioned. However, there are major differences in fluence requirements for the different responses both at the physiological and membrane levels, and it seems likely that in these various studies, we are looking at a minimum of two and perhaps more blue light-activated transduction chains. Considerable further experimentation will be required to sort these various systems out, identify their photoreceptors, and elucidate any possible interactions between them.

ACKNOWLEGEMENTS

The research from the author's laboratory was supported by National Science Foundation Grant DCB-88 19137 to WR Briggs; TW Short was supported by a National Science Foundation Predoctoral Fellowship. The authors gratefully acknowledge the able technical assistance of Anne McKillop.

This is Carnegie Institution of Washington Department of Plant Biology Publication No. 1090.

REFERENCES

Assmann SM, Simoncini L, Schröder JI (1985) Blue light activates electrogenic ion pumping in guard cell protoplasts. Nature 318:285-287

Baskin TI (1986) Redistribution of growth during phototropism and nutation in the pea epicotyl. Planta 169:406-414

Beebe SJ, Corbin JD (1986) Cyclic nucleotide-dependent protein kinases. In: The Enzymes Vol. 17, Boyer PD, Krebs EG (eds). Academic Press, New York, pp.44-100

Brain RD, Freeberg JA, Weiss CV, Briggs WR (1977) Blue light-induced absorbance changes in membrane fractions from corn and *Neurospora*. Plant Physiol 59:948-952

Briggs WR (1960) Light dosage and the phototropic responses of corn and oat coleoptiles. Plant Physiol 35:951-962

Briggs WR, Iino M (1983) Blue-light-absorbing photoreceptors in plants. Phil. Trans. Roy. Soc. Lond. B 303:347-379

Dharmawardhane S, Rubinstein B, Stern AI (1989) Regulation of transplasmalemma electron transport in oat mesophyll cells by sphingoid bases and blue light. Plant Physiol. 89:1345-1350

Gallagher S, Short TW, Ray PM, Pratt LH, Briggs WR (1988) Light-mediated changes in two proteins found associated with plasma membrane fractions from pea stem sections. Proc. Natl. Acad. Sci. U.S.A. 85:8003-8007

Galston AW (1959) Phototropism of stems, roots, & coleoptiles. In Handbuch der Pflanzenphysiologie, Ruhland W (ed) Vol. XVII 1, Springer-Verlag, Berlin, pp.492-529

Goldsmith MHM, Caubergs RJ, Briggs WR (1980) Light-inducible cytochrome reduction in membrane preparations from corn coleoptiles. Plant Physiol. 66:1067-1073

Hannun YA, Bell RM (1989) Function of sphingolipids and sphingolipid breakdown products in cellular regulation. Science 243:500-506

Heelis PF, Parsons BJ, Phillips GO, McKellar JF (1978) A laser flash photolysis study of the nature of flavin mononucleotide triplet states and the reactions of the neutral forms with amino acids. Photochem. Photobiol. 28:169-173

Hemmerich P, Masseym V, Weber G (1967) Photoinduced benzyl substitution of flavins by phenylacetate: a possible model for flavoprotein catalysis. Nature 213:728-730

Iino M (1988) Pulse-induced phototropisms in oat and maize coleoptiles. Plant Physiol. 88:823-828

Krust B, Galabru J, Hovanessian AG (1984) Further characterization of the protein kinase activity mediated by interferon in mouse and human cells. J. Biol. Chem. 259:8494-8498

Laskowski M, Briggs WR (1989) Regulation of pea epicotyl elongation by blue light. Plant Physiol. 89:293-298

Marrs KA, Kaufman LS (1989) Blue-light regulation of transcription for nuclear genes in pea. Proc. Natl. Acad. Sci. U.S.A. 86:4492-4495

Muñoz V, Butler WL (1975) Photoreceptor pigment for blue light in *Neurospora crassa*. Plant Physiol. 55:421-426

Palit A, Galland P, Lipson ED (1989) High- and low-intensity photosystems in *Phycomyces* phototropism: Effects of mutations in genes *madA*, *madB*, and *madC*. Planta 177:547-553

Rubinstein B, Stern AI (1990) The role of plasma membrane redox activity in light effects in plants. J. Bioenergetics Biomembranes (in press)

Schäfer E, Briggs WR (1986) Photomorphogenesis from signal perception to gene expression. Photobiochem. Photobiophys. 12:305-320

Schmidt W, Galland P, Senger H, Furuya M (1990). Microspectrophoto-metry of *Euglena gracilis*: Pterin- and flavin-like fluorescence in the paraflagellar body. Planta. (in press)

Schröder JI (1988) K^+ transport properties of K^+ channels in the plasma membrane of *Vicia faba* guard cells. J. Gen. Physiol. 92:667-683

Schröder JI, Hasegawa S (1989) Cytosolic calcium regulates ion channels in the plasma membrane of *Vicia faba* guard cells. Nature 338:427-430

Senger H. (ed) Blue light responses: Phenomena and occurrence in plants and microoganisms. Vol. I, pp. 1-169; Vol. II, pp. 1-260. CRC Press, Boca Raton, Florida

Shimizaki K, Iino M, Zeiger E (1986) Blue light-dependent proton extrusion by guard cell protoplasts of *Vicia faba*. Nature 31:324-326

Shinkle JR, Jones RL (1988) Inhibition of stem elongation in *Cucumis* seedlings by blue light requires calcium. Plant Physiol. 86:960-966

Short TW, Briggs WR (1990) Characterization of a rapid, blue light-mediated change in detectable phosphorylation of a plasma membrane protein from etiolated pea (*Pisum sativum*, L.) seedlings. Plant Physiol. 92:179-185

Spalding EP, Cosgrove DJ (1988) Large plasma-membrane depolarization precedes rapid blue-light-induced growth inhibition in cucumber. Planta 178:407-410

Thompson WF (1988) Photoregulation: diverse gene responses in greening seedlings. Plant Cell Environ. 11:319-328

Warpeha KMF, Hamm HE, Kaufman LS (1990) A blue-light-induced plasma-membrane-associated GTP-binding protein in pea. Plant Physiol. (suppl.) 93:30 (Abstr. 164)

Warpeha KMF, Kaufman LS (1989) Blue-light regulation of epicotyl elongation in *Pisum sativum*. Plant Physiol. 89:544-548

Warpeha KMF, Kaufman LS (1990) The distinct blue-light responses regulate epicotyl elongation in pea. Plant Physiol. 92:495-499

Warpeha KMF, Marrs KA, Kaufman LS (1989) Blue-light regulation of specific transcript levels in *Pisum sativum*. Plant Physiol. 91:1030-1035

Widdell S, Larsson C. (1987) Plasma membrane purification. In: Blue Light Responses: Phenomena and Occurrence in Plants and Microorganisms. Senger H (ed) Vol. II. CRC Press, Boca Raton, Florida. pp.99-107

Wilson E, Olcott MC, Bell RM, Merrill AH Jr, Lambeth JD (1986) Inhibition of the oxidative burst in human neutrophils by sphingoid long-chain bases. J. Biol. Chem. 261:12616-12623

Zeiger E (1983) The biology of stomatal guard cells. Annu. Rev. Plant Physiol. 34: 441-475

Appendix.

ROUND TABLE REPORTS

Round Table I

POST-TRANSLATIONAL MODIFICATION OF PHYTOCHROME

G.C. Whitelam,
Department of Botany,
University of Leicester,
Adrian Building, University Road,
Leicester, LE1 7RH,
UK.

INTRODUCTION

The biogenesis of holophytochrome within the cell involves several processing/ modification events which occur after, or during, translation of the phytochrome mRNA. The most obvious of these modifications is the conjugation of the bilin prosthetic group. From studies on the properties of isolated phytochrome (obtained almost exclusively from etiolated oat seedlings) the following modifications have been identified: cleavage of the N-terminal methionone, acetylation of the now N-terminal serine, dimerization of the phytochrome monomers, insertion of the bilin chromophore, phosphorylation, glycosylation and ubiquitination.

Many of these modifications, such as cleavage of N-terminal methionyl residues and acetylation of N-terminal serine residues, are fairly common events, and were not covered during the discussion. Ubiquitination of phytochrome appears tc be associated with light-induced phytochrome destruction and is dealt with in a separate discussion session. This particular discussion focussed on glycosylation and chromophore attachment.

GLYCOSYLATION

The idea that phytochrome is glycosylated is not new. There has been at least one report that isolated oat phytochrome contained carbohydrate, detectable by periodic acid Schiff staining. However, there have also been reports that this observation proved impossible to replicate (Pratt, 1982). Recently, Rudi Grimm (Freiburg) has obtained fresh evidence for glycosylation of phytochromes. This new evidence was initially based on the

NATO ASI Series, Vol. H 50
Phytochrome Properties and Biological Action
Edited by B. Thomas and C. B. Johnson

observation that a number of lectins will bind to highly purified oat phytochrome. By assessing the binding of a range of lectins, with different sugar specificies, it appears that phytochrome contains a complex type glycan moiety. For example, *Ulex europaeus* agglutinin I (UEA I), which is specific for fucosyl residues, will bind to oat phytochrome. In addition to lectin recognition, oat phytochrome also shows binding to an antibody raised to the potato storage glycoprotein, patatin. This antibody is known to recognise the *N*-linked complex glycan moiety of patatin. Mapping studies with oat phytochrome indicate that the glycan is attached at a site(s) located between amino acids 465-530 from the N-terminus. This region of the phytochrome molecule contains a conserved stretch of eight amino acids, rich in serine and threonine residues, that represents a potential glycan attachment site. Grimm also reported that the sugar residues had been analysed by GC-MS, following methanolysis of the attached glycan from highly purified oat phytochrome. A similar analysis had been undertaken, some years ago, by Ken Keegstra and Peter Quail, for immunopurified 118/114 kDa oat phytochrome (reported in Pratt, 1982). Although this analysis did detect some sugar, the levels obtained were lower than would be anticipated for sugars derived from a glycan moiety.

Lectin-binding based detection of a glycan moeity is not restricted to oat phytochrome. Collaborations between Rudi Grimm, Masaki Furuya and Brian Thomas, have resulted in evidence for similar glycosylation of pea phytochrome (both Type I and Type II, and Type I pea phytochrome expressed in yeast and *Escherichia coli*) and mustard (Type I) phytochrome. In the case of pea Type I phytochrome, expressed in yeast, it seems that the glycan is attached at a different site compared to native pea phytochrome. This conclusion is based on the observation that, although the full length pea phytochrome product from transgenic yeast is recognised by UEA I, an immuno-affinity isolated, 107 kDa product, believed to be missing about 120 amino acids from the the N-terminus, is not. The 107 kDa product contains the site to which the glycan is attached in phytochrome isolated from peas.

The possibility that the sugar residues, being detected in these studies, become attached to phytochrome during extraction was discussed. However, this seems an unlikely explanation for the observations. It is clear that the sugars are strongly attached to phytochrome and the enzymes that mediate glycan attachment and modification *in vivo* are unlikely to operate at appreciable rates *in vitro*. Lectin-binding is detected for phytochrome that is rapidly immunoprecipitated from crude extracts. However, it is not known whether all of the phytochrome molecules under analysis are glycosylated.

The presence of a complex type glycan moiety on phytochrome raises questions about the type, and cellular site of, glycan attachment and about the function of the glycan. Grimm reported that incubations of oat phytochrome with endoglycosidases, specific for two different types of *N*-linked glycans, fail to eliminate lectin-binding activity or anti-patatin antibody binding. This would suggest that the glycan is attached by an *O*-linkage. However, attempts to cleave chemically such a linkage with sodium hydroxide and sodium borohydride have, as yet, been unsuccessful.

With the exception of nuclear glycoproteins, plant glycoproteins are generally vacuolar (this includes protein bodies), membrane located or extracellular. These proteins are co-translationally inserted into the endoplasmic reticulum, where glycosylation occurs and then transported *via* the golgi, where glycan modification occurs. Cytoplasmic glycoproteins are relatively rare in plant cells and in animal cells. The machinery for glycan attachment and modification, outside of the endomembrane system, is poorly characterised at present. A complex type of glycan will have a molecular weight of about 2 kDa. Since the product of *in vitro* translation of phytochrome mRNA has been reported to have the same electrophoretic mobility as *in vivo* synthesised phytochrome, the possibility that glycosylation occurs in the wheat germ and reticulocyte lysate systems must be considered. If this is the case, these translation systems may prove to be valuable tools for characterising the glycosylation process.

The question of the possible function(s) of the glycans attached to phytochromes was briefly considered. By analogy with the proposed roles of other glycans, it was suggested that the complex glycan on phytochrome could function to stabilize a particular protein conformation, or, it could fulfil a targeting-like function to ensure that phytochrome remains cytoplasmic.

CHROMOPHORE ATTACHMENT

Recent work from the laboratory of Clark Lagarias has established that phycocyanobilin (the chromophore of phycocyanin), which is chemically similar to phytochromobilin, will attach spontaneously to apophytochrome to yield a photoreversible product (Elich and Lagarias, 1989; Lagarias and Lagarias, 1989). This attachment can be demonstrated to occur for apophytochrome that has been translated *in vitro*. These observations have led to the suggestion that chromophore attachment *in vivo* may be an autocatalytic process i.e. apophytochrome may function as a bilin lyase. Wolfhart Rüdiger (Munich) summarised the results of similar chromophore attachment studies that have been carried out by Alex Glazer's group (see Arciero *et al.*, 1989a,b,c). These studies show that phycocyanobilin will spontaneously attach to apophycocyanin and apophycoerythrin. However in this case, non-natural adducts, in which a new double bond is introduced into the A-ring of the bilin, are formed. A similar dehydrogenation can occur on treatment of the bilin with methanol. The attachment of phycocyanobilin to apophycocyanin *in vitro* is considered to be a purely chemical reaction, whereas correct linkage of the chromophore *in vivo* is believed to require the participation of another factor (maybe an enzyme). It is not known whether the spontaneous attachment of phycocyanobilin to apophytochrome results in modification of the bilin, so the possibility that some other factor is required for correct attachment of phytochromobilin *in vivo* cannot be discounted. Masaki Furuya reported that phycocyanobilin attachment to pea apophytochrome is very rapid, and that attempts to demonstrate that the reaction can be altered by extracts of pea tissues have been negative. Thus, a genuine autocatalytic process mediating chromophore attachment also remains a possibility.

REFERENCES

Arciero DM, Bryant DA, Glazer AN (1989a) J Biol Chem 263:18343-18349
Arciero DM, Dallas J, Glazer AN (1989b) J Biol Chem 263:18350-18357
Arciero DM, Dallas J, Glazer AN (1989c) J Biol Chem 263:18358-18363
Elich TD, Lagarias JC (1989) J Biol Chem 264:12902-12908
Lagarias JC, Lagarias DM (1989) Proc Natl Acad Sci USA 86:5778-5780
Pratt LH (1982) Ann Rev Plant Physiol 33:557-582

Round Table II

PHYTOCHROME DESTRUCTION

B. Thomas,
Horticulture Research International,
Littlehampton,
West Sussex, BN17 6LP,
UK.

INTRODUCTION

One of the highly distinctive properties of phytochrome concerns its turnover. Phytochrome accumulates in plants growing in darkness and is rapidly destroyed when they are transferred to light. The superficial explanation is that phytochrome is synthesised as Pr, which is relatively stable and thus accumulates in darkness. On the other hand Pfr is much less stable and is degraded rapidly within the cell. Pfr destruction, in dark grown seedlings which receive a brief irradiation, is rapid with a half-life of 1-2 hours. Destruction is a complex process involving biochemical modification and subcellular relocation. From being cytoplasmically dispersed immunostained phytochrome becomes aggregated in discrete areas of the cell before subsequently disappearing. Labile Pfr is a feature of Type I phytochrome, whereas Type II phytochrome is relatively stable as Pfr. The discussion in the round table centred on the role of ubiquitination and the possible involvement of PEST sequences in phytochrome destruction, plus an attempt to reconcile immunocytochemical with biochemical observations.

UBIQUITINATION

The properties of ubiquitin and the ubiquitination mechanism was introduced by R. Vierstra. Ubiquitin is a 76-amino acid protein with a globular core and flexible carboxy-terminus. It attaches to proteins *via* C-terminus to lysine residues, of which there are 64 in oat phytochrome. The ubiquitination pathway is complex, involving 35 proteins. Ligation is accomplished by an ATP-coupled series of reactions and occurs through the formation of an isopeptide bond between the C-terminus of ubiquitin and free amino

NATO ASI Series, Vol. H 50
Phytochrome Properties and Biological Action
Edited by B. Thomas and C. B. Johnson

groups on the target protein. Part of the specificity involves the formation of ubiquitin chains linked *via* the C-terminus and lys-48. Proteolysis then follows. This involves ubiquitin protein lyase, a specific protease which detaches ubiquitin. A second protease, the proteosome, recognises and completely degrades proteins which have ubiquitins attached.

Phytochrome-ubiquitin conjugates can be observed by immunoprecipitating partially purified phytochrome with antiphytochrome antibodies and then staining with anti-ubiquitin. By this technique high molecular conjugates, which are not present in dark-grown tissues can be seen to be formed and then disappear when oats are transferred from dark to light. There is an apparent correlation between conjugation and loss of phytochrome. Confirmation of the requirement for ubiquitination for Pfr degradation will require the usc of transgenic plants. These studies are in progress.

Evidence for the site of ubiquitination attachment in phytochrome comes from proteolytic mapping of the binding sites of monoclonal antibodies which recognise phytochrome but not phytochrome-ubiquitin conjugates. These data indicate a region between amino acids 747 and 790 in oat phytochrome. This region appears to be more exposed as Pfr than Pr, based on proteolytic mapping, suggesting a possible mechanism for selective degradation of Pfr.

In the resulting discussion E. Tobin and W. Briggs queried whether specific ubiquitination sites or more than one attachment site were involved in protein degradation. Vierstra indicated that work with yeast suggested specific sites were required. In phytochrome there are 3 conserved lysines in the presumptive ubiquitination area but some tentative proteolytic mapping of ubiquitinated phytochrome fragments suggests one attachment site. E. Schäfer pointed out that the kinetics of conjugate formation and phytochrome loss were difficult to resolve because the number of conjugates increase with time and then decrease, but destruction occurs most rapidly when the phytochrome content is highest i.e. at the beginning of the destruction curve. A. Batschauer brought up the relationship between sequestering and ubiquitination, both of which require ATP. He asked whether the ATP-requiring step was the same in both cases. Vierstra thought not, because pelletability occurred much faster than ubiquitination. He suggested that aggregation might be the signal for ubiquitination. In response to a question by G. Jenkins, Vierstra reported that there was no evidence for ubiquitination in plant organelles; none of the ubiquitin cycle genes they had sequenced had transit peptides. There was also no good evidence for phytochrome being in organelles. The final point for discussion, raised by P. Quail concerned the evidence for high molecular weight conjugates containing both phytochrome and ubiquitin. J. Cherry confirmed that there was coincidence between staining with anti-ubiquitin and anti-phytochrome on Western blots but further analysis by 2-D electrophoresis not been performed.

THE PEST HYPOTHESIS

The discussion was introduced by W. Rüdiger who outlined the essentials of the PEST hypothesis and the evidence that it might be relevant to phytochrome destruction. The hypothesis was originally put forward by M. Rechensteiner' group. A "PEST" sequence consists of two positively charged groups which might be arginine, lysine or histidine, between which are located proline, glutamic acid, serine and threonine residues in high molar proportion. The hypothesis is empirical, based on a correlation from a wide range of organisms between the presence of the PEST sequence in a protein and a half-life of less than one hour. In phytochrome, Pr is stable and Pfr unstable even though both have the same sequence. The hypothesis would have to be modified in some way to accommodate this. Analysis of the *Avena* sequence suggests two PEST regions. There is a conserved PEST region of about 25 amino acids beginning at the histidine (his-323) to the carboxy-side of the cysteine through which the chromophore is attached. In addition there is a PEST region at residues 537-547 which is not conserved in dicots. Interestingly, proteolytic studies with V8 protease suggests that the conserved PEST region becomes exposed on transformation from Pr to Pfr. In addition, if the acidic cluster is more exposed in Pfr, there should be a more negative charge in the Pfr form than in the Pr form. This can be confirmed experimentally by isoelectric focusing and native electrophoresis.

P. Quail presented comparisons of nine phytochrome sequences. If the PEST sequence is involved in the rapid degradation of phytochrome it might be expected that the sequence would be well-conserved in type I (*phyA*) sequences but not in type II phytochromes which are relatively stable as Pfr. Comparison of A sequences show conservation of the PEST region but with glutamate being replaced with aspartate, which is an allowable substitution. Sequences of B and C genes, on the other hand include arginine and histidine within the segment, which is not allowed in the hypothesis. In addition, proline is absent in B and C sequences. Analysis of the associated acid cluster gives a calculated net negative charge of -5 to -7 for A sequences, but -3 for B and C sequences. There is therefore a broad correlation between the stability of different phytochromes as Pfr and the presence of presumptive PEST regions.

In the resulting discussion it was agreed that the PEST region is hydrophilic and likely to be on the surface, although Rüdiger's data indicates differential exposure in Pr and Pfr, based on susceptibility to proteolysis. One problem with proteolytic mapping studies is that the presence of the acidic cluster might itself interfere with cleavage. Pratt reported that one of his MAbs, which binds near to C-side of the chromophore, shows differential binding, being greater with Pfr. This supports differential exposure of that region of the molecule. Schäfer suggested that, as the rate of Pr turnover after cycling is the same as Pfr, the difference could not simply be one of differential exposure in Pr and Pfr. There would have to be a modification of the protein to maintain the exposure in Pr. Rüdiger pointed out that the hypothesis is purely empirical and does not imply a mechanism. The PEST sequence could be a signal for attack, but not necessarily the site of attack. Photoconversion from Pr to Pfr and back is in the order of seconds, data from Schäfer's group indicating a minimum time of Pfr of 2-5 seconds for the recycled Pr to be destroyed. This would allow time for modification. There was some agreement that if the PEST sequence was in some way a signal, the PEST hypothesis was not inconsistent with the involvement of ubiquitination. Vierstra, however proposed caution as there are bacterial proteins with PEST sequences which are not ubiquitinated. Also there is as yet no evidence from *in vitro* mutagenesis studies that the PEST sequence is involved in determining protein stability.

SEQUESTERING

The redistribution of phytochrome from being cytoplasmically dispersed to being sequestered in discrete areas of the cell following irradiation appears to be an early stage in the destruction process. Schäfer confirmed that immunostaining showed enrichment for ubiquitin in sequestered particles over the cytoplasm if the appropriate antibody concentrations were used. Sequestered particles can also contain cycled Pr and this might help explain ubiquitination and destruction of Pr. Rüdiger asked whether sequestering occurred under conditions which allowed photoprotection of phytochrome, as has been described by Smith's group at Leicester. Although no attempt has been made to look at sequestering under those conditions, Vierstra reported that his group had attempted to study ubiquitination but did not have light sources of sufficiently high irradiance to repeat the original experiments. Smith pointed out that as photoprotection was observed at irradiances as low as 50-100 μmol m^{-2} s^{-1} the experiments should be possible.

Round Table III

PHYTOCHROME ACTION UNDER NATURAL CONDITIONS

D. Vince-Prue,
Department of Botany,
University of Reading,
Whiteknights,
Reading, RG5 2AT,
UK.

INTRODUCTION

The function of phytochrome is to signal information about the environment so that the plant can make appropriate responses. A great deal of the knowledge about phytochrome has, however, been obtained from physiological experiments carried out under highly unnatural environmental conditions. It is hard to imagine, for example, a more unnatural environmental signal than a brief pulse of red light. Consequently, it is often difficult to relate information obtained under such conditions to the operation of phytochrome in nature, where the light environment varies in duration, quality, irradiance and direction.

TERMINOLOGY

One of the problems that arises when attempting to correlate physiological studies of phytochrome-mediated responses with molecular and genetic approaches is that, at the different levels, the terminology used has not always been consistent. Introducing the subject, M. Furuya pointed out that, at the present moment, we are unable to relate particular phytochrome pools studied in the cell to particular genes and gene-products. In the cell, phytochrome behaviour can be studied by spectrophotometry and physiological responses. At this level, two 'pools' of phytochrome have been identified; one in which Pfr is stable for some time and one in which Pfr disappears more rapidly. Spectrophotometry has shown that the latter may involve loss of total measurable phytochrome (destruction) or reversion of Pfr to Pr. The relationship between spectrophotometrically and physiologically 'labile' and 'stable' pools is not at all clear and

NATO ASI Series, Vol. H 50
Phytochrome Properties and Biological Action
Edited by B. Thomas and C. B. Johnson

analysis is further complicated by the fact that Pfr may be present in the cell but not physiologically active. Thus, when using the terms stable and labile pools at this level of study, it is important to define clearly what is meant.

At the next level of analysis is the phytochrome protein. The use of antibodies has revealed phytochromes with different antigenicity. The terms phytochrome I and phytochrome II are used by many people for these different phytochromes. Since one phytochrome (Type I) predominates in etiolated seedlings, while the other (Type II) predominates in green plants, the terms 'etiolated' and 'green' phytochromes have sometimes been preferred. However, both types clearly co-exist in both green and etiolated plants. Antigenic studies have shown that phytochrome I accumulates in the dark but rapidly decreases to a low level in the light as it undergoes Pfr-mediated destruction. Type II phytochrome, on thc other hand, is present in similar amounts in light-grown and dark-grown seedlings and is stable for many hours in the light. Despite much discussion about the use of the terms phytochrome I and II, both in relation to physiological processes and as the products of specific genes, no general agreement was reached as to precise definitions.

Progress is rapidly being made at the gene level and an agreed terminology is needed for the phytochrome genes and gene products. It was proposed that the terminology should follow the usual rules, with the genes called, for example, *Arabidopsis phy A, phy B,* etc. The proteins coded by the gene would then carry the same name and be termed PHY A, PHY B, etc. The letters *A, B, C,* etc. would be based on homology of sequences within the genes. This proposal was generally supported but some felt that, since the protein has been known and studied for years, the gene products should carry the full name, phytochrome A, B, C, etc.

In conclusion, it was generally agreed that defined terms are needed at each level of study:- physiology, protein, gene - and some tentative recommendations were made. *At the gene level*, the conventions should be followed and the terminology *phy A, phy B,* etc. should be used, based on homology of the gene sequences. The proteins coded by the genes should carry the same name, PHY A, PHY B, etc. (or phytochrome A, phytochrome B could be acceptable, if preferred). Until the *phytochromes recognised and studied in the cell* can be unequivocally equated with the products of specific genes other terms are needed at this level. The terms phytochrome(s) I and II are useful but it is essential to define strictly what is meant when the term is used. Type I and Type II phytochromes have generally been determined and studied immunochemically. Etiolated and green phytochromes are also useful terms if they are clearly defined, i.e., as phytochromes which predominate in etiolated and light-grown plants respectively. *At the physiological level,* stable and labile phytochrome(s) are the preferred terms until the phytochrome associated with any particular physiological response can be identified more precisely. Physiologists should be careful to define what they mean by the terms 'stable' and 'labile', however, and say how this description has been arrived at, e.g., by spectrophotometry, deduced from physiological behaviour, etc. It is important, too, to remember that Pfr can revert to Pr, so that the protein can be stable while Pfr is labile.

RESPONSES TO LIGHT DURATION

Photoperiodism enables plants to make seasonal adjustments in their growth patterns. Flowering is the most studied response controlled by daylength but many others, e.g., dormancy, are equally important under natural conditions. P.J. Lumsden pointed out that phytochrome has several different functions in the photoperiodic control of flowering in short-day plants, but whether these are under the control of the same or different phytochromes is an unresolved question. Short-day plants flower when they are exposed to a sufficiently long night, the duration of which is measured by the operation of a circadian rhythm. The flowering response rhythm can be phase-shifted by a short pulse of red light, which is far-red reversible. One function of phytochrome is thus to control the phase of the flowering rhythm and it is assumed that this action of light is on the underlying oscillator. A second function of phytochrome is the red/far-red reversible inhibition of flowering by a night-break given at a particular phase of the rhythm.

A third function of phytochrome in the photoperiodic control of flowering is that (depending on the pre-treatment), floral induction may require the presence of Pfr during much or all of the inductive long night. This action of phytochrome appears to involve a highly stable pool of Pfr, since it has been found that flowering may be inhibited by a far-red pulse given after many hours (up to 50 in *Chenopodium*) in darkness. At certain times in the night, flowering can be inhibited either by removing Pfr with a far-red pulse, or by generating Pfr with a red pulse. At this time, in *Lemna*, the action spectrum corresponds to both Pfr and Pr. Physiologists are thus faced with an apparent paradox and it has been concluded that these two different actions of light must involve different pools of phytochrome. One is clearly a 'stable' Pfr. In the other pool, Pfr appears to be 'labile' since at the time of night-break sensitivity (i.e. after 8 or 9 hours of darkness) Pfr must be generated in order to inhibit flowering. It is stressed, however, that these are physiological descriptions and relate only to Pfr. The night-break pool could also be stable if it were a question of loss of sensitivity to Pfr already present in the cell.

The several actions of light in photoperiodism and the apparent complexity of phytochrome behaviour emphasise the need for careful definitions at the physiological level of analysis. Although different functions and apparently different kinetic pools have been proposed in the control of flowering in short-day plants, there is no evidence yet about the identity of the phytochrome(s) involved in any of the responses. However, the high stability of Pfr involved in the 'Pfr-requiring reaction' indicates that it may be a Type II phytochrome.

THE NATURAL ENVIRONMENT AND PHYSIOLOGICAL RESPONSES

In discussing this topic, H. Smith re-iterated the point that most physiological experiments are done under unnatural conditions - often of necessity - and stressed that when trying to interpret their relevance to natural conditions it is important to know what these conditions are. For example, the fluence rate for midsummer daylight is ~2 x 10^3 μmol m^{-2} s^{-1}. At this fluence rate, the VLFR would be saturated in 0.5 milliseconds, the LFR would be saturated in 2 seconds and even the HIR would be oversaturated. Under heavy vegetation shade, the VLFR would still be completed in about 50 milliseconds and the LFR within 3 minutes. However, under these conditions, the HIR would probably be within range. Very little information is available about light conditions under the soil, although obviously some light does penetrate with the development of a light gradient. Calculations for 10 mm down in a sandy soil showed that, even here, the VLFR would be completed within 2 seconds and it is difficult to understand the ecological significance of this physiological response. It is well to remember that the fluence-response curves for the VLFR have been determined under highly artificial conditions with seedlings grown in darkness for some time, illustrating the problems of relating such behaviour to the natural environment. On the other hand, the LFR is within the range of vegetation shade and under soil, where the response might be expected to be of ecological significance. The HIR is not only normally below the range under soil and above the range in midsummer sunlight but, in etiolated seedlings, has a wavelength dependence in the far-red. Pure far-red light does not occur in nature except under an entire, thick canopy of green leaves and so the ecological significance of the far-red HIR in etiolated seedlings is unclear.

In trying to relate observed physiological responses to natural conditions, Smith put forward the following scenario. Small seeds normally require red light to germinate and so do not germinate when deeply buried in the soil. Many can be inhibited by the far-red HIR and/or by a high FR/R ratio, thus preventing germination under a vegetation canopy. The VLF responses appear to be associated with the change from 'etiolated' to 'green' growth. Larger seeds germinate under the soil and accumulate large amounts of labile phytochrome (Type I?), allowing the seedlings to detect the presence of light as it nears the surface. This implies that the large amounts of labile phytochrome accumulating in etiolated seedlings acts as an antenna to detect the *presence* of light, not its quality. This contrasts with the situation later on when the other type of phytochrome, which is stable in light, operates through its photochromicity to monitor the R/FR ratio and detect competition from vegetation shading. In this way, a putative function for the different types of phytochrome is that 'labile' phytochrome accumulates in the dark to detect the presence of light very sensitively while 'stable' phytochrome monitors the R/FR ratio and controls acclimation reactions. Both may involve action at the level of gene expression.

In sunlight, cycling between the two forms may limit destruction and allow a considerable amount of labile phytochrome to remain in light-grown plants. This phenomenon is called photoprotection. In seedlings of *Amaranthus*, photoprotection of labile phytochrome began at about 30-100 μmol m^{-2} s^{-1} and continued to increase up to 1000 μmol m^{-2} s^{-1} (the maximum that was experimentally available). In etiolated oat seedlings, a significant amount of Type I phytochrome was immunochemically detectable at the end of a day in sunlight so that, under natural conditions, there may well be more Type I phytochrome present than is normally assumed. It was pointed out, however, that when seeds germinate under natural conditions, Type I phytochrome would not be expected to accumulate to be conserved in this way, while in normal light/dark cycles, Pfr would be destroyed during darkness. Thus, the amounts and functioning of Type I phytochrome in the plant once de-etiolation has occurred are still unresolved. One function that has been proposed is in the photoperiodic control of flowering in the long-day plant, wheat, where the flowering response appears to correlate with the amount of immunochemically-detectable Type I phytochrome in the leaf.

PHYTOCHROME DIMERS

W.J. Van der Woude pointed out that the physiological situation is additionally complicated by the fact the phytochrome is a dimer. For the HIR, calculations have predicted that the far-red peak at 710 nm is a function of the amount of heterodimer and the cycling rate. The peak in red, however, is a function of the homodimer. In far-red, there is a strong dependence on both fluence rate and total phytochrome, perhaps indicating the involvement of labile phytochrome. In red, fluence rate and total phytochrome are less important and a stable phytochrome may be involved. It was suggested that the heterodimer of labile phytochrome could have a function in plants growing in sunlight as cycling would allow the heterodimer to be constantly formed and re-formed in the cell, with the initiation of more signal transduction cascades. The question of conservation of labile phytochrome under these conditions was considered in the previous paragraph.

EVOLUTION OF PHYTOCHROMES

Phytochromes have obviously evolved under natural conditions and their properties and functions must relate to these. In a brief account of possible evolutionary patterns, H. Mohr emphasised that phytochrome is probably restricted to those plants which have arisen in the course of evolution from the green algae. However, in the paradigm of lower green algae, *Chlamydomonas*, a form of rhodopsin rather than phytochrome itself seems to be operating. Thus the decision between phytochrome and rhodopsin appears to have occurred at this stage of evolution. In the more highly developed green algae, such as *Mougeotia*, phytochrome is clearly present and active. This phytochrome appears to be stable and operates as a light sensor. There is, at present, no evidence that it operates on gene expression.

In ferns, phytochrome is also stable and is active in both the sporophyte and gametophyte. Most work has, however, been done on the gametophyte where phytochrome has a strong influence on development and so may act on gene expression. In mosses, spore germination and phototropism are under the control of a stable phytochrome, which differs from the phytochrome of higher plants.

Only a few studies have been made in Gymnosperms. In *Pinus*, the control by phytochrome of both seed germination and seedling photomorphogenesis is similar to that in higher plants. A low level of a stable phytochrome is detectable in cotyledons in white light; a labile phytochrome accumulates in darkness and is lost when plants are transferred to light. Thus, Gymnosperms, like Angiosperms, contain two pools of phytochrome with different properties. From physiological experiments with light/dark cycles followed by transfer to longer periods of darkness, it was concluded that a stable phytochrome controls the synthesis of the labile type. Thus, in this case, a stable phytochrome clearly acts on gene expression.

From the limited information available at present, it appears that a stable, PII-like phytochrome may have evolved before a labile, PI-like type. Stable phytochrome occurs in the green algae and apparently in all more highly-evolved groups. Labile phytochrome, on the other hand, has so far only been detected in Gymnosperms and Angiosperms. Information for lower plants is, however, extremely sketchy and further work may well reveal the presence of labile phytochrome in more primitive groups.

Round Table IV

Cis-ACTING DNA SEQUENCE ELEMENTS AND *trans*-ACTING FACTORS OF PHOTOREGULATED GENES

G.I. Jenkins,
Departments of Biochemistry and Botany,
University of Glasgow,
Glasgow, G12 8QQ,
UK.

This workshop was concerned with the molecular mechanisms involved in the photoregulation of gene expression. It is well established that light can have diverse effects on the expression of different genes within and between gene families and in different species; for example, light can cause either increases or decreases in expression, the responses of individual genes differ in their kinetics, different photoreceptors are involved, there are changes in photoregulation during development, and so on. This diversity in responses must be accounted for to a great extent by the regulatory DNA sequence elements associated with the genes and the sequence-specific DNA-binding proteins that interact with them. It is, therefore, not surprising that in recent years a corresponding complexity has been revealed in the nature and organisation of these sequence elements and in the number of *trans*-acting factors that may be required for photoregulation. The dissection of the *cis*-acting sequence elements of specific genes has involved studies of the expression of chimeric promoter-reporter gene constructs in transgenic plants and the identification of *trans*-acting factors has been achieved through gel retardation and footprinting studies. Table 1 shows the various factors that have been reported to date in terms of DNA-binding activity, and the sequences they recognise. In some cases, the activity has been shown to be present differentially in extracts from light- and dark-grown (or dark-adapted) leaf tissue. It should be emphasised that several of these factors may in fact be identical as insufficient experiments have been performed to demonstrate their uniqueness. The purpose of this workshop was to discuss some of the more recent findings in this area.

The workshop started with a consideration of some of the methodological aspects of the research and their inherent limitations. Although sequence homologies can be

NATO ASI Series, Vol. H 50
Phytochrome Properties and Biological Action
Edited by B. Thomas and C. B. Johnson

Table 1. A current list of *trans*-acting factors associated with photoregulated genes and the sequence elements they bind to. The factors listed have been identified in terms of DNA-binding activities. L and D refer to the presence of activity in either light-grown or illuminated leaf tissue (L) or dark-grown or dark-adapted leaf tissue (D). The question mark (?) indicates a lack of information. The asterisk (*) refers to data presented at this meeting.

Factor	Element	Activity	Reference
GT-1	boxes II and III and their homologues	L and D	Green *et al.*, 1987
GT-2	GT element in rice phytochrome promoter	Transcripts decrease in the light	K Dehesh and PH Quail*
GBF	G box	L and D	Guiliano *et al.*, 1988
AT-1	AT-1 box	controlled by phosphorylation	Datta and Cashmore, 1989
ASF-2/GATA BF	*as-2*/GATA box	L and D	Castresana *et al.*, 1987; Gidoni *et al.*, 1989; Lam and Chua, 1989; Gilmartin *et al.*, 1990
GAF-1	*ga-1*/I box	L > D	Gilmartin *et al.*, 1990
LRF-1 (GAF-1?)	X box	L > D	EM Tobin*; Buzby *et al., 1990*
3AF1	3*af*1 (pea *rbcS-3A* -50 to -31	L and D	Gilmartin *et al.*, 1990; Lam *et al.*, 1990
GC-1	GC element	?	Schindler and Cashmore 1990
ϕGC1	box 1 of rice phyto chrome promoter	?	S Kay*

identified in the promoters of genes from various species, that does not necessarily mean that the corresponding binding activities will be present in nuclear extracts. Moreover, a failure to observe binding activity *in vitro* need not necessarily imply that the factor is not functioning *in vivo*. The ability to detect factor binding is crucial and is obviously dependent on the techniques used; the ways nuclear extracts are prepared and the assays are carried out are clearly important. Tony Cashmore (Philadelphia) made the point that pea extracts appear to be deficient in G box binding activity (GBF; Giuliano *et al.*, 1988) although they have good GT-1 activity (the factor which binds box II and its homologues, originally identified in the pea *rbcS-3A* promoter; Green *et al.*, 1987). Elaine Tobin (Los Angeles) pointed out that the way gels are run is important and referred to some experiments in her own laboratory concerned with the investigation of different factors in *Arabidopsis*. The conditions used to identify one binding activity are not necessarily appropriate to see another and on a single gel one will probably not see all the factors that are present. The inability to detect a factor, therefore, does not necessarily mean that it is not there. Philip Gilmartin (New York) stated that the cloning of factors will ultimately enable antibodies to be produced which will permit direct investigation of the presence of particular factors in extracts from different species.

Steve Kay (New York) pointed out that the binding activities described by various laboratories are simply operational definitions; for example, in the New York laboratory GT-1 is defined as a set of binding activities which will compete with an oligonucleotide comprising four copies of the pea box II and not with the GG to CC mutant of box II. Until the factors are cloned there will not be a better definition. In response to a question from Harry Smith (Leicester), Kay explained that the four copies of box II are used to increase sensitivity in the *in vitro* assay. Although a tetramer has not been found *in vivo*, its use is valid *in vitro* because only one band is seen on a gel shift assay, suggesting strong co-operative binding. The synthetic nature of the assays is evident although it was agreed that they are very powerful. Elaine Tobin pointed out that we are trying to undertake a one dimensional analysis of a three dimensional problem and Steve Kay remarked that the structure of the DNA used in the *in vitro* assays is nothing like chromatin.

Nevertheless, it was agreed that the correlation between *in vitro* and *in vivo* observations to date has been very good. Philip Gilmartin mentioned that there is a good correlation between the ability of boxes to bind factors *in vitro* and their activity measured in *in vivo* experiments. For example, in the pea *rbcS-3A* promoter, substitution of box III (Green *et al*., 1987) by box II does not decrease binding *in vitro* nor the *in vivo* promoter activity, whereas substitution of box II by box III reduces the affinity for binding GT-1 *in vitro* and also the *in vivo* activity. Peter Quail (Berkeley and Albany) further observed that specific nucleotide substitution experiments show a good correlation between *in vitro* binding and *in vivo* activity, for example in the substitution of the two G residues critical for the binding of GT-1 to the box II cor e sequence by two C's, which abolishes activity (Green *et al*., 1988).

With regard to *in vivo* experiments, Kay mentioned some observations of E. Lam and N-H. Chua in which they linked four copies of box II to the -50 region of the cauliflower mosaic virus 35S promoter and saw no activity in transgenic plants; activity was observed with the -90 region of the promoter, presumably because of some sort of interaction with another factor binding to the additional promoter sequences. Gilmartin contributed a further observation. The -90 deletion of the 35S promoter is active in roots but not in leaves, but if four copies of the sequence that binds the factor ASF-1 are linked to the promoter, activity is then observed in leaves. The factor has been cloned and it is known that it is less abundant in leaf than in root tissue. Hence the appearance of activity in leaves in the above experiment can be explained simply by the presence of sufficient binding sites to compete for a limiting amount of the factor.

Bill Thompson (Raleigh) raised a question about GT-1. This factor has been shown to have an up-regulatory function in the *rbcS-3A* promoter, in that it increases expression in the light, whereas it has been found to have a negative effect in the context of the 35S promoter, in that three copies of box II caused a decrease in expression in darkness (Kuhlemeier *et al*., 1987; Green *et al*., 1988). Gilmartin remarked that the way the down-regulation is achieved in the 35S construct is unknown and is the subject of experimentation; there is no direct evidence that GT-1 is, or is not, involved. Cashmore pointed out that if GT-1 binding to box II is shown to be the responsible factor, then this would be a good example of a sequence element changing its activity when placed in the context of a different promoter.

Katy Dehesh and Peter Quail reported the cloning of a protein, designated GT-2, that binds to a regulatory sequence element in the rice phytochrome promoter. They constructed a cDNA library cloned into the expression vector λgt11 and screened duplicate nitrocellulose blots of the expressed proteins with a labelled oligonucleotide probe under conditions that minimise non-specific binding. This oligonucleotide probe was a multimer of a DNA sequence element (see below) found to have a role in controlling transcription of the rice phytochrome *phy A* gene. cDNA clones were selected which expressed a protein which bound to the oligonucleotide. Katy Dehesh noted that although this protein has been cloned on the basis of its sequence-specific binding activity, its function has yet to be demonstrated directly. In the rice phytochrome promoter there is a GT box element with two tandem sequences, GGTTAAT and GGTAATT (Fig. 1). The first of these is identical to the box II core sequence described

-250 TATCCCCCTA**GGTTAAT**TATTGGC**GGTAATT**AACTCCAGGTTGGCGTCGA<u>GGTAAA</u>TCCGC-190

Figure. 1. The -250 to -190 region of the rice phytochrome promoter (Kay *et al.*, 1989). The two GT-1 binding sequence elements identified by Kay (Kay *et al.*, 1989) and referred to by Dehesh are shown in bold type and the additional-sequence noted by Kay is underlined.

for the pea *rbcS-3A* gene and the oat phytochrome promoter, but the second, which lies 3' to the first in the rice phytochrome promoter, is different in two positions. The cloned factor binds only to the 3' sequence element and not to the other and is hence designated GT-2. It is, therefore, distinct on the basis of binding activity, to the rice nuclear GT-1 activity reported by Kay *et al.*, (1989) which binds to both elements (see below). The abundance of the transcripts encoding the GT-2 factor was investigated using northern blots of total RNA and the antisense RNA as a hybridisation probe. Exposure of dark-grown plants to 3 h white light decreased the abundance of the transcripts, but a corresponding decrease was not observed following red, far-red or ref/far-red illumination. GT-2 transcripts were observed in both oat and rice. Further experiments are required to determine whether there is a difference in the GT-2 binding activity in extracts from light- and dark-grown tissue.

Steve Kay presented some data which had a bearing on the observations of Dehesh and Quail on the binding of factors to the rice phytochrome promoter. He noted that the GGTAAT motif recognised by GT-2 is the core sequence of box II* previously identified in the pea *rbcS-3A* gene (Green *et al.*, 1987). Kay had carried out footprinting experiments using rice extracts and focused attention on the -220 region of the rice phytochrome promoter. The footprint showed 2 protected regions on both strands and addition of four copies of the pea box II core sequence competed with binding of the extract to both of these regions. These are the two GT-binding elements referred to by Dehesh. Kay's footprint additionally showed a third GT-binding sequence in the rice phytochrome promoter whose core is GGTAAA. These observations promoted the conclusion that there are 3 GT-1 binding sites in the promoter, but the cloning of GT-2, which, as mentioned above, binds to one of the boxes complicates the interpretation of these data. Kay's results clearly show that the box II core sequence can compete with binding to both of the sites mentioned by Dehesh. Furthermore, another footprinting experiment indicated that there is no apparent difference between the binding sites in their competition by the box II tetramer. Resolution of the data may well be achieved by consideration of the relative affinity of binding and the conditions for making the extracts and carrying out the footprinting experiments. The amount of competitor used, for instance, is important. However, it may be that there is a significant interaction between GT-1 and GT-2 in binding to this region. Kay raised questions of whether there is co-operativity of binding between GT-1 and GT-2, whether GT-1 and GT-2 compete for binding, and whether any such effects are significant *in vivo*. Although Dehesh has shown that GT-2 can bind without GT-1, its binding could be enhanced by GT-1. Quail made the point that in Dehesh's experiments the binding to the GT-1 target sequence was two orders of magnitude lower than to the GT-2 element. Further experiments are now required to resolve the nature of the factors that bind to these sites and the interaction between them.

Kay also reported that another factor, designated ϕGC1, is present in rice nuclear extracts. This binding activity is present in very small amounts even in fractionated extracts and there is insufficient to generate a footprint. Footprinting was carried out with factor obtained from the complex of protein bound to DNA. The footprint showed that the factor 14 nts of the rice promoter which are 100% conserved in oat and highly conserved in maize. At the 3' end of this sequence element in maize Quail and co-workers have found 8 bp of sequence that are also present in oat but not in rice. This factor therefore binds in the region referred to as box 1 by Quail which is important in photoregulation.

The possible roles of GT-1 and GT-2 in controlling transcription from the oat phytochrome promoter are not yet resolved. Quail mentioned that there is no evidence of a functional activity of these GT-1 elements in the pat *phy A* promoter at present on the basis of internal deletion experiments. Cashmore pointed out that these proteins are likely to be members of small gene families, as indicated by hybridisation of cloned DNA to Southern blots, and it may be difficult to resolve which factors bind which boxes *in vivo*. Although the best option would seem to be a genetic approach, this will be difficult because particular factors may be required for the expression of several genes and finding mutants of these factors will, therefore, be problematic because the mutations are likely to be pleiotropic.

There followed a discussion about the interactions between protein factors binding to regulatory *cis*-elements and the possible mechanisms of factor action. Philip Gilmartin presented data concerning the binding of GT-1 to boxes II and III and there homologues in the pea *rbcS-3A* promoter. Boxes II and III are on opposite strands of the DNA and face each other head to head. In contrast, boxes II* and III* and II** and III** are facing away from each other (Green *et al.*, 1987, 1988). The functional significance of this arrangement is not yet understood, but Gilmartin has carried out experiments to investigate the importance of the spacing between the box II and III elements (Gilmartin and Chua, 1990a). There are 33 bp between the G residues of boxes II and III which are critical for factor binding. Constructs were made which altered the spacing between the boxes in a -166 promoter deletion mutant and promoter activity was assayed through a reporter gene in transgenic plants. Removal of 5 bp (half a helical turn of DNA) places the boxes on opposite sides of the helix, which is likely to impair interaction; removal of 10 bp keeps the boxes on the same side of the helix but moves them even closer together. The spacing can be altered similarly by adding base pairs. Gilmartin observed that the removal of 5-7 bp did not affect activity whereas removal of a complete turn (10 bp) inactivated the promoter. Insertion of 2 bp decreased activity and activity was not restored by the insertion of a total of 10 bp, which keeps the boxes on the same side of the helix but moves them further apart. Thus there seems to be an absolute spacing requirement for activity of the boxes. The ability of the constructs with altered spacing to bind GT-1 *in vitro* was studied using competitive gel shift assays. A change in the spacing did not change the affinity of GT-1 to bind to the promoter in those constructs whose functional activity was altered. Thus GT-1 binding is necessary but not sufficient for functional activity; a spatial interaction of factors binding to the box II and III sites is evidently important. Similar experiments have not yet been undertaken with the box II and II homologues in the promoter, which are different in their relative orientations. The spacing between the critical G residues in the box II* and III* elements is retained, so it may be this spacing, rather than the orientation of the bound factors that is most important. Further experiments are required to investigate this point.

These findings, along with others discussed above, demonstrate the complexity of the interaction of *trans*-acting factors with specific DNA sequence elements in photoregulated genes. With the number of different factors and sequence elements described to date, and the possible interactions between like and different factors, there is clearly much to be done to understand the molecular mechanisms involved in the photoregulation of transcription. The way in which the binding and functional activity of the various factors is controlled by light is largely unknown. However, the interaction of the DNA-binding proteins with DNA can be investigated now that factors are being cloned. Furthermore, the way in which the synthesis of the factors themselves is controlled can be addressed. The availability of an *in vitro* transcription system will be important in assessing the activity of the factors, and that reported by Chua's group (Yamazaki *et al.*, 1990) may well be appropriate for this purpose.

Finally, Nigel Urwin (Glasgow) mentioned the importance of examining which photoreceptors mediated the activities observed with different constructs in transgenic plants. Most experiments have been carried out with white light and there is little evidence as yet to define the sequence elements concerned with responses mediated by different photoreceptors. Gilmartin said that the -166 promoter of the *rbcS-3A* gene is known to confer a response mediated by phytochrome (Gilmartin and Chua, 1990b), but this has not been investigated for the synthetic construct containing four copies of box II. Further work is needed to address this point.

REFERENCES

Buzby JS, Yamada T, Tobin EM (1990) The Plant Cell 2:805-814

Castresana C, Garcia-Luque I, Alonso E, Malik VS, Cashmore AR (1988) EMBO J. 7:1929-1936

Datta N, Cashmore AR (1989) The Plant Cell 1, 1069-1077

Gidoni D, Brosio P, Bond-Nutter D, Bedbrook J, Dunsmuir P (1989) Mol. Gen. Genet. 215:337-344.

Gilmartin PM, Chua N-H (1990a) The Plant Cell 2:447-455
Gilmartin PM, Chua N-H (1990b) Mol. Cell. Biol. 10:5565-5568
Gilmartin PM, Sarokin L, Memelink J, Chua N-H (1990) The Plant Cell 2:369-378
Guiliano G, Pichersky E, Malik VS, Timko MP, Scolnik PA, Cashmore AR (1988) Proc. Natl. Acad. Sci. USA. 85:7089-7093
Green PJ, Kay SA, Chua N-H (1987) EMBO J. 6:2543-2549
Green PJ, Yong M-H, Cuozzo M, Kano-Murakami Y, Silverstein P, Chua N-H (1988) EMBO J. 7:4035-4044
Kay SA, Keith B, Shinozaki K, Chye M-L, Chua N-H (1989) The Plant Cell. 1:351-360
Kuhlemeier C. Fluhr R, Green PJ, Chua N-H (1987) Genes and Development. 1:247-255
Lam E, Chua N-H (1989) The Plant Cell. 1:1147-1156
Lam E, Kano-Murakami Y, Gilmartin P, Niner B, Chua N-H (1990) The Plant Cell. 1:857-866
Schindler U, Cashmore AR (1990) EMBO J. 9:3415-3427
Yamazaki K, Chua N-H, Imaseki H (1990) Plant Mol. Biol. Reporter. 8:114-123

INDEX

A

acidic cluster 27
actin filaments 256
action spectrum 276
active phytochrome 39, 217
3AF-1 322
Agrobacterium tumefaciens 114, 170, 203
Amaranthus 317
amino acid
  sequences 19-23, 43, 63
  conservation 17
5-aminolevulinic acid 76
amphiphilic proteins 87, 101, 107
ancymidol 117
anthocyanin synthesis 243, 259
Antirrhinum 157
apical dominance 117
apophycocyanin 306
apophycoerythrin 306
Arabidopsis thaliana 2, 5, 14-34, 47, 62, 96, 157-164, 176-177, 202, 225, 298
ASF-1 148, 324
ASF-2 148, 322
Asplenium sp. 59
AT-1 322
ATPase 290
aurea mutant, *au* 7, 30, 224, 238-241
auroid mutant 241
auxin amide hydrolase 176
Avena sativa, oat 2, 14, 15, 39-54, 62, 96, 114-126, 182, 226, 291, 304
azide ions 297

B

B/UV-A photoreceptor 240, 267
b-type cytochrome 290
BC, *see* bound complex proteins
biliverdin 30
blue/UV-light response 185
blue-light responses 289
bound complex proteins, BC-1, BC-2 205-208
Bryophyta 59
bulk phytochrome 39, 217

C

cab
  genes 141, 191-198
  promoter 194-198
calcium 290
calcium-binding vesicles 256
CaMV 35S promoter 142, 170, 177, 209
carotenoid 297
carrot 298
cauliflower mosaic virus, *see* CaMV

CD spectra 72, 86-108
cell-specific gene expression 181-190
Ceratodon purpureus 62-65
Chaetomorpha linum 59
CHAPS 297
Chara hispida 59
Charaphytum 59
Chenopodium, 315
Chlamydomonas 59, 318
chlorophyll a/b binding protein, *see* cab
Chlorophyta 59
chloroplast reorientation 249
Chou-Fasman analysis 102
chromophore
  attachment 76-79, 305
  biosynthesis 3, 75-77
CHS 185
circadian rhythm 191-198
cis-acting elements 8, 129-138, 194, 202-204, 210, 321-328
competence 259
Corallina elongata 59
correlation plots 92-108
cryptochrome 267
cucumber, *see Cucumis sativa*
Cucumis sativa, cucumber 5, 292
Cucurbita 96
Cytoseira
  abies marina 59
  tamariscifolia 59
cytoskeleton 250

D

dark phytochrome 3
de-etiolation 223
deletion mutants 123, 194
difference spectra 124
DNA binding proteins 205-208
dwarfism 226

E

Edmunson wheel 89
ELISA 47, 281
end-of-day response 220, 223, 228
endogenous oscillator 198
enhancer region 195
Enteromorpha compressa 59
Escherichia coli 4, 45, 72
etiolated phytochrome 39-54, 315

F

fava beans 298
Fd-GOGAT, *see* ferredoxin-dependent glutamate synthase
Fe-protoporphyrin 76
Fed-1 201-214
ferredoxin-dependent glutamate synthase, *Fd*-GOGAT 267-270
ferredoxin gene 201-214
flavin 297
flowering 274-231
footprinting 131-134
Funaria hygrometrica 59

G

G box promoter 157-164
G proteins 294
GA, *see* gibberellin
gabaculin 76
GAF-1 145, 322
GATA BF 322
GBF 159-164, 322
GC-1 322
ϕGC-1 135, 322
Gelidium sp. 59
germination 239
gibberellin, GA 117, 238
Glycine max 253
glycosylation 75, 183, 303-305
green phytochrome 3, 39-54, 315
GT element 33
GT-1 129-138, 146-152, 322, 324, 326
GT-2 33, 322, 324, 326
GTP 293
GUS 171, 189, 195, 204, 209

H

HBP-1 160
heat shock element 144
high pigment mutants, *hp* 242-244
High Irradiance Response, HIR 7, 219, 226, 274, 316
HIR, *see* High Irradiance Response
HMG proteins 206
holoprotein assembly 3
hp, *see* high pigment mutants
HPLC 41, 91
hy mutants 30, 225
hydropathy 24
hydrophobic moment 87

I

immunoblot analysis 52, 115
immunoblotting 253
immunofluorescence 254
immunolocalization 249-256
immunoprecipitation 44
in situ hybridization 188
inclusion body 72

L

labile phytochrome 314
large phytochrome 123
LDP, *see* long-day plants
lectin-binding 304
Lemna gibba 167-176, 315
leucine zipper protein 159-164
LF, *see Low Fluence*
lh mutant 223
LIAC, *see* light-induced absorbance change
light-induced absorbance change 291
light-responsive element, LRE 129, 144
long hypocotyl mutants 176, 225
long-day plants, LDP 275-281
Low Fluence, LF 8, 219, 316

lower plants 57-69
LRE 129
LRE, *see* light responsive element
LRF-1 170, 322
Lycopersicon esculentum, tomato 5, 75, 226, 237-246, 298

M

MAb, *see* monoclonal antibody
maize, *see Zea mays*
Marchantia polymorpha 59
Mesotaenium 59, 249
N-methyl mesoporphyrin IX 76
microprojectile gene transfer 32
microsequencing 16, 41
monoclonal antibody, MAb 29, 44-53, 58-61, 114, 255, 276
Mougeotia 7, 59, 249-256
mRNA
  levels 192
  stability 213
mustard, *see Sinapis alba*
mutants 5, 30, 223-226, 237-245
  photoreceptor 237
  response 237
  transduction chain 237

N

N-linked glycan 304
N-terminus 96-102, 107-108, 123
natural environment 316
negative regulation 174-176
negatively regulated phytochrome genes, NPR 175
Nicotiana plumbaginifolia 144, 157
Nicotiana tabacum, tobacco 5, 75, 114-126, 157, 170, 205, 226-232, 267
NIR, *see* nitrite reductase
nitrate
  reductase, NR 262, 281-285
  reduction 67
  uptake 67
nitrite reductase, NIR 263-266
norflurazon 265
NPR, *see* negatively regulated phytochrome genes
NR, *see* nitrate reductase
nuclear run off experiments 192

O

oat, *see Avena sativa*
organ-specific expression 204
Oryza sativa, rice 5, 15, 19-23, 62, 96, 129-138, 226
overexpressing phenotype 115-118, 227

P

parsley cell culture 185
PCR, *see* polymerase chain reaction
pea, *see Pisum sativum*
pelletability 183
peptides 41

PEST region 27, 106, 309
Petunia 157
Phaeophyta 59
Phaseolus
  coccineus 253
  vulgaris 253
phenylacetic acid 297
phosphorylation 75
photoperiodism 220, 223, 315
photoprotection 311, 317
phototropism 290
phyA 14-38, 314
phyA promoter 130-138
phyB 14-38, 314
phyC 14-38, 314
α-C-phycocyanin 24
phycocyanobilin 305
Physcomitrella patens 59
phytochrome
  abundance 47
  accumulation kinetics 278
  apoprotein 45
  autoregulation 8, 32, 136
  biogenesis 71-81
  cDNA 2
  degradation 120
  destruction 4, 279, 307-311
  dimer 74, 106, 119, 318
  evolution 14, 318
  gene expression 31-32
  genes 2, 13-38
  half-life 279
  immunolocalisation 187
  intracellular location 181
  overexpression 5, 30, 75, 115-118, 227, 244
  pools 39, 313
  structural model 94
  structure 14
  terminology 313
  transcript abundance 31
  types 2
phytochrome-I 71-81, 244, 314
phytochrome-II 245, 314
phytochrome-L 218, 222
phytochrome-like proteins 57
phytochrome-S 218, 222
phytochromobilin 76, 305
PI, *see* phytochrome-I
PII, *see* phytochrome-II
Pinus sylvestris 267-270, 319
Pisum sativum, pea 3, 16, 19-23, 62, 96, 201-214, 290, 292
pizza test 151
plasma membrane 291
plastid factor 261
polyclonal antibodies 296
polymerase chain reaction, PCR 61
post-transcriptional control 202
post-translational
  control 284
  modification 75, 303-306
primers 60
procera mutant 241

protein
  kinase activity 25
  phosphorylation 294-298
protophorphyrinogen 76
protoporphyrin 76
Psilotum triquetrum 59
Pteridophyta 59

R

R/FR ratio 220
rbcs, *see* ribulose-1,5-bisphosphate carboxylase-oxygenase gene
Rhodophyta 59
rhodopsin 318
ribulose-1,5-bisphosphate carboxylase 261
ribulose-1,5-bisphosphate carboxylase-oxygenase gene, rbcs 141-152, 168-176
rice, *see Oryza sativa*
rRNA 193
run on transcription 211, 293

S

Saccharomyces cerevisiae, yeast 4
SAP, *see* sequestered areas of phytochrome
SDP, *see* short-day plants
secondary structure 72
seed germination 223
Selaginella martensii 59
sequestered areas of phytochrome, SAP 182
sequestering 308, 310
short-day plants, SDP 275, 315
Silene 201
Sinapis alba 259-266, 276, 281-286
Sphagnum auriculatum 59
sphingoid bases 291
SSU 193
stable phytochrome 314
stem elongation 116
stomatal guard cells 290
structural domain 93
  diagram 102
succinyl CoA 76
suicide gene selection 176
sunflower 298
supplementary Far-Red responses 229

T

tetrapyrrole synthesis 76
tms2 gene 176
tobacco, *see Nicotiana tabacum*
tomato, *see Lycopersicon esculentum*
trans-acting factors 9, 145-147, 321-328
transcription
  assays 193
  levels 174
  start sites 32
transduction chain 32
transgenic plants 5, 30, 113-126, 143-145, 205, 226-232

transient expression 172
Triticum aestivum, wheat 194-198, 274-281
Triton X-100 297
true phytochrome 57
type I phytochrome 3, 232, 276, 314
type II phytochrome 3, 183, 232, 314
type 2 phytochrome 16

U

ubiquitin 4, 119, 182
ubiquitination 307-309
Ulva rigida 59, 66-69

V

Very Low Fluence, VLF 7, 219, 316
VLF, *see* Very Low Fluence

W

western blot analysis 58, 72
wheat, *see Triticum aestivum*

Y

yeast, *see Saccharomyces cerevisiae*
yg mutants 241

Z

Zea mays, maize 19-23, 96, 131, 253, 298
Zn^{2+} -chelated phytochrome 85, 91, 95, 103-106
Zn^{2+} -induced
  fluorescence 41
  mobility shift 47
Zucchini 19-23, 62

NATO ASI Series H

Vol. 1: Biology and Molecular Biology of Plant-Pathogen Interactions.
Edited by J. A. Bailey. 415 pages. 1986.

Vol. 2: Glial-Neuronal Communication in Development and Regeneration.
Edited by H. H. Althaus and W. Seifert. 865 pages. 1987.

Vol. 3: Nicotinic Acetylcholine Receptor: Structure and Function.
Edited by A. Maelicke. 489 pages. 1986.

Vol. 4: Recognition in Microbe-Plant Symbiotic and Pathogenic Interactions.
Edited by B. Lugtenberg. 449 pages. 1986.

Vol. 5: Mesenchymal-Epithelial Interactions in Neural Development.
Edited by J. R. Wolff, J. Sievers, and M. Berry. 428 pages. 1987.

Vol. 6: Molecular Mechanisms of Desensitization to Signal Molecules.
Edited by T. M. Konijn, P. J. M. Van Haastert, H. Van der Starre, H. Van der Wel, and M. D. Houslay. 336 pages. 1987.

Vol. 7: Gangliosides and Modulation of Neuronal Functions.
Edited by H. Rahmann. 647 pages. 1987.

Vol. 8: Molecular and Cellular Aspects of Erythropoietin and Erythropoiesis.
Edited by I. N. Rich. 460 pages. 1987.

Vol. 9: Modification of Cell to Cell Signals During Normal and Pathological Aging.
Edited by S. Govoni and F. Battaini. 297 pages. 1987.

Vol. 10: Plant Hormone Receptors. Edited by D. Klämbt. 319 pages. 1987.

Vol. 11: Host-Parasite Cellular and Molecular Interactions in Protozoal Infections.
Edited by K.-P. Chang and D. Snary. 425 pages. 1987.

Vol. 12: The Cell Surface in Signal Transduction.
Edited by E. Wagner, H. Greppin, and B. Millet. 243 pages. 1987.

Vol. 13: Toxicology of Pesticides: Experimental, Clinical and Regulatory Perspectives.
Edited by L. G. Costa, C. L. Galli, and S. D. Murphy. 320 pages. 1987.

Vol. 14: Genetics of Translation. New Approaches.
Edited by M. F. Tuite, M. Picard, and M. Bolotin-Fukuhara. 524 pages. 1988.

Vol. 15: Photosensitisation. Molecular, Cellular and Medical Aspects.
Edited by G. Moreno, R. H. Pottier, and T. G. Truscott. 521 pages. 1988.

Vol. 16: Membrane Biogenesis. Edited by J. A. F. Op den Kamp. 477 pages. 1988.

Vol. 17: Cell to Cell Signals in Plant, Animal and Microbial Symbiosis.
Edited by S. Scannerini, D. Smith, P. Bonfante-Fasolo, and V. Gianinazzi-Pearson. 414 pages. 1988.

Vol. 18: Plant Cell Biotechnology.
Edited by M. S. S. Pais, F. Mavituna, and J. M. Novais. 500 pages. 1988.

Vol. 19: Modulation of Synaptic Transmission and Plasticity in Nervous Systems.
Edited by G. Hertting and H.-C. Spatz. 457 pages. 1988.

Vol. 20: Amino Acid Availability and Brain Function in Health and Disease.
Edited by G. Huether. 487 pages. 1988.

NATO ASI Series H

Vol. 21: **Cellular and Molecular Basis of Synaptic Transmission.**
Edited by H. Zimmermann. 547 pages. 1988.

Vol. 22: **Neural Development and Regeneration. Cellular and Molecular Aspects.**
Edited by A. Gorio, J. R. Perez-Polo, J. de Vellis, and B. Haber. 711 pages. 1988.

Vol. 23: **The Semiotics of Cellular Communication in the Immune System.**
Edited by E. E. Sercarz, F. Celada, N. A. Mitchison, and T. Tada. 326 pages. 1988.

Vol. 24: **Bacteria, Complement and the Phagocytic Cell.**
Edited by F. C. Cabello und C. Pruzzo. 372 pages. 1988.

Vol. 25: **Nicotinic Acetylcholine Receptors in the Nervous System.**
Edited by F. Clementi, C. Gotti, and E. Sher. 424 pages. 1988.

Vol. 26: **Cell to Cell Signals in Mammalian Development.**
Edited by S. W. de Laat, J. G. Bluemink, and C. L. Mummery. 322 pages. 1989.

Vol. 27: **Phytotoxins and Plant Pathogenesis.**
Edited by A. Graniti, R. D. Durbin, and A. Ballio. 508 pages. 1989.

Vol. 28: **Vascular Wilt Diseases of Plants. Basic Studies and Control.**
Edited by E. C. Tjamos and C. H. Beckman. 590 pages. 1989.

Vol. 29: **Receptors, Membrane Transport and Signal Transduction.**
Edited by A. E. Evangelopoulos, J. P. Changeux, L. Packer, T. G. Sotiroudis, and K. W. A. Wirtz. 387 pages. 1989.

Vol. 30: **Effects of Mineral Dusts on Cells.**
Edited by B. T. Mossman and R. O. Bégin. 470 pages. 1989.

Vol. 31: **Neurobiology of the Inner Retina.**
Edited by R. Weiler and N. N. Osborne. 529 pages. 1989.

Vol. 32: **Molecular Biology of Neuroreceptors and Ion Channels.**
Edited by A. Maelicke. 675 pages. 1989.

Vol. 33: **Regulatory Mechanisms of Neuron to Vessel Communication in Brain.**
Edited by F. Battaini, S. Govoni, M. S. Magnoni, and M. Trabucchi. 416 pages. 1989.

Vol. 34: **Vectors as Tools for the Study of Normal and Abnormal Growth and Differentiation.**
Edited by H. Lother, R. Dernick, and W. Ostertag. 477 pages. 1989.

Vol. 35: **Cell Separation in Plants: Physiology, Biochemistry and Molecular Biology.**
Edited by D. J. Osborne and M. B. Jackson. 449 pages. 1989.

Vol. 36: **Signal Molecules in Plants and Plant-Microbe Interactions.**
Edited by B. J. J. Lugtenberg. 425 pages. 1989.

Vol. 37: **Tin-Based Antitumour Drugs.** Edited by M. Gielen. 226 pages. 1990.

Vol. 38: **The Molecular Biology of Autoimmune Disease.**
Edited by A. G. Demaine, J-P. Banga, and A. M. McGregor. 404 pages. 1990.

Vol. 39: **Chemosensory Information Processing.** Edited by D. Schild. 403 pages. 1990.

Vol. 40: **Dynamics and Biogenesis of Membranes.**
Edited by J. A. F. Op den Kamp. 367 pages. 1990.

Vol. 41: **Recognition and Response in Plant-Virus Interactions.**
Edited by R. S. S. Fraser. 467 pages. 1990.

NATO ASI Series H

Vol. 42: Biomechanics of Active Movement and Deformation of Cells.
Edited by N. Akkaş. 524 pages. 1990.

Vol. 43: Cellular and Molecular Biology of Myelination.
Edited by G. Jeserich, H. H. Althaus, and T. V. Waehneldt. 565 pages. 1990.

Vol. 44: Activation and Desensitization of Transducing Pathways.
Edited by T. M. Konijn, M. D. Houslay, and P. J. M. Van Haastert. 336 pages. 1990.

Vol. 45: Mechanism of Fertilization: Plants to Humans.
Edited by B. Dale. 710 pages. 1990.

Vol. 46: Parallels in Cell to Cell Junctions in Plants and Animals.
Edited by A. W. Robards, W. J. Lucas, J. D. Pitts, H. J. Jongsma, and D. C. Spray. 296 pages. 1990.

Vol. 47: Signal Perception and Transduction in Higher Plants.
Edited by R. Ranjeva and A. M. Boudet. 357 pages. 1990.

Vol. 48: Calcium Transport and Intracellular Calcium Homeostasis.
Edited by D. Pansu and F. Bronner. 456 pages. 1990.

Vol. 49: Post-Transcriptional Control of Gene Expression.
Edited by J. E. G. McCarthy and M. F. Tuite. 671 pages. 1990.

Vol. 50: Phytochrome Properties and Biological Action.
Edited by B. Thomas and C. B. Johnson. 337 pages. 1991.

Vol. 51: Cell to Cell Signals in Plants and Animals.
Edited by V. Neuhoff and J. Friend. 404 pages. 1991.

Vol. 52: Biological Signal Transduction.
Edited by E. M. Ross and K. W. A. Wirtz. 560 pages. 1991.

Vol. 53: Fungal Cell Wall and Immune Response.
Edited by J. P. Latgé and D. Boucias. 472 pages. 1991.

Vol. 54: The Early Effects of Radiation on DNA.
Edited by E. M. Fielden and P. O'Neill. 448 pages. 1991.

Vol. 55: The Translational Apparatus of Photosynthetic Organelles.
Edited by R. Mache, E. Stutz, and A. R. Subramanian. 260 pages. 1991.

Vol. 56: Cellular Regulation by Protein Phosphorylation.
Edited by L. M. G. Heilmeyer, Jr. 520 pages. 1991.

Vol. 57: Molecular Techniques in Taxonomy.
Edited by G. M. Hewitt, A. Johnston, and J. P. W. Young. 420 pages. 1991.

Printing: Druckerei Zechner, Speyer
Binding: Buchbinderei Schäffer, Grünstadt